# NanoScience and Technology

Springer
Berlin
Heidelberg
New York
Barcelona
Hong Kong
London
Milan
Paris
Tokyo

Physics and Astronomy

ONLINE LIBRARY

http://www.springer.de/phys/

# NanoScience and Technology

*Series Editors:* P. Avouris   K. von Klitzing   H. Sakaki   R. Wiesendanger

The series NanoScience and Technology is focused on the fascinating nano-world, meso-scopic physics, analysis with atomic resolution, nano and quantum-effect devices, nano-mechanics and atomic-scale processes. All the basic aspects and technology-oriented developments in this emerging discipline are covered by comprehensive and timely books. The series constitutes a survey of the relevant special topics, which are presented by leading experts in the field. These books will appeal to researchers, engineers, and advanced students.

**Sliding Friction**
Physical Principles and Applications
By B. N. J. Persson

**Scanning Probe Microscopy**
Analytical Methods
Editor: R. Wiesendanger

**Mesoscopic Physics and Electronics**
Editors: T. Ando, Y. Arakawa, K. Furuya, S. Komiyama,
and H. Nakashima

**Biological Micro- and Nanotribology**
Nature's Solutions
By M. Scherge and S. N. Gorb

**Semiconductor Spintronics and Quantum Computation**
Editors: D. D. Awschalom, D. Loss, and N. Samarth

Series homepage – http://www.springer.de/phys/books/nst/

D. D. Awschalom    D. Loss    N. Samarth (Eds.)

# Semiconductor Spintronics and Quantum Computation

With 151 Figures

Springer

Prof. D. D. Awschalom
Department of Physics
University of California
Santa Barbara, CA 93106
USA

Prof. N. Samarth
Department of Physics
Pennsylvania State University
University Park, PA 16802
USA

Prof. D. Loss
Department of Physics and Astronomy
University of Basel
Klingelbergstr. 82
4056 Basel
Switzerland

*Series Editors:*

Professor Dr. Phaedon Avouris

IBM Research Division, Nanometer Scale Science & Technology
Thomas J. Watson Research Center, P.O. Box 218
Yorktown Heights, NY 10598, USA

Professor Dr., Dres. h. c. Klaus von Klitzing

Max-Planck-Institut für Festkörperforschung, Heisenbergstrasse 1
70569 Stuttgart, Germany

Professor Hiroyuki Sakaki

University of Tokyo, Institute of Industrial Science, 4-6-1 Komaba, Meguro-ku
Tokyo 153-8505, Japan

Professor Dr. Roland Wiesendanger

Institut für Angewandte Physik, Universität Hamburg, Jungiusstrasse 11
20355 Hamburg, Germany

Library of Congress Cataloging-in-Publication Data.
Semiconductor spintronics and quantum computation / D. Awschalom, N. Samarth, D. Loss (eds.) p. cm. –
(Nanoscience and technology) Includes bibliographical references and index. ISBN: 3540421769 (alk. paper)
1. Spintronics. 2. Quantum electronics. 3. Logic devices. I. Awschalom, D. (David) II. Samarth, N. (Nitin)
III. Loss, D. (Daniel) IV. Series.    TK7874.887.S45 2002    621.381–dc21    2002020925

ISSN 1434-4904
ISBN 3-540-42176-9 Springer-Verlag Berlin Heidelberg New York

Springer-Verlag Berlin Heidelberg New York
a member of BertelsmannSpringer Science+Business Media GmbH

http://www.springer.de

© Springer-Verlag Berlin Heidelberg 2002
Printed in Germany

Typesetting by the authors using a Springer TEXmacro package
Final processing by Steingräber Satztechnik GmbH Heidelberg
Cover design: *design& production*, Heidelberg

Printed on acid-free paper     57/3111 - 5 4 3 2

# Preface

The past few decades of research and development in solid-state semiconductor physics and electronics have witnessed a rapid growth in the drive to exploit quantum mechanics in the design and function of semiconductor devices. This has been fueled for instance by the remarkable advances in our ability to fabricate nanostructures such as quantum wells, quantum wires and quantum dots. Despite this contemporary focus on semiconductor "quantum devices," a principal quantum mechanical aspect of the electron – its spin has largely been ignored (except in as much as it accounts for an added quantum mechanical degeneracy). In recent years, however, a new paradigm of electronics based on the spin degree of freedom of the electron has begun to emerge. This field of semiconductor "spintronics" (spin transport electronics or spin-based electronics) places electron spin rather than charge at the very center of interest. The underlying basis for this new electronics is the intimate connection between the charge and spin degrees of freedom of the electron via the Pauli principle. A crucial implication of this relationship is that spin effects can often be accessed through the orbital properties of the electron in the solid state. Examples for this are optical measurements of the spin state based on the Faraday effect and spin-dependent transport measurements such as giant magneto-resistance (GMR). In this manner, information can be encoded in not only the electron's charge but also in its spin state, i.e. through alignment of a spin (either "up" or "down") relative to a reference (an applied magnetic field or magnetization orientation of the ferromagnetic film). This ability offers opportunities for a new generation of semiconductor devices combining standard microelectronics with spin-dependent effects that arise from the interaction between the spin of a charge carrier and the magnetic properties of the material. As mentioned above, the Pauli-based connection between spin and charge can be exploited to add a new level of functionality, where a device operates on both degrees of freedom together rather than one or the other. The advantages of these new devices would include non-volatility, increased data processing speed, decreased electric power consumption, and increased integration densities compared to conventional semiconductor devices.

Spin, moreover, offers the opportunity to store and manipulate phase coherence over length and time scales much larger than is typically possible in charge-based devices. Perhaps the most promising application of such control can be found in the potential for constructing quantum bits (qubits), which are comprised of superpositions of many spins leading to highly entangled quantum states with highly non-classical properties. Such quantum coherence and entanglement form the basis for quantum information processing, and, in particular, for building a spin-based quantum computer. The phase coherence of spins of localized and delocalized electrons plays a significant role in this field. In this sense, spintronics is a natural extension of mesoscopic physics, where the focus on phase coherence is shifted from the charge to the spin degrees of freedom of the electron.

The field of semiconductor spintronics is at an exciting stage since major fundamental problems (reviewed in this volume) are still being addressed by experiment and theory. These issues include the detection of spin coherence in nanoscale structures such as quantum wells and quantum dots, the optimization of electron spin lifetimes, transport of spin-polarized carriers across relevant length scales and hetero-interfaces, theoretical concepts for all-electrical spin control in quantum dots for quantum information processing, and the manipulation of both electron and nuclear spins on sufficiently fast time scales. Recent experiments suggest that the storage time of quantum information encoded in electron spins may be extended through their strong interplay with nuclear spins in the solid state. With their relatively long lifetimes (ranging from minutes to hours), future quantum technologies may utilize nuclear spins to store information, and electron spins to act as the information bus. Moreover, optical methods for spin injection, detection, and manipulation have been developed that exploit the ability to precisely engineer the coupling between electron spin and photons. It is envisioned that the merging of electronics, photonics, and magnetics may ultimately lead to spin-based multifunctional devices such as the spin-FET (Field Effect Transistor), spin-LED (Light-Emitting Diode), spin-RTD (Resonant Tunneling Device), as well as optical switches/modulators operating at terahertz frequencies. In addition, semiconductor spintronics may also allow us to develop the basic elements needed for quantum computation and communication (quantum bits, quantum gates, encoders, decoders, etc.). While there has been much recent progress in understanding spin dynamics in the solid state, the success of these ventures depends on a deeper understanding of the fundamental spin interactions in the relevant materials, the factors limiting coherence lifetimes, as well as the roles of materials fabrication, issues in spin transport and optical spin manipulation, dimensionality, defects, and semiconductor band structure in modifying the spin dynamics.

We hope that this volume of collected articles provides an up-to-date account of spintronics as seen by the contributing experts, and that this collection will help to define the central issues in this nascent area of research.

Santa Barbara, USA    *David D. Awschalom*
January 2002    *Daniel Loss*
    *Nitin Samarth*

# Contents

# Contributors

**David D. Awschalom**
Department of Physics
University of California
Santa Barbara, California 93106
USA
awsch@physics.ucsb.edu

**Jeremy Baumberg**
Department of Physics
and Astronomy
University of Southampton
Southampton, SO17 1BJ,
United Kingdom
J.J.Baumberg@soton.ac.uk

**Oliver Benson**
Quantum Entanglement Project,
ICORP, JST,
E. L. Ginzton Laboratory
Stanford University
Stanford, CA
USA

**Guido Burkard**
Department of Physics
and Astronomy
University of Basel
Klingelbergstrasse 82
4056 Basel
Switzerland
Guido.Burkard@unibas.ch

**Jeff M. Byers**
Naval Research Laboratory
Washington, D.C. 20375
USA
byers@foucault.nrl.navy.mil

**David DiVincenzo**
IBM
Yorktown Heights NY
USA
divince@watson.ibm.com

**Andrei Filip**
Applied Physics Department
Eindhoven University of Technology
P.O. Box 513
5600 MB Eindhoven
The Netherlands
filip@tue.nl

**Michael Flatté**
Dept. of Physics and Astronomy,
IATL 144,
University of Iowa
Iowa City, IA 52242
USA
michaelflatte@mailaps.org

**Friso Jedema**
Dept. of Applied Physics
and Materials Science Centre
University of Groningen
Groningen
The Netherlands
F.J.Jedema@phys.rug.nl

**Wayne H. Lau**
Department of Physics
and Astronomy and Optical Science
and Technology Center
University of Iowa
Iowa City, Iowa 52242
USA
wlau@ostc.physics.uiowa.edu


**Daniel Loss**
Department of Physics
and Astronomy
University of Basel
Klingelbergstrasse 82
4056 Basel
Switzerland
Daniel.Loss@unibas.ch


**Laurens W. Molenkamp**
Physikalisches Institut
Universität Würzburg
Am Hubland
97074 Würzburg
Germany
laurens.molenkamp
@physik.uni-wuerzburg.de


**F. G. Monzon**
California Institute of Technology
Condensed Matter Physics 114-36
Pasadena CA 91125
USA


**Hideo Ohno**
Laboratory for Electronic Intelligent
Systems, Research Institute
of Electrical Communication,
Tohoku University
Katahira 2-1-1, Aoba-ku
Sendai 980-8577
Japan
ohno@riec.tohoku.ac.jp

**Matthew Pelton**
Quantum Entanglement Project,
ICORP, JST,
E. L. Ginzton Laboratory
Stanford University
Stanford, CA
USA


**Nitin Samarth**
Department of Physics
Penn State University
104 Davey Lab, University Park
Pennsylvania 16802
USA
nsamarth@psu.edu


**Charles Santori**
Quantum Entanglement Project,
ICORP, JST,
E. L. Ginzton Laboratory
Stanford University
Stanford, CA
USA


**Glenn S. Solomon**
Quantum Entanglement Project,
ICORP, JST,
E. L. Ginzton Laboratory
Stanford University
Stanford, CA
USA


**Michael Roukes**
Condensed Matter Physics 114-36
California Institute of Technology
Pasadena CA 91125
USA
roukes@caltech.edu


**Axel Scherrer**
Dept. of Electrical Engineering
California Institute of Technology
Pasadena CA 91125
USA

**Georg Schmidt**
Physikalisches Institut
Universität Würzburg
Am Hubland
97074 Würzburg
Germany
georg.schmidt
@physik.uni-wuerzburg.de

**Hong Tang**
Condensed Matter Physics 114-36
California Institute of Technology
Pasadena, CA 91125
USA
htang@caltech.edu

**Bart Van Wees**
Dept. of Applied Physics
and Materials Science Centre
University of Groningen
Groningen
The Netherlands
B.J.van.Wees@phys.rug.nl

**Yelena Vuckovic**
Dept. of Electrical Engineering
California Institute of Technology
Pasadena, CA 91125
USA

**Yoshi Yamamoto**
Quantum Entanglement Project,
ICORP, JST,
E. L. Ginzton Laboratory
Stanford University
Stanford, CA
USA
yamamoto
@fastloki.Stanford.edu

# 1 Ferromagnetic III-V Semiconductors and Their Heterostructures

Hideo Ohno

## 1.1 Introduction

The unique interplay between semiconducting bulk properties and ferromagnetism via exchange interaction, first discovered in europium chalcogenides (e.g. EuO) and semiconducting spinels (e.g. $CdCr_2Se_4$), attracted much attention and studied extensively in the 1960's and 70's [1,2]. The interest in this first generation of ferromagnetic semiconductors gradually waned in the 80's, partly due to the difficulty associated with the preparation of single crystals and partly due to its low ferromagnetic transition temperatures, which made it difficult for practical room temperature applications.

The second generation of ferromagnetic semiconductors is based on nonmagnetic semiconductors alloyed with magnetic elements (mostly transition metals). Such alloys between a nonmagnetic semiconductor host and a magnetic element are called diluted magnetic semiconductors (DMS's), as small composition of magnetic elements ($x < 0.2$) is often used. The common choice of host semiconductor has been II-VI compounds (e.g. ZnSe and CdTe), because the valence of cation ($s^2$) is compatible with that of typical magnetic elements, e.g. Mn, $3d^5 4s^2$. These II-VI based DMS's have been extensively studied [3,4] and shown to exhibit pronounced change in their properties upon application of magnetic fields due to exchange interaction between the conduction carriers and the localized magnetic moments. The observed magnetism was limited to paramagnetism or spin-glass resulting from antiferromagnetic superexchange interactions among magnetic ions [5]. In addition to II-VI based DMS's, a IV-VI based DMS, (Pb,Sn,Mn)Te, was prepared in bulk form, in which carrier-induced ferromagnetism was studied [6].

The discovery of ferromagnetism in III-V based DMS's, (In,Mn)As and (Ga,Mn)As [7–9], has opened unique opportunities of introducing ferromagnetism in the semiconductors already in use in electronic applications (lasers and transistors) and integrating ferromagnetism in semiconductor heterostructures. The highest transition temperature so far reported in the III-V ferromagnetic semiconductors is 110 K [10], still below room temperature. However, a model for the ferromagnetism capable of describing a number of experimental observations suggests a route to above room temperature ferromagnetism in the family of tetrahedrally bonded semiconductors[11], which III-V DMS's are part of. Extensive research is directed toward above room temperature ferromagnetism and we begin to hear some successes [12,13]. Furthermore, the successful application of a II-VI based *paramagnetic* semiconductor (Cd,Hg,Mn)Te to the optical isolators used at room temperature for Er-doped fiber amplifiers, and growing interests in spin-manipulation in semiconductor heterostructures for quantum information technology (where operating temperature is *not* an issue), both suggest that the low ferromagnetic transition temperature alone is not an absolute obstacle for application of magnetic semiconductors any more.

In this chapter, we discuss first the preparation, properties and ferromagnetism of III-V based ferromagnetic semiconductors, mostly focusing on (Ga,Mn)As, and then we discuss the heterostructures made in combination with nonmagnetic III-V semiconductors showing a number of spin-dependent phenomena only accessible with ferromagnetic semiconductors.

## 1.2    Preparation of III-V Based Ferromagnetic Semiconductors

In order to make nonmagnetic III-V's such as GaAs and InAs magnetic, one needs to overcome the low solubility limit of the magnetic impurity in the host III-V's; the solubility limit is usually below the level of magnetic impurity concentration required for pronounced magnetic effects. Solubility was not an issue in II-VI based DMS's because of the valence compatibility of transition elements with group II elements. High level of doping realized by the use of low-temperature molecular-beam epitaxy (LT-MBE) allowed the first synthesis of an alloy between InAs and Mn, (In,Mn)As, in 1989 [14]. Reduction of the substrate temperature $T_S$ during growth from typical 500°C for InAs to 250°C was the key factor to suppress the formation of thermodynamically stable compounds such as MnAs and to realize a uniform alloy. This substrate temperature of 250°C was still high enough to provide thermal energy for epitaxial growth of single crystal on a substrate. The Mn in the alloy not only provides localized magnetic moment but also acts as an acceptor, leaving a hole in the valence band and making the host InAs $p$-type (this actually depends on growth conditions; $T_S$ below 200°C results in $n$-type conduction [8]). It was later discovered that thus provided holes mediate ferromagnetic interaction, leading to ferromagnetism in (In,Mn)As [15]. This was followed by the successful synthesis [16] and discovery of ferromagnetism in (Ga,Mn)As [17].

MBE growth of (Ga,Mn)As using solid source MBE with elemental sources Ga, Mn, and As, proceeds in the following way. (001) oriented GaAs is the common choice of substrate. After thermal removal of the surface oxide, a thin (> 100 nm) GaAs or (Al,Ga)As buffer layer is grown at $T_S = 580$–600°C (typical $T_S$ for GaAs growth) to make the surface smooth. Before growth of (Ga,Mn)As, $T_S$ is reduced to 250°C and epitaxial growth of (Ga,Mn)As is commenced by opening the Ga and Mn shutters at the same time [18]. Growth rate is in the range of 0.6–1.2 $\mu$m/h. Reflection high-energy electron diffraction (RHEED) patterns are used to monitor the surface reconstruction during growth. Growths are always carried out under As-stabilized conditions. RHEED pattern during and after (Ga,Mn)As growth is ($T_S = 250°C$) ($1 \times 2$). Although the properties of grown (Ga,Mn)As do depend on growth parameters such as As overpressure and $T_S$ [19,20], as long as the established growth procedure is followed, the properties of (Ga,Mn)As films are reproducible; for example, for a given Mn concentration $x$, the ferromagnetic transition temperature $T_C$ can be maintained in the range of $2000 \times x \pm 10$ K. Even if the growth condition is off, one can anneal the sample at around growth temperature to "restore" the highest possible $T_C$ that can be achieved by the given composition [21,22]. The process behind this annealing behavior is not fully understood, yet, but it is argued that the removal of excess antisite As (As in group III site), which increases the hole concentration and hence results in higher $T_C$ (see the discussion of the carrier-induced ferromagnetism). Recent calculation shows that local configuration of Mn and As in the zincblende lattice may be of importance in determining the magnetic interaction [23].

At lower $T_S$, the films tend to be insulating and even lower $T_S$ ($< 180°C$), only polycrystalline films are obtained. When the Mn flux or the substrate temperature or both were too high, an RHEED pattern indicative of the appearance of the MnAs (NiAs-structure) second phase on the surface emerges. The maximum $x$ achieved by LT-MBE without formation of second phases is a function of $T_S$, but for the most extensively studied (Ga,Mn)As, this appears to be less than 0.1 even under optimized conditions. One can instead insert submonolayer of MnAs sheets in the GaAs host instead of making an alloy [24]. It was shown that a single submonolayer MnAs results in ferromagnetism.

Mn composition $x$ in (Ga$_{1-x}$Mn$_x$)As films is determined by X-ray diffraction measurements (XRD), which is calibrated against electron probe microanalysis (EPMA) results. The lattice constant $a$ of (In,Mn)As and (Ga,Mn)As follows the Vegard's law (i.e. $a$ is a linear function of $x$). For (Ga,Mn)As it is $a = 0.566(1 - x) + 0.598x$ (nm) [7]. Here, $a = 0.598$ nm for $x = 1$ is the lattice constant of hypothetical zincblende MnAs. The free standing lattice constant $a$ (i.e. under no strain) is calculated from the diffraction peak position in the measured XRD spectra assuming that (Ga,Mn)As is fully strained and that (Ga,Mn)As has the same elastic constant as GaAs as,

$$a = \frac{1 - \nu}{1 + \nu} a_{XRD} + \frac{2\nu}{1 + \nu} a_{GaAs}.$$

Here, $a_{XRD}$ is the measured lattice spacing, $a_{GaAs}$ the lattice constant of the substrate, and $\nu$ the Poisson ratio. The above relationship between $a$ and $x$ contains some error due to the fact that the details of lattice constant depend on the growth conditions such as $T_S$ and As overpressure, presumably because of the As antisite incorporation [20,25]. Although (Ga,Mn)As has larger lattice constant than GaAs, no indication of strain relaxation was observed at least up to thickness of 2 $\mu$m with $x = 0.057$ [26].

Extended X-ray absorption fine structure measurements on (In,Mn)As and (Ga,Mn)As showed that the Mn atoms substitutes group III sublattice in both cases [27,28], in accordance with the observation of the Vegard's law.

As will be discussed later, strain in the film determines the direction of the magnetic easy axis and thus is an important parameter to control magnetism. When directly grown on GaAs substrate, (Ga,Mn)As epitaxial films are compressively strained. Tensile strain can be introduced into the magnetic layer by growing a lattice relaxed (In,Ga)As buffer layer having a larger lattice constant than the subsequent (Ga,Mn)As layer. (In,Mn)As grown on AlSb, GaSb or an alloy between the two is under tensile strain.

Preparation of systems other than (Ga,Mn)As and (In,Mn)As is also in progress. Ferromagnetism has been observed in (Ga,Mn)Sb ($T_C = 25$ K), but it appears difficult to completely eliminate the second phase MnSb in this system [29]. Among alloys of GaAs and Cr, Fe, and Co, only (Ga,Cr)As shows ferromagnetic interaction [30], whereas others exhibits paramagnetism down to 2 K [31,32]. Magnetic gallium nitride has also been prepared [33–36]. Very recent report suggests room temperature ferromagnetism in (Ga,Mn)N [37].

Among semiconductors other than magnetic III-V's, (Cd,Mn,Ge)P$_2$ [12] and TiO$_2$:Co [13] are reported to exhibit ferromagnetism at room temperature. Although not a semiconductor, zincblende CrAs [38] and CrSb [39] have also been prepared epitaxially on GaAs substrates, which do not exist in naturally in the zincblende form. Both of them are ferromagnetic at room temperature and the former has been predicted to be half-metallic by first principle calculation [40,38]. The first magnetic group VI semiconductor, Ge:Mn, has also been prepared [41].

## 1.3   Magnetic Properties

Square hysteresis in the magnetization curves (magnetization ($M$) versus magnetic field ($B$) curves) at low temperature is one of the evidences of the ferromagnetic order in the film. Figure 1.1 is a result of magnetization measurements by a superconducting quantum interference device (SQUID) magnetometer at 5 K of a 150 nm thick Ga$_{0.965}$Mn$_{0.035}$As layer grown on a (001) GaAs substrate. The temperature independent diamagnetic response of thick GaAs substrate has been subtracted from the as-measured curves. When $B$ is applied parallel to the magnetic layer, a clear hysteresis is observed as shown in the upper left inset; the magnetic easy axis is in the plane of the film and shows a weak four-fold symmetry within the plane. The direction perpendicular to the plane is a hard axis, as can be seen from the absence of hysteresis in the magnetization curve. The anisotropy energy $K$ calculated from the difference between the two curves ($B \perp$ plane and $B \parallel$ plane) is $K = 2.9 \times 10^3$ J/m$^3$. This anisotropy energy is strain dependent and can be

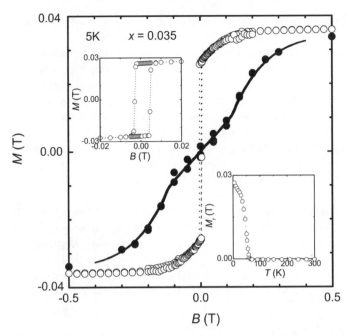

**Fig. 1.1.** The magnetization versus field curve of a 150 nm thick (Ga,Mn)As layer (Mn composition $x = 0.035$) grown on a low-temperature GaAs buffer layer. Substrate is (001) semi-insulating GaAs. Diamagnetic response from the GaAs substrate is subtracted. Open symbols are taken with the field in-plane and closed symbols with field perpendicular to the plane of the sample ($T$= 5 K). The data show that the magnetic easy axis is in-plane. The solid line is the Hall resistance $R_{\mathrm{Hall}}$ scaled to match the magnetization curve, showing that $R_{\mathrm{Hall}} \propto M$ perpendicular to the plane. The *upper left inset* shows the magnified view of the data at low fields, where a clear and square hysteresis is observed. The *lower right inset* shows the temperature dependence of remanence, indicating that the transition temperature is 60 K and negligible ferromagnetic MnAs contribution above 60 K.

made perpendicular to the plane by reversing the sign of the strain in the film [42]. The strain in Fig. 1.1 is $-2.4\%$ (compressive). The strain dependent easy axis has also been observed in (In,Mn)As [43].

Paramagnetic response is often observed after closure of hysteresis as seen in the curve of Fig. 1.1. This paramagnetic response appears to be correlated with the transport properties of the film; the more metallic the sample is (in terms of the metal-insulator transition) the less the portion of the paramagnetic response [44]. The total saturation magnetization ($M_{\mathrm{SAT}}$) after the saturation of the paramagnetic response is always consistent with Mn having five half spins ($S = 5/2$). For example, the curve shown in Fig. 1.1 yields $S = 2.2$ using $M_{\mathrm{SAT}} = xN_0 g\mu_B S$ and nominal $x$, where $N_0$ is the density of cation sites, $g$ (= 2) the $g$-factor of Mn, $\mu_B$ the Bohr magneton, and $S$

the spin of Mn. Note that accurate determination of $S$ from magnetization is difficult because of the error involved in determining $x$ ($\pm 10\%$).

Shown in the lower right inset of Fig. 1.1 is the temperature dependence of the remanence, $M_r$, which reveals that the ferromagnetic transition temperature $T_C$ of this particular film is approximately 60 K. The trace also shows that there is no indication of inclusion of NiAs-structure MnAs ($T_C \approx 310$ K), as only negligible $M_r$ is present above $T_C = 60$ K. This, however, is not always the case; field independent $M_r$ that persists up to around 300 K is occasionally observed, especially in samples grown at higher $T_S$, having high $x$, or annealed at temperatures well above the growth temperature. The highest $T_C$ so far obtained in the ferromagnetic III-V semiconductor family is 110 K with $x = 0.053$ [10].

Stripe shaped domain structure has been observed using scanning Hall microprobe in a (Ga,Mn)As sample with a perpendicular easy axis introduced by tensile strain ($x = 0.043$). The shape and its temperature dependence are in good agreement with the standard theory of domain structure [45]. Scanning SQUID microscopy revealed random domain sizes from below 30 $\mu$m to 100 $\mu$m in (Ga,Mn)As samples ($x = 0.047$) having in-plane axis [46].

## 1.4   Transport Properties

### 1.4.1   The Hall Effect

The Hall effect of magnetic materials is a sum of the ordinary Hall effect and the anomalous Hall [47]. The sheet Hall resistivity $R_{\text{Hall}}$ is expressed as,

$$R_{\text{Hall}} = \frac{R_0}{d} B + \frac{R_M}{d} M \tag{1.1}$$

where $R_0$ is the ordinary Hall coefficient, $B$ the magnetic field, $R_M$ the anomalous Hall coefficient, $M$ the magnetization perpendicular to the film, and $d$ the thickness of the layer. $R_M$ is proportional to $R_{\text{sheet}}^\gamma$ with temperature independent proportionality constant $c$. Normally $\gamma$ is either 1 or 2 depending on the origin of the scattering; the skew-scattering results in $\gamma = 1$, whereas the side-jump $\gamma = 2$ [47]. The solid line in the main panel of Fig. 1.1 that traces closely $M$ measured with $B \perp$ plane is the trace of the Hall voltage of the same sample. As in this case, very often, especially in (Ga,Mn)As, the anomalous Hall effect dominates the Hall effect and $R_{\text{Hall}} \approx cR_{\text{sheet}}M$ (or $c'R_{\text{sheet}}^2 M$ depending on the dominant scattering mechanism), and $R_{\text{Hall}}/R_{\text{sheet}}$ is a often good measure of $M$ perpendicular to the plane of the film.

Figure 1.2a shows $R_{\text{Hall}}$ versus $B$ curves measured at different temperatures of a 200 nm thick $\text{Ga}_{0.947}\text{Mn}_{0.053}\text{As}$ grown on an $\text{Al}_{0.9}\text{Ga}_{0.1}\text{As}$ buffer layer. The inset shows the temperature dependence of the sheet resistivity $R_{\text{sheet}}$ of the same sample. Since $R_{\text{Hall}}/R_{\text{sheet}} \propto M$, temperature dependence of spontaneous magnetization ($[R_{\text{Hall}}/R_{\text{sheet}}]_s$) can be obtained from such

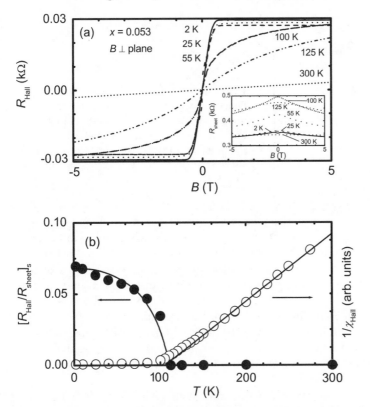

**Fig. 1.2.** (a) Temperature dependence of the sheet Hall resistivity $R_{\mathrm{Hall}}$ versus field curve taken on a 200 nm thick (Ga,Mn)As ($x = 0.053$). Inset shows the corresponding sheet resistivity $R_{\mathrm{sheet}}$ of the sample. Note that $R_{\mathrm{Hall}}/R_{\mathrm{sheet}}$ is proportional to $M$. No hysteresis is present because the field is applied perpendicular to the plane along the hard axis. (b) Temperature dependence of $[R_{\mathrm{Hall}}/R_{\mathrm{sheet}}]_s$ proportional to spontaneous magnetization using Arrott plots (closed symbols) and inverse susceptibility $1/\chi$, both obtained from the transport data shown in (a). The transition temperature is 110 K.

transport data using Arrott plots, which minimize the effect of magnetic anisotropy and domain rotation [48]. Note that even though it is difficult to accurately determine whether $R_{\mathrm{Hall}} \approx c' R_{\mathrm{sheet}}^2 M$ or $R_{\mathrm{Hall}} \approx c R_{\mathrm{sheet}} M$, these two cases result in only about 20% difference, due to moderate temperature dependence of $R_{\mathrm{sheet}}$ of the particular sample studied here. The result is shown in Fig. 1.2b (solid circles). $T_{\mathrm{C}}$ obtained this way (110 K) is in good agreement with $T_{\mathrm{C}}$ form the direct magnetization measurements. Although the temperature dependence of spontaneous magnetization in Fig. 1.2b can be fitted rather well by a mean-field theory assuming a Brillouin function form as is done in the solid line [10], it can be quite different from what is determined from the direct magnetization measurements. One possible explanation for

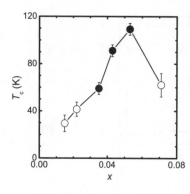

**Fig. 1.3.** Manganese composition dependence of transition temperature $T_C$ determined from transport. Closed symbol shows that the sample is on the metal side of the metal-insulator transition and open symbol show that it is on the insulator side (see Fig. 1.4).

the difference is the way the magnetization is measured; a magnetometer measures all magnetic moment equally, whereas the anomalous Hall effect is weighed by the conductivity. The non-uniform conductivity implied by this explanation could be due to a result of localization, as the sample is at near the metal-insulator boundary. By repeating the transport measurements for different samples, a $T_C$ vs. $x$ curve is generated as shown in Fig. 1.3.

Parenthetically, no clear indication of the presence of MnAs clusters has been observed by transport, even in the cases where the direct magnetization measurements detect a small MnAs contribution. This may be due to the Schottky barrier formation around MnAs clusters, which prevents the clusters to interact with carriers.

Because of the dominance of the anomalous Hall effect, we expect $\chi \propto \{d(R_{\text{Hall}}/R_{\text{sheet}})/dB\}_{B=0}$ and a Curie-Weiss plot may be generated using the results of magnetotransport. The straight line obtained in the $\{d(R_{\text{Hall}}/R_{\text{sheet}})/dB\}^{-1}$ vs. $T$ plot shown in Fig. 1.2b clearly indicates that the Curie-Weiss law describes the temperature dependence of the susceptibility of (Ga,Mn)As. It also shows that carrier concentration of these samples cannot be obtained from the Hall measurements at room temperature, as the contribution of the ordinary Hall effect is still negligible near room temperature. The intercept of the extrapolation from the high temperature part shows that $\theta$ is 107 K, in good agreement with the direct measurement.

### 1.4.2   Temperature and Magnetic Field Dependence of Resistivity

Due most probably to the disorder introduced by the high impurity concentration, the ferromagnetic III-V semiconductors reside close to the metal-insulator transition (MIT) boundary. Figure 1.4 shows the temperature dependence of resistivity $\rho$ of six 200 nm (Ga,Mn)As samples with different $x$. Low and high Mn concentration samples ($x < 0.03$, $x > 0.06$) are on the insulator side of MIT, whereas the samples with intermediate Mn concentration ($0.06 \geq x \geq 0.03$) are on the metal side. All the samples exhibit negative magnetoresistance at low temperatures.

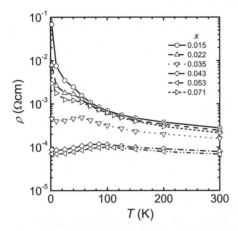

**Fig. 1.4.** Temperature dependence of resistivity $\rho$ measured on six different samples with different Mn concentrations. Low and high Mn concentration samples are on the insulator side of the metal-insulator transition, whereas the intermediate ones are on the metal side. Note also the presence of local maximum around the transition temperature on each curve, which is attributed to spin-disorder scattering.

Figure 1.5 shows temperature dependence of $\rho$ of one of the metallic (Ga,Mn)As samples ($x = 0.053$). $\rho$ shows a peak around $T_\mathrm{C}$, where negative magnetoresistance also peaks; $\rho$ decreases by 20% at most at $B = 7$ T (at 100 K). The moderate magnitudes of negative magnetoresistance and its temperature dependence suggest that the spin disorder scattering by thermodynamic fluctuations of the magnetic spins is involved in the resistance peaks in the metallic samples; *i.e.* the peak around $T_\mathrm{C}$ can be interpreted as a result of critical scattering by packets of the magnetic spins with a ferromagnetic short-range order characterized by the correlation length comparable to the wavelength of the carriers at the Fermi level. The negative magnetoresistance occurs because the spin-disorder scattering is reduced with magnetic-field induced spin-alignment. The corresponding contribution to the resistivity is given by the following formula, which takes into account the presence of correlation between the neighboring spins, $\langle \boldsymbol{S}_i \boldsymbol{S}_j \rangle \neq \langle \boldsymbol{S}_i^2 \rangle \delta_{ij}$ [4],

$$\rho = 2\pi^2 \frac{k_\mathrm{F}}{pe^2} \frac{m^2 \beta^2}{h^3} \frac{k_\mathrm{B}T}{g^2 \mu_\mathrm{B}^2} (2\chi_\perp(T,B) + \chi_\parallel(T,B)). \tag{1.2}$$

Here $k_\mathrm{F}$ is the Fermi wave vector determined from $p$ assuming a spherical Fermi surface, $m$ is the hole effective mass taken as $0.5m_0$ ($m_0$: the mass of free electron), $\beta$ the exchange integral between holes and magnetic spins, and $h$ the Planck constant. $\chi_\perp$ and $\chi_\parallel$ are the transverse and longitudinal magnetic susceptibilities, respectively, which can be determined from the magnetotransport data assuming no anisotropy; $\chi_\perp = \partial M/\partial B$ and $\chi_\parallel = M/B$.

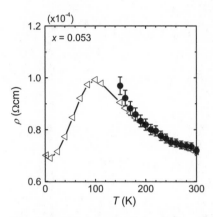

**Fig. 1.5.** Temperature dependence of resistivity $\rho$ of a sample with Mn concentration of 0.053 (*open triangles*). Closed symbols are a fit from which the exchange is determined (see text).

Equation (1.2) reproduces the high temperature part of the temperature dependence of $\rho$ as shown in Fig. 1.5 [49]; the measured data can be fitted by $\chi T$, characteristic temperature dependence of critical scattering. Deviation of $\chi T$ fit from experiment as $T_C$ approaches occurs because $q = 0$ value of $\chi$ ($q$ the wave vector) is used in the place of $q$ dependent $\chi$ responsible for the scattering. $\chi(q)$ is normally a decreasing function of $q$ in ferromagnetic materials and thus the fit overestimates the scattering near $T_C$, where $\chi(q = 0)$ diverges.

From the fit of (1.2) to $T$ and $B$ dependence of $\rho$, $|N_0\beta|$ is obtained, where $N_0$ is the density of cation sites. Both $T$ and $B$ fits yield $|N_0\beta|$ of $1.5\pm0.2$ eV. Previous estimate of $|N_0\beta| = 3.3$ eV from the expression for the high temperature limit of (1.2) gives the upper limit of the interaction [10], because the correlation of neighboring spins is ignored. The exchange constant obtained from transport compares favorably with that determined by photoemission experiments on (Ga,Mn)As, 1.0–1.2 eV (negative) [50]. However, large $p$-$d$ exchange (2.5 eV, positive) was deduced from the magneto-optical study of GaAs doped with Mn [51]. Note that typical $p$-$d$ exchange ($N_0\beta$) in II-VI DMS's is about $-1$ eV.

There is a number of factors one needs to take into account to properly understand $|N_0\beta|$ obtained from transport, however. One is the effect of localization induced by disorder, which enhances the ferromagnetic interaction through hole-hole interaction. Another possible source of ambiguity is the complex valence band structure, which may not allow simple usage of constant effective mass in (1.2).

Very large negative magnetoresistance, with a large anisotropy, has been observed in "reentrant" insulating samples at high $x$ [52,53]. Formation of bound magnetic polarons as invoked in an earlier work on (In,Mn)As [15] and increase of Fermi energy in a spin split band, which reduces the localization length by reducing the energy difference between the mobility edge and the Fermi level are suggested for the mechanism.

Because of the presence of the anomalous Hall effect and the negative magnetoresistance, one needs to go to low temperature high field to determine the conduction type and the carrier concentration of (Ga,Mn)As. This has been done for a (Ga,Mn)As sample with $x = 0.053$ as shown in Figs. 1.6a,b.

Note that although $M$ saturates at relatively low $B$, the negative magnetoresistance needs to be saturated as it determines the anomalous Hall coefficient. Measurement of $R_{Hall}$ at 50 mK in the field range of 22–27 T shown in Fig. 1.6 revealed that the conduction type is $p$-type, consistent with the acceptor character of Mn, and the hole concentration being $p = 3.5 \times 10^{20}$ cm$^{-3}$, about 30% of the nominal concentration of Mn [49]. This method is only applicable for the metallic samples as the insulating nature of the samples in the insulator side makes it difficult to measure the Hall effect at low temperatures.

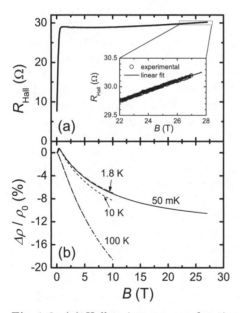

**Fig. 1.6.** (a) Hall resistance as a function of magnetic field. The inset shows a magnified voew of the high-field part, where the Hall resistance is proportional to the field. (b) Negative magnetoresistance of the same 200 nm sample with $x = 0.053$, which gradually saturates at higher fields. $\rho_0$ is the resistivity at zero field.

## 1.5   Carrier-Induced Ferromagnetism

Since the magnetic interaction between Mn has been shown to be antiferro-magnetic in $n$-type paramagnetic (In,Mn)As [54] and in semi-insulating fully carrier-compensated (Ga,Mn)As using Sn as a donor [55,56], the ferromagnetic interaction in magnetic III-V's is most likely carrier induced. Here, we present a mean field model of the ferromagnetism in magnetic III-V's, which can account for a number of experimental results reported to date, including the magnitude of $T_C$, and appears to present a good starting point for the understanding of the ferromagnetism [11,57,58].

The mean-field model starts by considering two interacting spin subsystem; the spins of the delocalized carriers and the localized spins of the magnetic elements. These two subsystems interact with one another via exchange interaction, which is parameterized as $N_0\beta$ for the case of valence band as the ferromagnetic III-V's are $p$-type. $N_0\beta$ is often called the $p$-$d$ exchange because the valence band top is made of $p$-orbitals, whereas $N_0\alpha$ is used for the $s$-$d$ exchange between conduction band electrons ($s$-orbitals) and localized magnetic spins. This parameterization scheme has been very successful in explaining numerous optical and magnetic experiments on II-VI based paramagnetic semiconductors [3], and is believed to be valid for III-V magnetic semiconductors, especially those with Mn, as shown by a theoretical appraisal [59] of the magneto-spectroscopic experiments on Mn acceptor in GaAs [60].

The ferromagnetism is described in the following way. A finite $M$ produces a band splitting in the valence band and reduces the carrier energy. The same $M$ increases the free energy of the localized spin subsystem, but this free energy penalty reduces as temperature is lowered. At the ferromagnetic transition temperature, the two energies balances and further reduction of temperature gives rise to spontaneous spin splitting and polarization of spins, i.e. ferromagnetism. It is useful to look into this in a parameterized way. The total free energy of the system $\Delta F_{\text{total}}$ is a sum of the free energies of the localized spin system $\Delta F_{\text{Mn}}$ (here we use Mn as a symbol for the localized spin system), and the carrier spin system $\Delta F_c$. If we consider a small but finite $M$ and the mean field $H$ acting on the system, these two free energies can be expressed in terms of the susceptibility of the localized spin system $\chi_{\text{Mn}}$ and the carrier system $\chi_c$, respectively. Here we assume that the entropy term of the carrier free energy is small, which is true as long as the Fermi energy is much greater than $kT$. The expression for the free energy becomes,

$$
\begin{aligned}
\Delta F_{\text{total}} &= \Delta F_{\text{Mn}} + \Delta F_c \\
&\approx \frac{M^2}{2\chi_{\text{Mn}}} - \frac{\chi_c}{2} H^2 \\
&= \left( \frac{1}{2\chi_{\text{Mn}}} - \frac{\chi_c}{2\chi_{\text{Mn}}^2} \right) M^2.
\end{aligned}
\tag{1.3}
$$

The minus sign of the carrier energy term indicates that the carrier system always gains energy. The temperature at which inside the parenthesis becomes zero is the ferromagnetic transition temperature, $T_C$. $\chi_c$ can be expressed using the effective $g$-factor $g^*$ of the carrier system as, $\chi_c = (g^*\mu_B)^2 \rho_s(E_F)/4$, where $\rho_s(E_F)$ is the spin-dependent density of states at the Fermi energy. The spin-splitting of the valence band $\Delta E$ (between $J_z = 3/2$ and $-3/2$) in terms of $g^*$ is $\Delta E = g^*\mu_B H$. The same $\Delta E$ expressed in terms of $N_0\beta$ is $\Delta E = xN_0\beta S$. By equating the two, and using that $M = xN_0g\mu_B S$, we obtain expression for $g^*$ as $g^* = \chi_{Mn}\beta/(g\mu_B^2)$. Here, $g$ is the $g$-factor of the localized spins ($g = 2$ for Mn) and $N_0$ the density of cation sites. Using the standard Curie form for $\chi_{Mn}$, we obtain $T_C$ as,

$$T_C = \frac{xN_0S(S+1)A_F\rho_s(E_F)\beta^2}{12k_B},  \tag{1.4}$$

where $k_B$ is the Boltzmann constant. We have included the enhancement due to electron-electron interaction as $A_F$ ($= 1.2$ calculated by the local spin density approximation [57]).

As $x$ (magnetic impurity molefraction) and $S$ ($= 5/2$ for Mn) are known, we need to determine $\beta$ and $\rho_s(E_F)$ to compute $T_C$ of such materials like (Ga,Mn)As and (In,Mn)As. Because the magneto-luminescence spectroscopy commonly used to determine $\beta$ in II-VI magnetic semiconductors is not available for III-V magnetic semiconductors as no photoluminescence is observed, we here adopt a value for (Ga,Mn)As deduced from photoemission spectroscopy combined with a modeling based on configuration-interaction, $N_0\beta = -1.2$ eV [50]. For alloys other than (Ga,Mn)As, we use a chemical trend $N_0\beta = N_0\beta(\text{GaMnAs}) \times (a/a_{\text{GaMnAs}})^{-3}$ proposed in ref [58]. The spin-dependent density of states $\rho_s(E_F)$ needs to be calculated from a realistic band structure model including the effect of exchange. The calculation performed here uses the $6 \times 6$ $k \cdot p$ model to obtain the valence band structure. The spin density of states can be obtained noting that the carrier energy reduction $\Delta E$ is proportional to $M^2$ and is expressed as,

$$\Delta E = -\frac{A_F\rho_s(E_F)\beta^2}{2(2g\mu_B)^2}M^2,$$

from which we obtain $\rho_s(E_F)$.

$T_C$'s calculated for (Ga,Mn)As and (In,Mn)As as a function of hole concentration $p$ are shown in Figs. 1.7a and 1.7b, respectively. All the Luttinger parameters were assumed to be identical to the host binary compound. The Mn concentration is chosen to make it possible to compare the calculated results with the related experiments. Figure 1.7a shows that $T_C$ is 130 K at the hole concentration $p = 3.5 \times 10^{20}$ cm$^{-3}$, which is in good agreement with the experimental value of 110 K (circle) [10]. As discussed later, the results for (In,Mn)As shown in Fig. 1.7b agree well with the experimental observations. The same mean field picture can also reproduce the $T_C$'s of hole-induced ferromagnetism observed recently in II-VI DMS (Zn,Mn)Te [61]. These agreements between theory and experiment suggest that the chemical

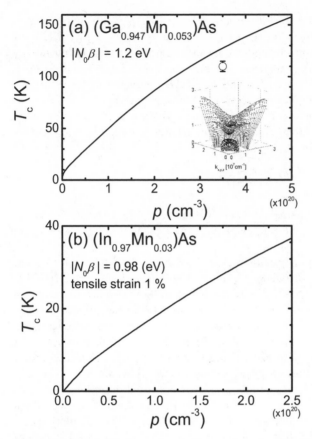

**Fig. 1.7.** Calculated $T_C$ versus hole concentration for (**a**) $Ga_{0.947}Mn_{0.053}As$ with p-d exchange ($|N_0\beta| = 1.2$ eV) without strain and (**b**) $(In_{0.97}Mn_{0.03})As$ ($|N_0\beta| = 0.98$ eV) with 1% tesile strain. Electron-electron interaction enhancement factor of 1.2 is assumed. The open symbol in (**a**) is the experimental data, which is in good agreement with theory. *Inset* shows the valence band structure with exchange.

trend is correctly accounted for in the present model. Using the reported $k \cdot p$ parameters together with the mean-field model, it has been suggested that compounds made of light elements (like ZnO, GaN, and diamond) would show room temperature ferromagnetism if one can dope these compounds to $x_{Mn} = 0.05$ and $p = 3.5 \times 10^{20}$ cm$^{-3}$ [58]. The factors contribute to the trend are a smaller spin-orbit interaction, which allows spins to align easier, a larger $N_0\beta$ as the cahemical shift suggests that it is inversely proportional to $a^3$, and to some extent a heavier mass due to the opening of the gap.

The present mean field model can explain a number of experimental resulats other than $T_C$ [58]. First, it can explain the absence of ferromagnetism in $n$-type materials; in $n$-type materials, the small density of states together with the small $N_0\alpha$ (smaller than $N_0\beta$ by about a factor of 5 ($N_0\alpha \approx 0.2$ eV), which is established in II-VI DMS's) makes it difficult for the carrier-induced ferromagnetic interaction to overcome the direct antiferromagnetic coupling among Mn.

Second, inclusion of strain terms in the Luttinger-Kohn Hamiltonian predicts the correct direction of the magnetic anisotropy [11,58,62]. Although the theory predicts that the easy axis direction and the anisotropy energy is a function of carrier concentration and strain, for the experimentally relevant hole concentration and strain, the experimental observations are well accounted for by the theory. Experiments show that the magnetic easy-axis of (Ga,Mn)As lies in the plane of the film [17] when grown on (001) GaAs substrates, whereas insertion of a (In,Ga)As buffer layer having larger lattice constant than the subsequent (Ga,Mn)As layer (introduces tensile strain in the (Ga,Mn)As layer) results in the easy axis perpendicular to the plane of the film [42]. For (In,Mn)As films grown on a (Al,Ga)Sb buffer layer, the strain is tensile and the easy axis is perpendicular to the plane [43].

Third, the model can reproduce an anomalous temperature dependence observed in a narrow spectral region of magnetic circular dichroism of (Ga, Mn)As [58,63]. This is understood as a result of carrier redistribution among the spin-split heavy and light hole states. Ferromagnetism in insulating samples can be understood within the same framework. The length-scale involved in the transport is considerably longer than the one in the magnetism; in insulating samples, the localization length is smaller than the sample size but still be significantly larger than the length scale of magnetic interactions. This length scale difference is believed to make the mean field approach a reasonable starting point for the insulating samples.

The model presented here has its origin in a model proposed by Zener (see ref. [11]). The Zener model was later abandoned, because in metals the carrier concentration is usually much higher than the concentration of doped magnetic elements, and one needs to take into account the Friedel oscillation to describe the exchange interaction (the Ruderman-Kittel-Kasuya-Yosida (RKKY) interaction). In magnetic III-V's, the carrier concentration is usually smaller than the magnetic impurity density. In this case, the mean distance between the carriers is greater than that between the spins, and the Friedel oscillation of the carrier spin-polarization around the localized spins tends to average out [64].

The limitations of the present mean-field model are (1) it breaks down at the high carrier concentration regime, (2) it does not take into account the low energy excitation, and (3) the effect of localization is neglected. The dynamic aspect of the mean-field theory and its relation to the RKKY regime has been discussed in ref. [65]. The effect of positional disorder is discussed in the recent studies [66,67].

Other approaches to the ferromagnetism in magnetic III-V's have also been reported [68,69]. Bloembergen-Rowland type indirect interaction has been put forward to explain ferromagnetism in insulating samples [69], where a gap between an impurity band and the valence band of 0.1 eV, which corresponds to the ionization energy of isolated Mn in GaAs, is assumed. Infrared absorption spectroscopy results on (Ga,Mn)As [70] and on (In,Mn)As [71] show, however, no clear gap in the spectrum region of interest.

First principle calculations have also shown that the ground state of (Ga,Mn)As is ferromagnetic [72–77]. An approach among them by Akai emphasizes the role of $d$-band formed by the transition metal impurities. Akai stressed the partial $d$-character of the holes and invoked a double-exchange picture for the observed ferromagnetism [73]. It remains to be seen whether this approach is capable of explaining the reported experimental findings. It is worth pointing out that the partial $d$-character of the holes is the source of enhancement of $|N_0\beta|$ over $N_0\alpha$. Material systems other than Mn alloyed GaAs and InAs have also been investigated by the first principle calculations. Theoretical appraisal of ZnO with V, Cr, Mn, Fe, Co, and Ni doping have suggested that they lead to ferromagnetism except for the case of Mn [78,79]. This is interpreted as once the $3d^5$ configuration is destabilized by adding or subtracting electrons, a double exchange channel opens leading to ferromagnetism. Essentially the same argument holds for the results of GaN with V, Cr, Mn, Fe, Co, or Ni doping [80]. The end compounds, from zincblende TiAs to NiAs, have also been calculated [40]. It is shown that among compounds studied ferromagnetic phase is stable for VAs, CrAs and MnAs, with the highest energy gain obtained in the CrAs compound, which was realized experimentally [38].

## 1.6 Basic Properties of Ferromagnetic III-V Semiconductor Heterostructures

Because of the close lattice matching between the ferromagnetic III-V alloys and its host III-V compounds, the ferromagnetic III-V's allow us to grow ferromagnetic/nonmagnetic semiconductor heterostructures. These heterostructures are used to investigate and to explore the spin-dependent phenomena in semiconductors. X-ray diffraction study has shown that (Ga,Mn)As/GaAs superlattices (23.5 nm period) grown by low-temperature MBE are of high crystalline quality [81]. Thickness dependence of magnetization curves of (Ga,Mn)As films measured by the Hall effect [19] has indicated that the films are ferromagnetic and no significant change in $T_C$ is observe from 1 $\mu$m film thickness down to 10 nm, apart from the increase in anisotropy. Below 5 nm, however, we have observed that the (Ga,Mn)As films become often insulating with considerably lower $T_C$ or paramagnetic down to the lowest temperature of measurement (usually 4.2 K). This disappearance of ferromagnetism in thin (Ga,Mn)As was also observed in thin (Ga,Mn)As quantum wells (QW's) of less than 5 nm [82]. In contrast to the QW and thin film results, ferromagnetism was observed in GaAs layers with submonolayer planes of MnAs [24]. Further study is apparently necessary to understand the thickness dependence of magnetic properties in these materials.

The alignment of the bands at the interface is one of the most fundamental heterojunction properties. However, the band alignment (or band offset) has not been established for the Mn-doped III-V's. The difficulty in determining the band offset arises from a very high doping level in Mn-doped III-V's and relatively small Mn concentration that makes the offset small. Close examination of current-voltage ($I$-$V$) characteristics of $p$-$i$-$n$ and $p$-$p$ diodes made of (Ga,Mn)As and GaAs shows that a thermoionic emission current across a barrier flows at temperature higher than the transition temperature [83,84]. Analysis of the temperature dependence of the $I$-$V$ characteristics has revealed that a barrier exists between (Ga,Mn)As and GaAs; holes in (Ga,Mn)As have to overcome a barrier of about 100 meV to reach the GaAs valence band [9,85]. To obtain the band offset between (Ga,Mn)As and GaAs from this, the (Ga,Mn)As Fermi energy (usually of the order of 100 meV) has to be added resulting in total of 200 meV offset, which appears to be rather insensitive to $x$. Band gap renormalization can only account for 20 meV at most and thus the rest should be a result of alloying Mn with GaAs. The formation Mn impurity band together with its merging with the host valence band may have to be considered to interpret the results.

## 1.7   Spin-Dependent Scattering and Tunnel Magnetoresistance in Trilayer Structures

Ferromagnet/nonmagnet/ferromagnet trilayer structure constitutes the most fundamental building block of modern magnetic sensors and storage elements and is useful for evaluation of various magnetotransport processes. The spin-dependent scattering in the structures, where the nonmagnetic part is conducting, is the process that gives rise to the giant magnetoresistance (GMR) effect. The tunneling magnetoresistance (TMR) effect in the structures, where the nonmagnetic layer is a thin insulator, is used for nonvolatile magnetic random access memories. In order to investigate the presence and the nature of such spin-dependent transport processes in trilayer structures made of semiconductors alone, (Ga,Mn)As/(Al,Ga)As/(Ga,Mn)As structures have been prepared and studied [86,87].

Figure 1.8 shows the magnetic field dependence of $R_{\mathrm{Hall}}$ and $R_{\mathrm{sheet}}$ of a trilayer structure at 30 K, where current is flown in-plane. The structure consists of 30 nm (Ga$_{0.95}$Mn$_{0.05}$)As/ 2.8 nm (Al$_{0.14}$Ga$_{0.86}$)As/ 30 nm (Ga$_{0.97}$Mn$_{0.03}$)As and is grown on a 50 nm (Al$_{0.30}$Ga$_{0.70}$)As/lattice relaxed 1 $\mu$m (In$_{0.15}$Ga$_{0.85}$)As buffer layer structure. The (In$_{0.15}$Ga$_{0.85}$)As buffer layer introduces tensile strain and makes the magnetic easy axis perpendicular to the plane; with this easy axis direction, $R_{\mathrm{Hall}}$ can be used to monitor $M$. The plateau structure seen in $R_{\mathrm{Hall}}$ in Fig. 1.8 indicates that $M$ of the two (Ga,Mn)As are anti-parallel in this magnetic field range (note that $R_{\mathrm{Hall}}$

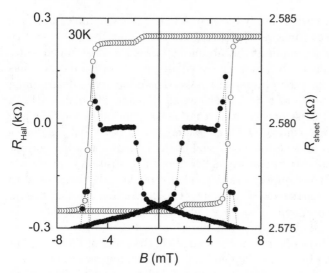

**Fig. 1.8.** Hall resistance, $R_{Hall}$, and sheet resistance, $R_{sheet}$, versus $B$ at 30 K for a $(Ga_{0.95}Mn_{0.05})As/(Al_{0.14}Ga_{0.86})As/(Ga_{0.97}Mn_{0.03})As$ trilayer structure. The increase in $R_{sheet}$ shown by *closed circles* is clearly observed in the plateau region of 2–6 mT in $R_{Hall}$ (open circles, proportional to magnetization), where the magnetizations of the two (Ga,Mn)As layers are anti-parallel. This increase is attributed to the presence of spin-dependent scattering.

is a weighted average of $M$ of the two (Ga,Mn)As layers). A clear increase of sheet resistance $R_{sheet}$ was observed at this plateau region of 2–6 mT, which demonstrates the presence of spin-dependent scattering in a spin-polarized hole system in a trilayer structure made of semiconductors alone.

Vertical transport measurements of trilayer structures revealed the presence of TMR. Figure 1.9a shows the $B$ dependence of $M$ and (b) that of magnetoresistance, both at 20K, of a trilayer structure with a 3 nm AlAs barrier layer [87]. The trilayer structure is grown on a lattice relaxed (In,Ga)As buffer layer structure for perpendicular easy axis to restrict $M$ in the direction perpendicular to the plane to avoid ambiguities arising from in-plane magnetostatic structures. For TMR measurements, one electrode was formed on the top (Ga,Mn)As and the other on the bottom (Ga,Mn)As. The difference in coercive fields produces the plateau structure observed in Fig. 1.9a, where $M$ of the two ferromagnetic layers are antiparallel. The resistance increase is observed between 8 mT and 16 mT, in the field region of antiparallel configuration of $M$. The MR ratio is about 5.5% at 20 K. This is TMR, because the AlAs valence band barrier is high (about 550 meV from the GaAs valence band), all the holes tunnel across the AlAs layer. Recently TMR ratio in over 70% was reported in a (Ga,Mn)As / AlAs / (Ga,Mn)As structure with the AlAs thickness of 5 monolayers (1.5 nm) [88], which indicates that spin polarization of (Ga,Mn)As is quite high.

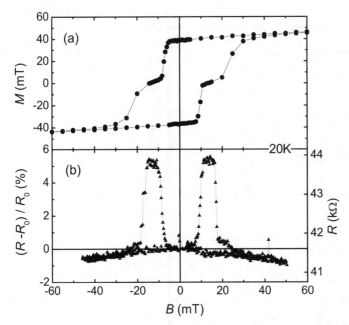

**Fig. 1.9.** (a) Magnetization and (b) tunneling magnetoresistance curve of a 30 nm $(Ga_{0.95}Mn_{0.05})As/$ 3 nm AlAs/ 30 nm $(Ga_{0.97}Mn_{0.03})As$ tunnel junction at 20 K. A clear increase in tunnel resistance is observed whenever the magnetizations of the two (Ga,Mn)As layers are anti-parallel.

## 1.8   Ferromagnetic Emitter Resonant Tunneling Diodes

Spontaneous magnetization in ferromagnetic semiconductors gives rise to spin splitting in the conduction and valence bands due to the exchange interaction. This spin splitting of bands should manifest itself in the current-voltage $(I$-$V)$ characteristics of resonant-tunneling diodes (RTD's) having a ferromagnetic emitter. Nonmagnetic double barrier AlAs/GaAs/AlAs RTD's with a (Ga,Mn)As emitter has reported to show spontaneous splittings of resonant peaks below the ferromagnetic transition temperature of (Ga,Mn)As in the absence of magnetic fields [89,90]. Figure 1.10a shows the temperature dependence of $dI/dV$ versus $V$ curves with no magnetic fields of an RTD having a (Ga,Mn)As emitter. Spontaneous splitting of resonant peaks labeled HH2 and LH1 is observed below the ferromagnetic transition temperature of 60 K. The structure studied here consists of (from the surface side) 150 nm $(Ga_{0.97}Mn_{0.03})As/$ 15 nm undoped GaAs spacer / 5 nm undoped AlAs barrier / 5 nm undoped GaAs quantum well / 5 nm undoped AlAs barrier / 5 nm undoped GaAs spacer / 150 nm Be doped GaAs $(p = 5 \times 10^{17} \, cm^{-3})$ / 150 nm Be doped GaAs $(p = 5 \times 10^{18} \, cm^{-3})$ / $p^+$ GaAs substrates. Each label in Fig. 1.10 indicates the resonant state in the GaAs well, where total of 6 states are present in the 5 nm GaAs well.

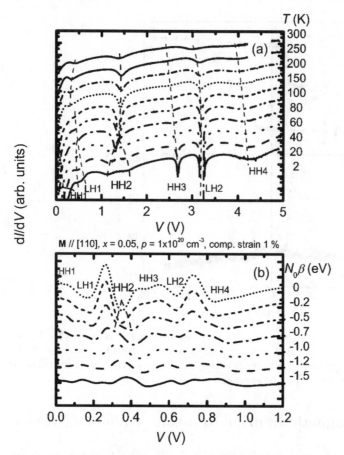

**Fig. 1.10.** (a) Temperature dependence of $dI/dV$ versus $V$ curves of an RTD having a (Ga,Mn)As emitter. No magnetic field is applied. Spontaneous splitting of resonant peaks labeled HH2 and LH1 is observed below the ferromagnetic transition temperature of (Ga,Mn)As, 60 K. (b) Calculated resonant tunneling spectra as a function of $N_0\beta$, which is used to represent the spontaneous magnetization. The HH2 resonant peak shows a splitting as $N_0\beta$ (magnetization) increases, whereas other peaks (except LH1) show no pronounced splitting. At high $N_0\beta$, all the features smear out.

We have calculated the corresponding $dI/dV$-$V$ characteristics of an RTD with a ferromagnetic emitter, taking into account the strong $k$-dependent mixing with the presence of exchange [91]. The calculation is based on previous theoretical approaches for $p$-type RTD's [92,93] and the $p$-$d$ exchange interaction is included in the $6 \times 6$ Luttinger-Kohn Hamiltonian. The resonant states in the GaAs well is calculated using the $4 \times 4$ Hamiltonian. The current density for a given hole concentration and voltage was then obtained by the following expression.

$$J_z = \frac{e}{(2\pi)^3 \hbar} \sum_{i,j,l=1}^{4} \iiint_{E \leq E_F} d^3k \cdot \left(\frac{\partial E}{\partial k_z}\right) \cdot T^* T_{ij} \cdot [f(E) - f(E - eV)] \cdot c_{il}$$

Here, $T^* T_{ij}$ is the transmission coefficient from the heavy or light hole states in the emitter to the heavy and light hole states in the collector, $f(E)$ the Fermi-Dirac distribution function, $c_{il}$ is the probability amplitude of the $j$th hole state in the $l$th band. Figure 1.10b shows the calculated $dI/dV$-$V$ characteristics with $|N_0\beta|$ as a parameter, which represents magnetization here. A 1 nm AlAs barrier is assumed to avoid numerical instabilities. A 5% compressive strain is included in the (Ga,Mn)As ($x = 0.05$) layer with $p = 1 \times 10^{20}$ cm$^{-3}$. The magnetization is assumed to be in the in-plane [110] direction. As can be seen in Figure 1.10b, the HH2 resonant peak shows a clear splitting as $|N_0\beta|$ (magnetization) increases, whereas other peaks (except probably LH1) do not show such pronounced splitting, both of which are in accordance with the experiment. The features are smeared out at high $|N_0\beta|$ as the spin polarization of the carrier system increases. The HH2 peak splitting is pronounced because the dispersion of the HH2 resonant state is similar to that of the dominant emitter valence band state. Other resonant states either show very different dispersion and/or lie high in energy where the transmission coefficient peaks become broad. Note that no clear cut-off of currents is observed in experiments nor in theory because of the rather high hole concentration. The present $k \cdot p$ calculation indicates that it is necessary to go through rather elaborate calculation to understand the origin of the peaks and their splitting when $k$-dependent spin splittings are involved.

In spite of the successful observation of spin splitting in nonmagnetic RTD's with a ferromagnetic emitter, efforts to make the quantum well ferromagnetic have met considerable difficulty. This is due to difficulty in the growth of high quality RTD structure at reduced temperature by MBE, as one needs to keep the growth temperature low once the ferromagnetic layer is grown to prevent precipitation of second phases. Calculation similar to the one described above has been done for the case of a ferromagnetic well [94].

## 1.9   Spin-Injection in Ferromagnetic Semiconductor Heterostructures

Electrical spin-injection (spin-polarized current flow) is one of the building blocks for transport and manipulation of spins in semiconductors; carrier spins themselves may be used to transport spin information, or a spin polarization of carrier system may be used to initialize (polarize) another spin subsystem. The spontaneous magnetization in ferromagnetic (Ga,Mn)As introduces imbalance of spin population in the carrier system. Thus it can be used as a spin polarized carrier source (no magnetic field needed) for electrical spin injection in GaAs based structures. (Ga,Mn)As also offers an epitaxial interface with the AlAs-GaAs system minimizing the spin depolarization that may occur in less ideal interfaces. It also has smaller "conductivity mis-match" that prevents diffusive spin injection from ferromagnetic metals into semiconductors, which is discussed in detail in Chaps. 2 and 3.

An electrical spin injection from (Ga,Mn)As into GaAs has been demon-strated by the use of a light emitting diode (LED) structure [95]. Partially spin polarized holes are injected from a $p$-type (Ga,Mn)As layer through an intrinsic GaAs into an (In,Ga)As quantum well in an $p$-$i$-$n$ structure, where they recombine with spin unpolarized electrons from nonmagnetic $n$-type GaAs. Spin polarization of the recombining holes and hence spin injection is demonstrated by the observation of electroluminescence (EL) polarization. Figure 1.11b shows the relative polarization change $\Delta P$ as a function of magnetic field at various temperatures. Two polarization states are clearly observed in the absence of an external magnetic field at low temperatures. A series of experiments was done to exclude the effect of fringing fields and magnetic circular dichroism from the adjacent (Ga,Mn)As. Because of the in-plane easy axis of (Ga,Mn)As grown on GaAs, EL was collected from the direction parallel to the QW plane. Since the selection rule at the very bottom of the band in a QW does not allow circular polarization in this direction, the effect of band filling (i.e. nonzero $k$) may need to be taken into account to

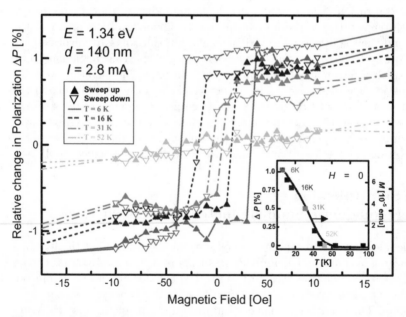

**Fig. 1.11.** Relative change in polarization $\Delta P$ as a function of magnetic field at four different temperatures. Clear hysteresis is observed at low temperatures, where (Ga,Mn)As is ferromagnetic. Inset compares the remanent magnetization with the remanent $\Delta P$, where virtually identical temperature dependence is observed. After [95].

fully understand the results. Recent experiments on "vertical" spin injection LED's, where the structures are grown on a lattice relaxed (In,Ga)As buffer layer to orient the easy axis perpendicular to the sample plane, have shown higher EL polarization in the direction normal to the quantum well plane; here circular polarization is fully allowed by the selection rule [96].

Injection of electron spins is preferable from the application point of view as electrons usually exhibit longer spin lifetime. This has been demonstrated using paramagnetic semiconductor based on II-VI compounds [97,98]. Because of the lack of established $n$-type ferromagnetic semiconductor, the most probable scheme for electrical *electron* spin injection using ferromagnetic semiconductors is through band-to-band tunneling. This can be realized by using a broken gap heterojunction system such as InAs/GaSb, where the valence band edge of GaSb is energetically higher than the conduction band edge of InAs. The successful growth of ferromagnetic (Ga,Mn)Sb [29] may allow us to inject spin polarized electrons into nonmagnetic InAs. The other scheme is to use an Esaki tunnel diode made of $p^+$-(Ga,Mn)As/$n$-GaAs [99]. Here the spontaneous spin splitting in the hole system of the (Ga,Mn)As side creates a spin-polarized hole current and is transformed into a spin-polarized electron current on the $n$-GaAs side via tunneling.

## 1.10   Electric-Field Control of Hole-Induced Ferromagnetism

The mean field model of the hole-induced ferromagnetism in Mn-doped III-V's indicates that the change in the hole concentration results in modification of ferromagnetic interaction among Mn spins and leads to increase or decrease of the ferromagnetic transition temperature. Since carrier concentration in semiconductors can be modified by the field-effect, the control of magnetic phase transition may be done isothermally and reversibly in a field-effect transistor (FET) structure, as schematically depicted in Fig. 1.12a. This has been demonstrated using an insulating-gate FET structure with an (In,Mn)As channel [100]. The 5 nm thick channel layer ($x = 0.03$) was grown on a buffer layer structure designed in such a way to relax almost all the lattice mismatch between the epitaxial structure and the GaAs substrate. Figure 1.12b shows the magnetization curves at three different gate voltages $V_{\mathrm{G}}$ ($+125$, $0$, $-125V$) applied to the gate of the FET, having a 0.8 $\mu$m gate insulator. Here, magnetization is measured by the Hall resistance, which is proportional to $M$ perpendicular to the film, as the magnetization of a small channel area (60 $\mu$m $\times$ 270 $\mu$m) is not easily accessible with an ordinary magnetometer. Due to the lattice mismatch between (In,Mn)As and AlSb, tensile strain is introduced in the (In,Mn)As layer making the easy axis perpendicular to the plane. At zero gate bias ($V_{\mathrm{G}} = 0$ V), the channel is weakly ferromagnetic at 22.5 K as can be seen from the presence of soft hysteresis. Application of positive gate voltage $V_{\mathrm{G}} = +125$ V partially depletes the holes

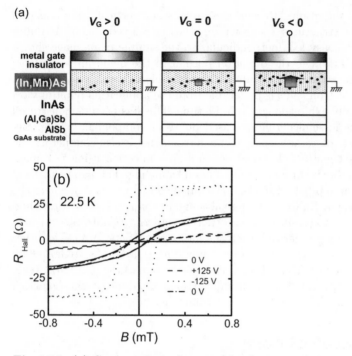

Fig. 1.12. (a) Cross-sections of a metal-insulator-magnetic semiconductor structure under gate biases $V_G$, which controls the hole concentration in the magnetic
semiconductor channel (through which the ferromagnetic interaction is mediated).
(b) Hall resistance, $R_{Hall}$, of an insulated gate (In,Mn)As field-effect transistor as a
function of magnetic field under three different gate voltages. $R_{Hall}$ is proportional
to the magnetization of the (In,Mn)As channel. Positive gate voltage of 125 V
partially depletes holes and results in weaker ferromagnetic interaction and a paramagnetic response, whereas negative gate voltage produces square hysteresis. Zero
gate voltage data before and after application of positive and negative gate voltages
are virtually identical.

and reduces the ferromagnetic interaction mediated by them resulting in a
paramagnetic magnetization curve without hysteresis. When holes are accu-
mulated by applying negative gate voltage ($V_G = -125$ V), a clear hysteresis
appears. The magnetization curve resumes its original curve as the gate voltage returns to 0 V. The 125 V swing gives rise to ±6% change in the total hole
concentration ($p = 0.9 \times 10^{20} cm^{-3}$ from the Hall effect at room temperature)
and results in the transition temperature change of ±4%, which agrees well,
together with the magnitude of $T_C$ (about 20 K), with the results calculated
by the mean-field model (see Fig. 1.7b). The mean-field model assumes a
three-dimensional carrier system, whereas the modulation of $T_C$ implies that
the carrier system is two-dimensional-like (the carrier concentration of the
undepleted channel region remains the same in the three-dimensional case).

Multi-subband occupation of the hole system due to the high hole concentration and the effect of disorder that broadens the subbands are among the few that probably need to be taken into account in order to understand the electric-field effect further. Additional contribution to the modulation of $T_C$ may have come from the distortion of wavefunction [101]. For example, application of positive bias distorts the relevant 2D wavefunctions away from the (In,Mn)As layer, resulting in reduction of carrier-magnetic-impurity exchange interaction and hence in reduction of $T_C$. It has been pointed out theoretically that the magnetization reversal and quantum confinement effects are closely linked and the field-effect modulation should result in a dramatic change in the magnetization reversal process when the hole system occupies only the ground subband [102]. It is worth noting that photogenerated carriers has also been used to modify the properties of (In,Mn)As [103,104].

The electric-field control of ferromagnetic phase transition can be used to define the ferromagnetic wires and dots by lithography allowing us to explore the man-made magnetic nanostructures with controllable exchange.

## 1.11   Summary and Outlook

The hole-mediated ferromagnetism in magnetic III-V's allows us to explore the spin-dependent phenomena not available in structures made of nonmagnetic III-V semiconductors alone, because ferromagnetic III-V's can be incorporated into nonmagnetic III-V heterostructures.

As to the understanding of its ferromagnetism, it is intriguing to observe that the mean-field theory presented here, which has its origin in the early Zener's idea, works so well at all despite its number of limitations. If the Zener model is the correct model for the present system, then we have found a ferromagnetic material system, where the material parameters are such that the original Zener's idea works for the first time. This means that we should be able to examine the limit of such formalism experimentally by increasing or decreasing magnetic impurity concentration and/or doping of the semiconductor. The gate-control of ferromagnetism should eventually allow us to do this in a single sample once a good insulator with high breakdown electric field is established. It will also enable gate-defined magnetic nanostructures, magnetic wires and dots, with variable ferromagnetic interaction. These would certainly advance our understanding of magnetism in the nanometer scale.

As to the semiconductor heterostructures with ferromagnetism, there are two directions for the exploration of spin-dependent phenomena. One is "classical", in which the direction of magnetization is critical, such as optical isolators, magnetic sensors and memories. This direction requires above room temperature $T_C$. Apart from a recent report on (Ga,Mn)N [37], the above room temperature $T_C$ has not been established in III-V based ferromagnetic semiconductors yet. However, the mean-field model [11,58] shows a chemical trend towards higher $T_C$, and envision such above room temperature ferromagnetism in some of the compound semiconductors if certain conditions are met. Metallic compounds compatible with III-V's such as CrAs and CrSb have already been shown to exhibit room temperature ferromagnetism for possible inclusion in III-V heterostructures. Although not III-V's, it is encouraging to learn that above room temperature ferromagnetism has been realized in semiconductors such as $CdMnGeP_2$ [12] and Co doped $TiO_2$ [13]. The freedom offered by semiconductor heterostructures together with ferromagnetism at room temperature will certainly enable a new family of magnetic sensors and memories, which may result in breakthroughs of semiconductor electronics and magnetic recording industries.

The other direction is "quantum": new developments in III-V ferromagnetic heterostructures reviewed here together with the recent progress in manipulation of carrier spin coherence in bulk [105] and across interface [106], and manipulation of nuclear spins via carrier spins in a quantum well [107,108] will pave the way to future quantum information technologies using spins in semiconductors [109–112].

Ferromagnetic III-V heterostructures are thus an excellent tool to explore a new field of semiconductor physics and technology, where both charge and spin of electrons play critical roles.

## Acknowledgements

The authors thank T. Dietl, F. Matsukura, Y. Ohno, and D. D. Awschalom for illuminating discussion and fruitful collaboration. The assistance as well as discussion with D. Chiba is also acknowledged. The work at Tohoku University was supported in part by the "Research for the Future" Program (# JSPS-RFTF97P00202) from the Japan Society for the Promotion of Science and by Grant-in-Aids (#12305001) from the Ministry of Education, Culture, Science, Sport, and Technology, Japan.

## References

1. T. Kasuya and A. Yanase, Rev. Mod. Physics, 40, 684 (1968).
2. A. Mauger and C. Godart, Physics Reports, 141, 51 (1986).
3. J. K. Furdyna and J. Kossut: Semiconductor and Semimetals, Vol.25 (Academic, New York, 1988).
4. T. Dietl, (Diluted) Magnetic Semiconductors, in Handbook of Semiconductors, (ed. S. Mahajan) Vol.3B (North-Holland, Amsterdam, 1994), p. 1251.
5. Recent progress in the $p$-type doping technology resulted in observation of carrier-induced ferromagnetism in II-VI DMS's. For a review of ferromagnetism in II-VI compounds see T. Dietl, J. Cibert, P. Kossacki, D. Ferrand, S. Tatarenko, A. Wasiela, Y. Merle d'Aubigne, F. Matsukura, N. Akiba, and H. Ohno, Physica E, 7, 967 (2000).

6. T. Story, R. R. Galazka, R. B. Frankel, and P.A. Wolff, Phys. Rev. Lett. 56, 777 (1986).
7. H. Ohno, Science, 281, 951 (1998).
8. H. Ohno, J. Mag. Mag. Materials, 200, 110 (1999).
9. H. Ohno, F. Matsukura, and Y. Ohno, Solid State Commun., 117, 179 (2001).
10. F. Matsukura, H. Ohno, A. Shen, and Y. Sugawara: Phys. Rev. B, 57, R2037 (1998).
11. T. Dietl, H. Ohno, F. Matsukura, J. Cibert, and D. Ferrand, Science, 287, 1019 (2000).
12. G. A. Medvedkin, T. Ishibashi, T. Nishi, K. Hayata, Y. Hasegawa, and K. Sato, Jpn. J. Appl. Phys., 39, L949 (2000).
13. Y. Matsumoto, M. Murakami, T. Shono, T. Hasegawa, T. Fukumura, M. Kawasaki, P. Ahmet, T. Chikyow, S. Koshihara, and H. Koinuma, Science, 291, 854 (2001).
14. H. Munekata, H. Ohno, S. von Molnar, A. Segmuller, L. L. Chang, and L. Esaki, Phys. Rev. Lett., 63, 1849 (1989).
15. H. Ohno, H. Munekata, T. Penney, S. von Molnar, and L. L. Chang: Phys. Rev. Lett., 68, 2664 (1992).
16. J. De Boeck, R. Oesterholt, A. Van Esch, H. Bender, C. Bruynseraede, C. Van Hoof, and G. Borghs, Appl. Phys. Lett. 68 2744 (1996).
17. H. Ohno, A. Shen, F. Matsukura, A. Oiwa, A. Endo, S. Katsumoto, Y. Iye, Appl. Phys. Lett. 69 363 (1996).
18. For the result of alternating shutter sequence see, J. Sadowski, R. Mathiu, P. Svedlindh, J. Z. Domagala, J. Bak-Misiuk, K. Swiatek, M. Karlsteen, J. Kanski, L. Ilver, H. Asklund, and U. Sodervall, Appl. Phys. Lett., 78, 3271 (2001).
19. F. Matsukura, A. Shen, Y. Sugawara, T. Omiya, Y. Ohno, and H. Ohno, Proc. 25th Int. Symp. Compound Semiconductors, Institute of Physics Conference Series, No. 162 (IOP Publishing Ltd, Bristol, 1999) p. 547.
20. H. Shimizu, T. Hayashi, T. Nishinaga, and M. Tanaka, Appl. Phys. Lett. 74, 398 (1999).
21. T. Hayashi, Y. Hashimoto, S. Katsumoto, and Y. Iye, Appl. Phys. Lett., 78, 1691 (2001).
22. S. J. Potashnik, K. C. Ku, S. H. Chun, J. J. Berry, N. Samarth and P. Schiffer, Appl. Phys. Lett. **79** 1495 (2001).
23. S. Sanvito and N. A. Hill, Appl. Phys. Lett., 78, 3493 (2001).
24. R. K. Kawakami, E. Johnston-Halperin, L. F. Chen, M. Hanson, N. Guebels, J. S. Speck, A. C. Gossard, and D. D. Awschalom, Appl. Phys. Lett. 77, 2379 (2000).
25. G.M. Schott, W. Faschinger, L.W. Molenkamp, Appl. Phys. Lett. **79**, 1807 (2001).
26. A. Shen, F. Matsukura, S. P. Guo, Y. Sugawara, H. Ohno, M. Tani, H. Abe, and H. C. Liu, Journal of Crystal Growth, 201/202, 679 (1999)
27. Y. L. Soo, S. W. Huang, Z. H. Ming, Y. H. Kao, and H. Munekata, Phys. Rev. B, 53, 4905 (1996).
28. R. Shioda, K. Ando, T. Hayashi, and M. Tanaka, Phys. Rev. B, 58, 1100 (1998).
29. E. Abe, F. Matsukura, H. Yasuda, Y. Ohno, and H. Ohno, Physica E, 7, 981 (2000).

30. H. Saito, W. Zaets, R. Akimoto, K. Ando, Y. Mishima, and M. Tanaka, J. Appl. Phys., 89, 7392 (2001).
31. S. Haneda, M. Yamura, Y. Takatani, K. Hara, S. Harigae, and H. Munekata, Jpn. J. Appl. Phys., 39, L9 (2000).
32. D. Okazawa, K. Yamamoto, A. Nagashima, and J. Yoshino, Physica E, 10, 229 (2001).
33. H. Akinaga, S. Nemeth, J. De Boeck, L. Nistor, H. Bender, G. Borghs, H. Ofuchi, and M. Oshima, Appl. Phys. Lett., 77, 4377 (2000).
34. M. Zajac, R. Doradzinski, J. Gosk, J. Szczytko, M. Lefeld-Sosnowska, M. Kaminska, A. Twardowski, M. Palczewska, E. Grzanka, and W. Gebicki, Appl. Phys. Lett., 78, 1276 (2001).
35. S. Kuwabara, K. Ishii, S. Haneda, T. Kondo, and H. Munekata, Physica E, 10, 233 (2001).
36. N. Theodoropoulou, A. F. Hebard, M. E. Overberg, C. R. Abernathy, S. J. Pearton, S. N. G. Chu, and R. G. Wilson, Appl. Phys. Lett. 78, 3475 (2001).
37. S. Sonoda, S. Shimizu, T. Sasaki, Y. Yamamoto, H. Hori, presented at Int. Conf. Crystal Growth (July 30–August 4, 2001, Kyoto, Japan) to appear in J. Crystal Growth.
38. H. Akinaga, T. Manago, and M. Shirai, Jpn. J. Appl. Phys., 39, L1118 (2000).
39. J. H. Zhao, F. Matsukura, K. Takamura, E. Abe, D. Chiba, and H. Ohno, Appl. Phys. Lett., 79, 2776 (2001).
40. M. Shirai, Physica E, 10, 143 (2001).
41. Y. D. Park, A. Wilson, A. T. Hanbicki, J. E. Mattson, T. Ambrose, G. Spanos, and B. T. Jonker, Appl. Phys. Lett., 78, 2739 (2001).
42. H. Ohno, F. Matsukura, A. Shen, Y. Sugawara, A. Oiwa, A. Endo, S. Katsumoto, and Y. Iye, Proc. of the 23rd Intl. Conf. on Physics of Semiconductors (Eds., M. Scheffler and R. Zimmermann, World Scientific, Singapore, 1996), pp. 405-408.
43. H. Munekata, A. Zaslavsky, P. Fumagalli, and R. J. Gambino, Appl. Phys. Lett. 63, 2929 (1993).
44. A. Oiwa, S. Katsumoto, A. Endo, M. Hirasawa, Y. Iye, H. Ohno, F. Matsukura, A. Shen, and Y. Sugawara, Solid State Commun., 103, 209 (1997).
45. T. Shono, T. Hasegawa, T. Fukumura, F. Matsukura, and H. Ohno, Appl. Phys. Lett., 77, 1363 (2000).
46. T. Fukumura, T. Shono, K. Inaba, T. Hasegawa, H. Koinuma, F. Matsukura, and H. Ohno, Physica E, 10, 135 (2001).
47. C. L. Chien and C. W. Westgate: The Hall Effect and Its Applications, (Plenum, 1980).
48. A. Arrot, Phys. Rev., 108, 1394 (1957).
49. T. Omiya, F. Matsukura, T. Dietl, Y. Ohno, T. Sakon, M. Motokawa, and H. Ohno, Physica E, 7, 976 (2000).
50. J. Okabayashi, A. Kimura, O. Rader, T. Mizokawa, A. Fujimori, T. Hayashi, and M. Tanaka, Phys. Rev. B., 58, R4211 (1998).
51. J. Szczytko, W. Mac, A. Stachow, A. Twardowski, P. Becla, and J. Tworzydlo, Solid State Commun., 99, 927, (1996).
52. A. Oiwa, S. Katsumoto, A. Endo, M. Hirasawa, Y. Iye, H. Ohno, F. Matsukura, A. Shen and Y. Sugawara, phys. status solidi (b), 205, 167 (1998).
53. S. Katsumoto, A. Oiwa, Y. Iye, H. Ohno, F. Matsukura, A. Shen, and Y. Sugawara, phys. status solidi (b), 205, 115 (1998).

54. S. von Molnar, H. Munekata, H. Ohno, and L. L. Chang, J. Magn. Magn. Mater., 93, 356 (1991).
55. Y. Satoh, N. Inoue, Y. Nishikawa, and J. Yoshino: 3rd. Symp. on Physics and Application of Spin Related Phenomena in Semiconductors, pp. 23–25 (Nov. 17-18, 1997, Sendai, Japan).
56. Y. Satoh, D. Okazawa, A. Nagashima, and J. Yoshino, Physica E, 10, 196 (2001).
57. J. Köning, H.-H. Lin, and A. H. MacDinald, Phys. Rev. Lett. 84, 5628 (2000).
58. T. Dietl, H. Ohno and F. Matsukura, Phys. Rev. B, 63, 195205 (2001).
59. A. K. Bhattacharjee and C. B. a la Guillaume, Solid State Commun., 113, 17 (2000).
60. M. Linnarsson, E. Janzen, B. Monemar, M. Kleverman, and A. Thilderkvist, Phys. Rev. B, 55, 6938 (1997).
61. D. Ferrand, J. Cibert, C. Bourgognon, S. Tatarenko, A. Wasiela, G. Fishman, A. Bonanni, H. Sitter, S. Kolesnik, J. Jaroszynski, A. Barcz, and T. Dietl, J. Crystal Growth, 214/215, 387 (2000).
62. M. Abolfath, T. Jungwirth, J. Brum, and A. H. MacDonald, Phys. Rev. B, 63, 054418 (2001).
63. B. Beschoten, P. A. Crowell, I. Malajovich, D. D. Awschalom, F. Matsukura, A. Shen, and H. Ohno, Phys. Rev. Lett., 83, 3073 (1999).
64. T. Dietl, A. Haury and Y. Merle d'Aubigne, Phys. Rev. B, 55, R3347 (1997).
65. J. Schliemann, J. König, H.-H. Lin, and A. H. MacDonald, Appl. Phys. Lett., 78, 1550 (2001).
66. O. Sakai, S. Suzuki, and K. Nishizawa, J. Phys. Soc. Japan, 70, 1105 (2001).
67. R. N. Bhatt and M. Berciu, xxx.lanl.gov/abs/cond-mat/0011319 (2000); M. Berciu and R. N. Bhatt, Phys. Rev. Lett. **87**, 107203 (2001).
68. J. Inoue, S. Nonoyama, and H. Itoh, Phys. Rev. Lett., 85, 4611 (2000).
69. V. I. Litvinov and V. K. Dugaev, Phys. Rev. Lett., 86, 5593 (2001).
70. Y. Nagai, T. Kunimoto, K. Nagasaka, H. Nojiri, M. Motokawa, F. Matsukura, T. Dietl, and H. Ohno, Jpn. J. Appl. Phys., 40, 6231 (2001).
71. K. Hirakawa, A. Oiwa, and H. Munekata, Physica E, 10, 215 (2001).
72. M. Shirai, T. Ogawa, I. Kitagawa, and N. Suzuki, J. Magn. Magn. Mater., 177-181, 1383 (1998).
73. H. Akai, Phys. Rev. Lett., 81, 3002 (1998).
74. T. Ogawa, M. Shirai, N. Suzuki, and I. Kitagawa, J. Magn. Magn. Mater., 196-197, 428 (1999).
75. S. Sanvito, P. Ordejon, and N. A. Hill, Phys. Rev. B, 63, 165206 (2001)
76. M. van Schlifgaarde and O. N. Mryasov, Phys. Rev. B 63, 233206 (2001).
77. T. C. Schulthess and W. H. Butler, J. Appl. Phys. 89, 7021 (2001).
78. K. Sato and H. Katayama-Yoshida, Jpn. J. Appl. Phys., 39, L555 (2001).
79. K. Sato and H. Katayama-Yoshida, Jpn. J. Appl. Phys., 40, L334 (2001).
80. K. Sato and H. Katayama-Yoshida, Jpn. J. Appl. Phys., 40, L485 (2001).
81. A. Shen, H. Ohno, F. Matsukura, Y. Sugawara, Y. Ohno, N. Akiba, and T. Kuroiwa, Jpn. J. Appl. Phys., 36, L73 (1997).
82. T. Hayashi, M. Tanaka, K. Seto, T. Nishinaga, and K. Ando, Appl. Phys. Lett., 71, 1825 (1997).
83. Y. Ohno, I. Arata, F. Matsukura, K. Ohtani, S. Wang, and H. Ohno, Appl. Surf. Sci., 159/160, 308 (2000).
84. I. Arata, Y. Ohno, F. Matsukura and H. Ohno, Physica E, 10, 288 (2001).

85. Y. Ohno, I. Arata, F. Matsukura, and H. Ohno, Appl. Surf. Sci., 159, 308 (2000).
86. N. Akiba, D. Chiba, K. Nakata, F. Matsukura, Y. Ohno, and H. Ohno, J. Appl. Phys., 87, 6436 (2000).
87. D. Chiba, N. Akiba, F. Matsukura, Y. Ohno, and H. Ohno, Appl. Phys. Lett., 77, 1873 (2000).
88. Y. Higo, H. Shimizu, and M. Tanaka, Physica E, 10, 292 (2001). See also M. Tanaka and Y. Higo, Phys. Rev. Lett., 87, 026602 (2001).
89. H. Ohno, N. Akiba, F. Matsukura, A. Shen, K. Ohtani, and Y. Ohno, Appl. Phys. Lett., 73, 363 (1998).
90. N. Akiba, F. Matsukura, Y. Ohno, A. Shen, K. Ohtani, T. Sakon, M. Motokawa, and H. Ohno, Physica B, 256-258, 561 (1998).
91. F. Matsukura, unpublished.
92. C. Y.-P. Chao and S. L. Chuang, Phys. Rev. B, 43, 7027 (1991).
93. A. C. R. Bittencourt, A. M. Cohen, and G. E. Marques, Phys. Rev. B, 57, 4525 (1998).
94. A. G. Petukhov, D. O. Demchenko and A. N. Chantis, J. Vac. Sci. Technol. B 18, 2109 (2000).
95. Y. Ohno, D. K. Young, B. Beschoten, F. Matsukura, H. Ohno, and D. D. Awschalom, Nature, 402, 790 (1999).
96. D. K. Young, E. Johnston-Halperin, D. D. Awschalom, Y. Ohno, and H. Ohno, Appl. Phys. Lett., 80, 1598 (2002).
97. R. Fiederling, M. Keim, G. Reuscher, W. Ossau, G. Schmidt, A. Waag, and L. W. Molenkamp, Nature, 402, 787 (1999).
98. Y. D. Park, B. T. Jonker, B. R. Bennett, G. ltskos, M. Furis, G. Kioseoglou, and A. Petrou, Appl. Phys. Lett., 77, 3989 (2000).
99. M. Kohda, Y. Ohno, K. Takamura, F. Matsukura, and H. Ohno, Jpn. J. Appl. Phys., 40, L1274 (2001); E. Johnston-Halperin, D. Lofgreen, R. K. Kawakami, D. K. Young, L. Coldren, A.C. Gossard, and D. D. Awschalom, Phys. Rev. B, 45, 041306 (2002).
100. H. Ohno, D. Chiba, F. Matsukura, T. Omiya, E. Abe, T. Dietl, Y. Ohno, and K. Ohtani, Nature, 408, 944 (2000).
101. B. Lee, T. Jungwirth, A. H. MacDonald, Phys. Rev. B, 61, 15606 (2000).
102. B. Lee, T. Jungwirth, A. H. MacDonald, xxx.lanl.gov/abs/cond-mat/0106536 (2001).
103. S. Koshihara, A. Oiwa, M. Hirasawa, S. Katsumoto, Y. Iye, C. Urano, H. Takagi, and H. Munekata, Phys. Rev. Lett., 78, 4617 (1997).
104. A. Oiwa, T. Slupinski, and H. Munekata, Appl. Phys. Lett., 78, 518 (2001).
105. J. M. Kikkawa and D. D. Awschalom, Nature, 397, 139 (1999).
106. I. Malajovich, J. M. Kikkawa, D. D. Awschalom, J. J. Berry, and N. Samarth, Phys. Rev. Lett., 84, 1014 (2000).
107. J. M. Kikkawa and D. D. Awschalom, Science, 287, 473 (2000).
108. G. Salis, D. T. Fuchs, J. M. Kikkawa, D. D. Awschalom, Y. Ohno, and H. Ohno, Phys. Rev. Lett., 86, 2677 (2001).
109. B. E. Kane, Nature, 393, 133 (1998).
110. D. Loss and D. P. DiVincenzo, Phys. Rev. A, 57, 120 (1998).
111. S. D. Sarma, J. Fabian, X. Hu, and I. Zutic, Superlattice and Microst. 27, 285 (2000).
112. R. Vrijen, E. Yablonovitch, K. Wang, H. W. Jiang, A. Balandin, V. Roychowdhury, T. Mor, and D. DiVincenzo, Phys. Rev. A, 62, 012306 (10) (2000).

# 2 Spin Injection and Transport in Micro- and Nanoscale Devices

Hong X. Tang, F.G. Monzon, Friso J. Jedema, Andrei T. Filip,
Bart J. van Wees, and Michael L. Roukes

## 2.1 Overview

Experiments to explore the transfer of a spin-polarized electric current within small devices have been ongoing for nearly 30 years. But attaining the same level of exquisite control over the transport of *spin* in micro- or nanoscale devices, as currently exists for the flow of charge in conventional electronic devices, remains elusive. Much has been learned since the time of the first demonstrations of spin polarized tunneling by Tedrow and Meservey. During this period we have witnessed the transformation of spin-based electronic devices from laboratory experiments to the realm of commercially available products. This has been driven especially, just in this past decade, by the robust phenomena of giant magnetoresistance (GMR) [1]. Even more recently, magnetic tunnel junction devices, involving transport of spin polarized electrons across interfaces, have proceeded to commercial development [2]. Meanwhile, spin injection devices – and by "injection" we here denote transferal of spin-polarized carriers into an otherwise *nonmagnetic conductor* (or semiconductor) – have not reached a similar, commercially viable, state of maturation. In fact, it is fair to say that, at present, even the fundamental physics and materials science of the spin injection process remains in need of significant elucidation.

Yet the problem of spin injection continues to capture the focus of many researchers including ourselves. Ongoing interest in spin electronics – especially in semiconductors of late – is, in part, motivated by the expectation that, in the near term, this field may lead to the large scale integration of semiconductor micro- and nanodevices capable of performing very high speed logic and memory operations, such as performed via conventional charge-based electronics, but at a fraction of the power. In the long term, surveying the spectrum of other possible solid-state embodiments, many researchers anticipate that the spin degree of freedom may provide the most robust foundation upon which practical realizations of qubits and quantum computers may ultimately be constructed. All of these prospects clearly require us to understand how to transfer electron spins across interfaces, and how to preserve their polarization during this traversal.

In this chapter we shall review some of the most important developments in the field of spin injection that have emerged since the earliest experiments.

With apologies at the outset to many important contributors to this area whose work may not be included, we strive herein more to develop a coherent overall perspective, rather than for absolute completeness of coverage. Below we shall attempt to summarize the evolution of thinking about spin injection by describing, in part, the succession of research that has ensued.

## 2.2  Background

### 2.2.1  Spin Polarized Tunneling

In 1971 Tedrow and Meservey (TM) first studied spin polarized tunneling using superconductor/insulator/ferromagnet devices based upon Aluminum/Aluminum Oxide junctions [3], [4]. Their experiments took advantage of the fact that application of a magnetic field splits the superconductor quasiparticle density of states into spin up and spin down bands. This splitting, equal to the Zeeman energy of electronic spins in the magnetic field, allows the superconductor to be used as a spin analyzer. Varying the voltage between the ferromagnet and superconductor produces a tunneling current that is sensitive to the detailed combined density of states in both the ferromagnet and the superconductor. The ferromagnet's density of states can then be obtained by deconvolving the tunneling conductance. Polarization values obtained for some of the elemental ferromagnets by TM were 44% for Fe, 34% for Co, and 11% for Ni. These are in direct proportion to the saturation magnetization of the bulk materials: $1.714 \times 10^3$, $1.422 \times 10^3$, and $0.484 \times 10^3$ emu/cm$^3$ respectively [5].

This correspondence may be intuitively sensible, but is not particularly straightforward to understand quantitatively. Even the ostensibly simpler question as to whether it is majority or minority spins that comprise the larger percentage of the tunneling current is complicated. The sign of the polarization observed by TM in their tunneling experiments was, in fact, the opposite expected for majority spins. Attempts at resolving this apparent discrepancy took several years. Investigation into the precise nature of the injected electrons continues today, but first discussions were first provided by Stearns [6], later by Sloncezwski [7], and, subsequently, several other groups [8],[9],[10]. Collectively the picture that has emerged from this work is that the tunneling conductance can vary dramatically from band to band, and the contribution from one particular minority channel band at the Fermi energy may be predominant in the overall majority spin conductance. A detailed analysis of the electronic band structure of materials at the interface is therefore required to understand the complex behavior observed in experiments.

In 1975, Julliere extended this class of tunneling experiments to a system with two ferromagnets (Fe/Ge/Co); these were carried out at 4.2K where the Ge became semi-insulating [11]. In this work the magnetization of Fe and Co could be varied independently, which resulted in a variation of the tunneling conductance for the parallel and antiparallel magnetization configurations. Based upon TM's previous analysis, Julliere expressed this magnetoconductance (the difference in conductance values for the parallel and antiparallel magnetization configurations) as $\Delta G/G = 2P_1P_2/(1 + P_1P_2)$, where $P_1$ and $P_2$ represent the conduction electron spin polarizations in the Fe and Co. The maximum measured value was 14%, somewhat below the anticipated value based on the 26% polarization previously deduced by the measurements of TM. Importantly, Julliere pointed out that both coupling between the ferromagnets (resulting in less-than-saturated magnetization), and spin-flip scattering in the Ge or at the interfaces, could reduce the overall effect. This first demonstration of spin-filtering, without the use of a superconducting analyzer film, opened up concrete experimental prospects for spin injection at higher temperatures.

## 2.2.2   Spin Injection in Clean Bulk Metals

In 1976, A.G. Aronov predicted that electron spin resonance (ESR) signals should change in the presence of a spin-polarized current injected directly from a ferromagnet into a normal metal [12]. The injected spins, he proposed, would also induce a polarization of the nuclei as the spins scattered. His proposal implies that one should be able to quantify the injected spin density through a local measurement of the spin polarization of either the conducting electrons or the nuclei.

Direct evidence of spin-polarized transport within an "Ohmic", non-tunneling configuration was first provided in 1985 by Johnson and Silsbee (JS). They carried out experiments where a non-ferromagnetic, paramagnetic metal (**P**) was placed between two metallic ferromagnets [13]. Fig. 2.1 provides a conceptual illustration of the experimental configuration for this class of experiments. A current sourced through a ferromagnet, **F1**, acquires a spin polarization due to the remnant magnetization of the ferromagnetic material. Injection of this current into a nonmagnetic metal, **P**, induces a net spin polarization within it (represented as a shaded cloud of magnetization in Fig. 2.1); this decays spatially with a characteristic length scale $\lambda_{\mathrm{sf}}$. If the separation, $d$, between the two ferromagnets **F1** and **F2**, is less than this spin diffusion length, one would expect that the second ferromagnet, **F2** should "interact" in some characteristic way with this nonequilibrium spin polarization. As described below, for conditions of open-circuited output terminals this interaction results in an induced voltage, $V_s$, as shown in Fig. 2.1. The initial ideas and experiments involved *diffusive* electron transport; this is valid in the limit where both $d$ and $\lambda_{\mathrm{sf}}$ greatly exceed the mean free path for momentum scattering.

For such experiments it is crucial to confirm that the induced voltage, $V_s$, is truly representative of spin accumulation within the paramagnet, rather than simply being the result of spurious, uncontrolled potential drops within the device. In the spin injection work to date, there are two principal ways in

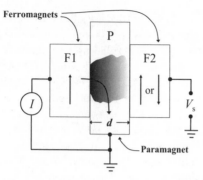

**Fig. 2.1.** Sketch illustrating the basic components of a three-terminal spin injection device. An external magnetic field is employed to controllably switch the relative orientation of the ferromagnetic electrodes (**F1**, **F2**). The ground connection (current return) for the paramagnetic conductor (**P**) is attached many spin diffusion lengths away from the region of spin injection, and is therefore far from the cloud of induced magnetization (*shaded region*). The ground connection for the detected voltage need not necessarily be the same as the ground for current return, allowing for nonlocal four-terminal measurements, as described in the text.

which this has been pursued. The first, and most unambiguous of these techniques was employed by JS in their 1985 experiments; it is based on what is called the Hanle effect. (Below, we shall describe the second technique, which we term "polarize/analyze" experiments.) The Hanle effect experiments involve application of an external magnetic field to induce precession of the injected spins as they traverse from injector to collector across the device. Changing the magnitude of this field changes the rate of precession and, hence, the total precession angle accrued by each carrier during its traversal of the device. (Special care must be taken to insure the magnetization of both the ferromagnetic injector and collector remain unaffected by the applied field – which is usually applied perpendicular to the plane of the sample). The total spin current from injector to collector obviously involves the entire ensemble of diffusing carriers, hence there can be a rather wide variation in the path lengths that are traversed. This leads to a wide variation in transit times, and hence also in the total precession accrued by the individual spins during their traversal. When the external field is sufficient to result in an *average* total precession during transit of order $\pi$, the net effect sums incoherently over the ensemble, and the spin-induced signal at the collector $V_s$ becomes suppressed. These arguments directly lead to a field scale for the decay of the Hanle effect in a diffusive system [14].

Figure 2.2a provides a schematic of the actual device geometry employed by JS. In their first experiments **P** comprised a $\sim$50 μm thick aluminum foil; on top of this a thin permalloy (a NiFe alloy) film was lithographically patterned to form both the ferromagnetic injector and detector electrodes. The device dimensions were quite large by contemporary scales, but this

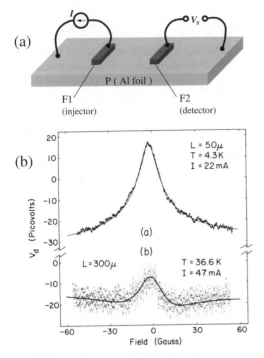

Fig. 2.2. (a) Schematic representations of the spin devices of Johnson and Silsbee employing Al foil, which was 50 μm thick and 100 μm wide [13]. (b) Representative data from all-metallic spin devices: Hanle effect experiment [13],[14].

was compensated by the high purity of the single crystal Al foil employed, which in some devices was reported to yield an electron mean free path of 17 μm. The spin diffusion lengths, $\lambda_{sf}$ (discussed below), were therefore also correspondingly large, reportedly approaching 500 μm. In these experiments the separation between injector(**F1**) and detector (**F2**) was varied between 50 μm and 300 μm.

Representative Hanle effect data obtained at liquid helium temperatures from these devices are shown in Fig. 2.2b. The signal levels observed in these first experiments were exceedingly low – at best of order tens of picovolts, even for current drive levels up to ~ 30 mA. The data were obtained by sourcing current through **F1** and into the Al bar, then extracting it from the near end, i.e. away from **F2**. A SQUID picovoltmeter was used to measure the induced voltage $V_s$ at **F2** with respect to opposite end of the Al foil. From this a spin transresistance, $R_s = V_s/I_s$, could be deduced. This nonlocal, four-terminal measurement configuration (sometimes termed a "potentiometric" measurement) is conceptually equivalent to that pictured in the three-terminal device of Fig. 2.1. Here it allows further decoupling of the "spin component" of transport, arising from diffusion of spin-polarized electrons in the Al from one ferromagnet to the other, without obfuscation from potentially large background voltage offsets associated potential drops associated with the regular unpolarized components of transport. However, the measurements can still be plagued by small background offsets due to asymmetries in electric field gradients (arising from device asymmetries) and from Hall voltages in the paramagnetic conductor. The latter can be especially vexatious in semiconductors; we shall discuss them in more detail below.

Particularly noteworthy is that – despite the clear evidence that the Hanle effect provides for spin injection – to date, the data of Fig. 2.2b are still, to our knowledge, the only such traces that can be found in the entire literature of spin injection.

### 2.2.3   Conceptual Picture of Spin Injection

A conceptual understanding of the basic origin of spin-induced signal is provided in Fig. 2.3 for the all-metallic case, based upon the qualitative picture developed by JS in their early work [14]. A series of density of states diagrams representing the device components are shown, each with spin down on the left of the vertical axis and spin up on the right. This represents the case in which the magnetization of the ferromagnets is parallel. At zero current bias, as shown in Fig. 2.4a, the Fermi levels in the injector (**F1**), paramagnet (**P**), and detector (**F2**) all line up and the system is in equilibrium. When a current is sourced into **F1**, as shown in Fig. 2.4b, spin polarized electrons are injected from **F1** into **P**; as a result its down-spin population increases, while its up-spin population decreases as the initially unpolarized carriers depart from the current return lead. The steady-state balance between spin injection and spin relaxation in **P** directly determines the amount by which the electrochemical potentials of its two spin subbands are offset. The open-circuit boundary condition on **F2**, which corresponds to an ideal voltage measurement, implies that (in steady-state) there is no net current flow into **F2**. For this to hold, the down-spin electrochemical potential of **F2** must rise to match that of the down-spins in **P**. In this very simple picture it is assumed that the device is formed from ideal (Stoner) ferromagnets, **F1**

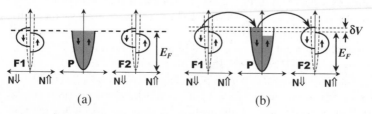

**Fig. 2.3.** Schematic diagrams of the densities of states (*horizontal axis*) versus energy (*vertical axis*) in the ferromagnets and paramagnets. In (**a**) the zero current bias case is shown while in (**b**) bias current from **F1** to **P** preferentially builds up one spin population (*shaded*).

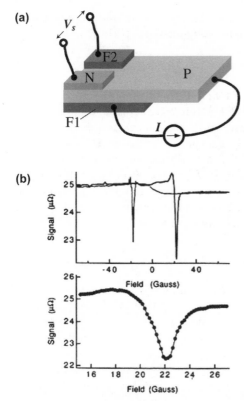

**Fig. 2.4.** (a) The geometry of Johnson's all-metal "spin transistor". (b) Data obtained from a polarize/analyze experiment with such a device [17]

and **F2**, possessing 100% spin polarization at the Fermi energy, and that **P** is a simple free-electron paramagnetic metal.

Based on this simple picture of spin injection it is possible to arrive at a crude estimate of the spin-induced voltage [14]. Magnetic moments are injected into **P** at the rate $J_\mathrm{m}$ per unit area, this magnetization current density injected into the paramagnet can be written as,

$$J_\mathrm{m} = \eta \frac{\mu_B J_e}{e} \tag{2.1}$$

where $J_e$ is the electric current density; $\mu_B$ is the Bohr magneton; $e$ is the electron charge, and $\eta$ is a phenomenological parameter that subsumes all "nonidealities" associated with the injection process – such as partial polarization at the Fermi energy, differences in Fermi velocities, and partial or inefficient spin transfer across the interface. We shall discuss it further below.

In steady state these injected spins are relaxed at the rate $1/\tau_\mathrm{sf}$, where $\tau_\mathrm{sf}$ represents a simple relaxation time approximation for the spin flip scattering rate (i.e. some form of $k$-space average over the Fermi surface electrons). Therefore the resulting nonequilibrium magnetization, $M$, in **P** is

$$M = \frac{J_\mathrm{m}\tau_\mathrm{sf}}{d} = \frac{\eta_1 \mu_\mathrm{B} J_e \tau_\mathrm{sf}}{ed} \tag{2.2}$$

where $d$ is the separation between **F1** and **F2** and $\eta_1$ is the "non-ideality parameter" for injection across the **F1**–**P** interface. Again, we stress that this result applies to the diffusive case which holds for $d \gg \ell_0 = v_\mathrm{F}\tau$, where $v_\mathrm{F}$ and $\tau$ are the Fermi velocity and momentum lifetime in **P**, respectively. We also assume that the separation $d$ is smaller than (or at least comparable to) the spin diffusion length $\lambda_\mathrm{sf} = \sqrt{v_\mathrm{F}^2\tau\tau_\mathrm{sf}/3}$, so the effect is not diminished by spin relaxation. The spin-induced voltage detected at **F2** is proportional to the rise in chemical potential in **F2** associated with the injected spins in **P**. Since $M/\mu_\mathrm{B}$ is the number of nonequilibrium spins:

$$\eta_2 \frac{M}{\mu_\mathrm{B}} = N(E_\mathrm{F})eV_\mathrm{s} \tag{2.3}$$

where $N(E_\mathrm{F})$ is the density of states at the Fermi level. This is valid assuming linear response, i.e., that $V_\mathrm{s}$ is sufficiently small so that $N(E)$ has a roughly constant value $\sim N(E_F)$ over the energy range of the injected electrons. Assuming **P** is a simple, free electron metal, $N(E_\mathrm{F}) = 3n/2E_\mathrm{F}$, where $n$ is the volume density of electrons, we arrive at

$$R_\mathrm{s} = \frac{2\eta_1\eta_2}{3e^2} \frac{E_\mathrm{F}\tau_\mathrm{sf}}{nAd}. \tag{2.4}$$

Apparently the success of the simple description above hinges upon the phenomenological parameters, $\eta_1$ and $\eta_2$. (In the simplest picture used here, one assumes that the interfacial properties are uniform and identical over the entire contact area $A$, hence $\eta \equiv \eta_1 = \eta_2$. Buried within these "fudge factors" are the real physics of spin injection from the ferromagnet into the paramagnet (and vice-versa).

With knowledge of all parameters on the right side of expression Eq. 2.4, values for $R_\mathrm{s}$ can be extracted. However, it is clear that there is significant ambiguity regarding the values for $\tau_\mathrm{sf}$ that should be applicable to such experiments, and almost complete lack of *a priori* knowledge about the appropriate values for the spin injection efficiencies $\eta_1$ and $\eta_2$. If we assume $\tau_\mathrm{sf}$ is equal to $T_2\sim10$ns, as measured in CESR experiments on Aluminum at 4K [15], and furthermore assume the ideal situation holds where $\eta_1 = \eta_2 = 1$, the maximum expected spin transresistance should be of order $R_\mathrm{s} \sim 1\ \mu\Omega$. The value found in the JS experiments is significantly smaller, $R_s \sim 2$ n$\Omega$. Alternatively, one can turn the problem around by assuming the interfaces behave identically, and deduce values for $\eta_1 = \eta_2 \equiv \eta$ from the measured values of $R_\mathrm{s}$ and values of $\tau_\mathrm{sf}$ (assumed equal to $T_2$) obtained from the literature. For these early Al spin injection devices, the values for $\eta$ deduced range from 0.043 to 0.075. Given that Tedrow and Meservey observed Fermi surface spin polarization for metallic ferromagnets in the range of tens of percent, it initially appears that an additional order-of-magnitude suppression of the spin injection occurred. Presumably this may have arisen from uncontrolled interfacial effects. But the results of this qualitative and rather incomplete picture should not be interpreted too literally. Soon after the publication of the JS experiments, van Son [16] et al. provided a more detailed model of transport across the metal-metal interface that provides deeper insight to the problem. After careful matching of the chemical potential for different spin bands, they found that the spin splitting can be significantly suppressed due to the high conductivity of metals. This shall be described in further detail in Sect. 2.5 of this chapter.

### 2.2.4   Spin Injection in Impure Metal Films

In 1993 Johnson announced the realization of a spin (injection) transistor based upon thin metallic films [17]. From the results obtained, prospects for non-volatile transistor-like memory elements and spin injection logic elements were raised [19]. The devices formed a sandwich structure, as depicted in Fig. 2.4a, in which the two thin metal film ferromagnets, **F1** and **F2**, were positioned on opposite sides of a 1.6-μm-thick disordered, polycrystalline, Au film; (labeled **P**). A second normal metal contact, **N**, (which sensed the average local, spin-averaged electrochemical potential in **P**) was positioned near the detector ferromagnet **F2** to provide a local "ground" reference for the spin-dependent voltage that was detected.

Evidence for spin injection in these devices was based solely through the second of the two aforementioned techniques, namely polarize/analyze experiments. These are somewhat easier to carry out, compared to spin precession experiments; the measurements involve application of a magnetic field in the plane of the sample, along the easy axis of the ferromagnets. This is swept to cause their relative magnetizations to switch. A voltage jump at the detection terminals, i.e. in $V_s$, is expected when the injector and detector magnetizations change from parallel to antiparallel. In other words, these devices yield a two-state output depending on the relative orientation of the magnetization in the two ferromagnetic contacts.

Representative data from these experiments are shown in Fig. 2.4b. The pronounced dips in voltage correspond to antiparallel orientation of the magnetization in the ferromagnetic contacts. An injected current of several mA resulted in a spin-induced voltage at the output of order several μV – *immensely* larger than in the first experiments. The possible explanation advanced at the time was that the much smaller size of the paramagnet yielded a profound increase in the efficiency of spin accumulation. The magnitude was much greater than should be expected, however, even for 100% polarization of the injected electrons. This profound signal enhancement occurred despite the known, and dramatic reduction in $\lambda_{sf}$, (deduced to be ~1.5 μm), suffered as a direct result of the much lower quality of the impure, poly-crystalline Au films employed for **P**. The data could only be explained by values of $\eta$ exceeding unity. This is obviously unphysical and reflects an in-complete understanding of the nature of the data obtained. Furthermore the device geometry of [17] suffers from the fact that the current flows not only perpendicular, but also parallel, to the F layers. This means that any mag-netoresitance in the layers (in particular AMR) may contribute to, or even completely dominate the measured signal.

Eight years elapsed before further confirmation of spin injection phenom-ena in all-metal systems was obtained. These latter experiments, in conjunc-tion with significant theoretical development that occurred in the intervening period, have significantly enhanced our understanding of the complexities of spin injection. These more recent experimental and theoretical developments in diffusive spin injection are described in Sect. 2.5.

## 2.3   Toward a Semiconducting "Spin Transistor"

### 2.3.1   Why a Spin Transistor?

Johnson motivated his work on the all-metal spin transistor by proposing that such devices would offer new routes for realizing nonvolatile memory and spin-based logic elements [19]. However, since that time, as mentioned, spin injection has continued to be a topic of fundamental research while the much more robust phenomena of GMR and magnetic tunneling have al-ready evolved into commercial products. Although the push to develop spin-injection based *memory elements* has evaporated in the intervening period, interest in low power logic may remain [20]. In fact, recent concepts for novel, all-semiconductor spintronic devices have emerged, and these are generating new enthusiasm for "spin transistor" research [23], [24], [25]. More specifi-cally, motivation for realizing robust electrical spin injection now arises from recent developments in "spin transfer" [21] and "spin imprinting" [22]. The former refers the use of spin polarized currents to achieve magnetic actuation (e.g. domain reversal in nanomagnets), and the latter refers to prospects for electrical control of the polarization of localized nuclear or electronic mo-ments. Both phenomena may ultimately play important roles in future spin-tronic systems and in the longer-term quest for realizing solid-state quantum logic.

### 2.3.2   Why Semiconductors?

Semiconducting materials offer the possibility of new device functionalities not realizable in metallic systems. Equilibrium carrier densities can be var-ied through a wide range by doping. Furthermore, because the typical carrier densities in semiconductors are low compared to metals, electronic properties are easily tunable by gate potentials. There is, of course, a vast body of knowledge concerning semiconductor materials and processing; these are amongst the most pure materials available commercially. All these attributes converge to allow definition of microelectronic devices with power gain, enabling the fan-out necessary to create massively integrated systems. Additionally, recent advances have allowed optimization of interfaces, at the level of atomic-scale control, between different epitaxial materials. In fact, many of these processes have already been scaled up to commercial production lines. These factors, in concert with recent advances in materials science of high-quality magnetic semiconductors, now make semiconductor materials perhaps the first, and natural choice for future spintronic applications – especially those involving large scale integration of spintronic devices.

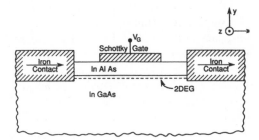

**Fig. 2.5.** A spin-FET proposed by Datta and Das, conceptually similar in operation to an electro-optic modulator. Two micromagnets serve as polarizer and analyzer. The propagation medium between them – capable of inducing a gate-controllable net rotation of spin orientation – is a two-dimensional electron gas (2DEG).

### 2.3.3   Concept

In 1990, special attention was drawn to the possibilities of spin injection in semiconducting systems by the proposal of Datta and Das for a "spin transistor." Their idea was based upon the manipulation of the spin state of carrier via controlled spin precession [26]. This device is at first glance much like a conventional FET; it has a current source, a current drain, and a channel in between with a tunable conductance. However, the spin transistor was based upon contacts envisaged as spin selective, i.e. capable of injecting and accepting only one spin component of the carrier distribution. Of course, polarize/analyze experiments are configured precisely this manner. However, the similarity with previous spin injection experiments carried out with diffusively propagating carriers, ends here.

The Datta/Das proposal was couched in terms of spin-coherent *transmission* through a hypothetical transistor. Their idea was drawn from direct analogy to an electro-optic modulator (EOM), a device that provides electrically-tunable optical transmittance. An EOM operates by first polarizing the incident light, then rotating the resulting beam's polarization by propagating it through an optical medium with birefringence that can be tuned electrically. Finally, modulation of the beam's intensity is realized when this transmitted beam is "analyzed" by a second polarizer just before exiting the device. In the spin transistor, by analogy, the proposed method of achieving controlled rotation of the transmitted "beam's" *spin* orientation was through gate-tunable spin-orbit coupling for the electrons within the device channel. This will be described in more detail below.

Initially it might seem that the same conditions required for observation of the Hanle effect in a diffusive system – namely that the spin diffusion length greatly exceed the device dimensions – should be sufficient for the operation of spin transistor. This is not the case; the requirements for operation of the spin transistor are far subtler. The substantial additional complexity arises from the fact that the magnetic field that induces spin precession is here not simply an externally applied uniform magnetic field as in the Hanle case. Instead, as described below, for the spin transistor it is an effective "Rashba field" that is relevant. Its dependence upon carrier momentum introduces new ways in which spin coherence can be scrambled as the carriers propagate from the injector to collector within a spin transistor.

For a completely ballistic device, these issues become somewhat simplified as is described in Sect. 2.6 of this chapter. But even in this limit, if the injected electron distribution is characterized by finite occupancy of *multiple* momentum states (subbands), significant complexity persists. Hence the simplest case, considered by Datta and Das, is that of a ballistic device involving occupancy of a *single* transverse subband (Sect. 2.6). At this juncture even for *unpolarized* electrons it remains quite difficult to realize ballistic, single-subband transport within devices having any appreciable length. Since the original demonstration of single-subband transport in a quantum point contact [28], [29], such transport over longer "device scale" dimensions has been demonstrated solely, to our knowledge, in quantum wires fabricated by cleaved-edge overgrowth [30] and in nanotubes [31].

### 2.3.4 Prerequisites for Realizing a Spin Transistor

There are four fundamental requirements for successful implementation of a spin transistor: *injection* of a spin polarized current, spin coherent *propagation*, induction of controlled spin *precession*, and spin-selective *collection*. Below we shall first briefly review the status quo of our knowledge regarding spin coherent propagation and the gate control of spin precession. The state-of-the-art of electrical spin injection and spin-selective collection will emerge in subsequent sections of this chapter.

### 2.3.5    Spin Lifetime in the Conduction Channel

In the past four years, it has been proven experimentally that spin coherence can persist in doped semiconductors on very long temporal (>100 ns at low temperatures) and spatial scales (> 100 μm) [32–36]. A detailed introduction and discussion is found in Chap. 5 of this book. All of these experiments were carried out using optical techniques, involving time resolved pump-probe method. This involves excitation of spin-polarized electron-hole pairs by short pulses of circularly polarized light from an ultrafast laser. This excitation results in carriers far from equilibrium. In general the spin polarized holes are found to relax quickly, but the spin polarized electrons persist for long times. At present, the extent to which the spin dynamics of these hot electrons differ from those of spin-polarized electrons at Fermi surface is unclear. Since spin scattering is closely related to momentum scattering, one might expect the spin relaxation rate at the Fermi surface to be *even slower* than that observed for hot carriers. But, to date, difficulties in manipulating near-equilibrium electrons, have precluded measurement of spin lifetime at the Fermi surface. But the picture that emerges is extremely promising for electrical spin injection devices.

### 2.3.6    Gate Control of the Spin Orbit Interaction (Theory)

Most III-V semiconductors have zincblende lattice structure, which is asymmetric with respect to inversion. The intrinsic crystal fields lead to a conduction band spin splitting proportional to $k^3$, even at zero magnetic field. Spin-orbit coupling can also be induced by an *interfacial electric field* within a heterostructure. Carriers confined to moving in an asymmetric quantum well will experience an effective magnetic field, called the Rashba field, that may induce spin precession [18]. The rate of this Rashba-field-induced precession should be tunable through an external gate voltage, which can serve to alter the pre-existing, "built-in" confinement potential. The Rashba Hamiltonian is usually written as,

$$H_{\mathrm{R}} = \alpha\left[\boldsymbol{\sigma} \times \boldsymbol{k}\right] \cdot \hat{z} \tag{2.5}$$

where the spin-orbit interaction parameter $\alpha$ is linearly dependent on $\langle E_z \rangle$ through the energy gap and the effective mass, $\boldsymbol{\sigma}$ represents the Pauli spin matrices, and $\hat{z}$ is the direction of the electric field. The total Hamiltonian, assuming that the Rashba effect dominates all other spin-coupled factors, is $H_{\mathrm{tot}} = H_{\mathrm{k}} + H_{\mathrm{R}}$, where $H_{\mathrm{k}} = \hbar^2 k^2 / 2m^*$ is the kinetic energy part of the Hamiltonian. The eigenstates for up and down spins are then [18],

$$E^{\pm}(\mathrm{k}) = \frac{\hbar^2 k^2}{2m^*} \pm \alpha \left|k\right| \tag{2.6}$$

the spin splitting energy, $\Delta_{\mathrm{R}}$, at zero magnetic field at the Fermi energy is $\Delta_{\mathrm{R}} = 2\alpha k_{\mathrm{F}}$. In traversing a distance $L$ in the quantum well, an electron will precess an amount $\Delta\theta = \omega_L L/v_F$, where $\omega_L = \Delta_R m^* L/\hbar^2 k_F$. The angle through which an electron precesses in traversing a distance $L$ is thus [26],

$$\Delta\theta = 2m^*\alpha L/\hbar^2 \tag{2.7}$$

Values for $\alpha$ given by Nitta et al. [37] and Heida et al. [38] range between 0.5 and $1 \times 10^{-11}$ eVm, corresponding to an energy splitting $\Delta_R = 1.5$ to 6 meV. From the expression above, the tunability of spin-orbit coupling is derived through tuning the value of the parameter $\alpha$ by an external gate. We now examine the evidence for such tunability.

### 2.3.7   Gate Control of the Spin Orbit Interaction (Experiment)

Gate control of spin splitting in quantum wells has been demonstrated in various two dimensional electron gas systems. There have been considerable theoretical studies on the spin-splitting of the conduction band in zincblende compounds [45], [46], [47], [48]. Experimental estimates have been presented, mostly through extrapolations to zero field of magnetotransport data and (with reduced accuracy) through electron-spin resonance experiments. Other experimental measurements of this splitting have been obtained from spin relaxation [49], spin precession [50], weak-localization magnetotransport measurements [51], and zero-field Raman scattering experiments [52].

a) **AlGaAs/GaAs systems.** There has been extensive study of both two-dimensional electron gases (2DEG) and two-dimensional hole gases (2DHG) in GaAs heterostructures. Using electron-spin-resonance techniques, the spin splitting at zero magnetic field for an $Al_x Ga_{1-x}As/GaAs$ 2DEG was found by extrapolation to be about 30 µeV [53]. This result, however, was later argued to be an artifact of extrapolation from the small magnetic field range over which the experiments were performed, namely in the regime where the Fermi energy located between spin-split Landau levels [54]. Through Raman scattering a spin splitting energy of 0.74 meV was measured at zero magnetic field for an 18nm-thick quantum well. By detecting the beating pattern in the Shubnikov-de Haas oscillations, Ramvall [55] et al. measured an increase in the spin splitting from 0.46 to 0.61 meV with fixed electron concentration. Given their control of the density, this increase was attributed to gate-controlled increase in the value of $\alpha$. These measured zero-field spin splittings are quite small compared to those reported for 2DEG's in narrow-gap systems (usually 1.5 to 3.5 meV). The dominant mechanism inducing such zero-field splitting is generally considered to originate from interfacial potential. Nevertheless, in most cases direct modulation of $\alpha$ through an external gate does not seem to be unequivocally demonstrated. Magneto-transport measurements of 2DHG's in both triangular and square quantum wells of GaAs/AlGaAs were made by Störmer [56] and Eisenstein et al. [57]. Their results reveal a lifting of the spin degeneracy in triangular quantum wells, but the spin degeneracy remains in structures with symmetric wells.

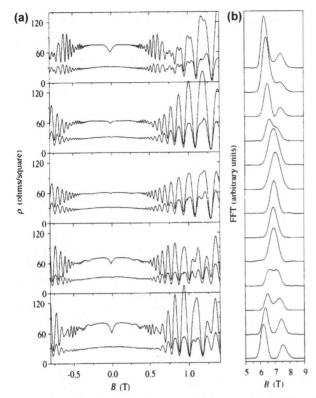

**Fig. 2.6.** Shubnikov de Haas oscillations in a two dimensional hole gas providing an especially clear demonstration of the beating characteristic of spin split bands. (After Ref. [44]). Spin splitting in these experiments was changed by applying an electric field perpendicular to the 2D plane, $E_\perp$. Through simultaneous use of both front and back gate electrodes the spin splitting could be tuned while the electron density was held fixed. (**a**) Magnetoresistance traces, all at a density of $3.3 \times 10^{11}$ cm$^{-2}$ but at different values of $E_\perp$. The data shown are from the low-mobility [00$\bar{1}$] (*top trace in* each panel) and high-mobility [$\bar{2}$33] (*bottom trace* in each panel) directions. (**b**) Fast Fourier transforms (FFT) of the Shubnikov de Haas oscillations, showing that the spin splitting is being tuned through a minimum.

Recently, very convincing evidence has been provided for wide tunability of $\alpha$ in both classes of quantum wells, via the application of both substrate and surface gate biases [39], [44].

**b) InGaAs system.** This is the system originally motivated the idea of the spin transistor, and it enables achievement of both high mobility and high zero-field spin-splitting. Extremely clear MR oscillations were first reported by Das et al. [58], and the 1.5–2.4 meV spin splittings deduced from them were attributed to the Rashba mechanism. Nitta et al. subsequently measured the spin-orbit interaction in InGaAs in an inverted InGaAs/InAlAs quantum

**Table 2.1.** Observed spin band splitting and Rashba coupling parameter in various InAs quantum wells

| Quantum Well type | $\Delta_R$ (meV) | $\alpha$ $\times 10^{-11}$eVm | Tunability | Reference |
|---|---|---|---|---|
| GaSb/InAs/GaSb | 3.7 | 0.9 | – | Luo [60] |
| AlSb/InAs/AlSb | 3.3–4.5 | 0.6 | No | Heida [38] |
| AlSb/InAs/AlSb | 0 | 0 | – | Brosig [61] |
| AlSb/InAs/AlSb | 0 | 0 | – | Sasa [62] |
| AlGaSb/InAs/AlSb | 5.6–13 | 1.2–2.8 | – | Sasa [62] |
| InGaAs/InAs/InGaAs | 5.1–6.8 | 0.6–1.1 | Yes | Nitta [41] |
| InGaAs/InAs/InGaAs | 9–15 | 2–4 | Yes | Grundler [42] |
| SiO2/InAs(p–type) | 5.5–23 | 1–3 | Yes | Matsuyama [43] |

well [37]. In this case the spin-orbit coupling was tunable from 0.6 to 1 × $10^{-11}$ eVm by a gate potential. The corresponding changes in the spin-splitting energy were from 6 to 4.5 meV. Later, in experiments performed on a quantum well with higher In concentration, anisotropic spin-orbit coupling was found [59] with values as high as 7.8 × $10^{-11}$ eVm in certain crystallographic directions. On the other hand, Engels et al. [40] also found that significant tuning of $\alpha$ is possible in a modulation doped InP/InGaAs/InP quantum well, by biasing with a top gate.

**c) InAs quantum well.** Because of its small effective mass and narrow bandgap, this system has received much attention for its potential application as a medium for realizing the spin transistor. Spin splitting is generally predicted to increase in smaller band gap materials [47]. These materials also have the additional advantage that ohmic contacts can be easily be formed between them and ferromagnetic metals since there is no Schottky barrier at the interface. Conflicting results have been presented in InAs 2DEGs over the past decade. Zero-field spin splittings as large as 3.7 meV were first observed by Luo et al. [60] in InAs/GaSb quantum wells, via magnetotransport measurements. The Rashba mechanism was purported to be dominant over intrinsic, crystal field effects and a spin-orbit coupling constant $\alpha_s = 0.9 \times 10^{-11}$ eVm was deduced for a 7.5 nm well. In the InAs/AlSb system, tunability was not initially demonstrated [38], [61], [62], although many attempts were made to realize spin injection devices in this system. Even the spin splitting itself remains controversial in this materials system. Heida et al. [38] first reported a spin splitting of 3.3–4.5 meV at zero magnetic field. The value of $\alpha$ derived from their measurements did not change significantly with electron density. Subsequent experiments indicated that the expected beat pattern in the magnetoresistance is completely absent in square quantum wells made from this materials system [61], [62]. These difficulties subsequently induced more careful heterostructure designs involving asymmetric quantum wells. These efforts have now proven to be quite successful. One method that was developed is to replace AlSb on one side of the quantum well by AlGaSb [62], which gives a fairly large band splitting (13 meV) and (values as high as $3\times10^{-11}$ eVm). The other trend is to asymmetrically insert an InAs quantum well into an inverted InGaAs/InAlAs heterostructure (InGaAs system). Nitta et al. reported that the spin splitting parameter, (for such an InAs inserted InGaAs/InAlAs quantum well) is of order of $10^{-11}$ eVm, and was directly controllable by an applied gate voltage [41]. Very recently, Grundler revealed a very pronounced band splitting on a similar structure tuned by a positive back-gate voltage [42]. An astonishingly high value for the resulting Rashba parameter, $\alpha$, was reported, and it was tunable over a factor of about 2 using an additional front gate without perturbing the well. On the other hand, comparably large tuning of $\alpha$ has been achieved in a 2DES confined in inversion layers of metal-oxide-semiconductor field-effect transistors on p-type bulk InAs [43]. A potential of triangular shape with a rather steep and high barrier is formed at the InAs/SiO$_2$ interface.

In summary, zero field spin splitting has now been observed in all of the above systems. It is generally agreed that in heterostructures such as 2DEGs, 2DHGs, and quantum wells, this is dominated by inversion asymmetry arising not from the crystal lattice, but from the artificially grown interfacial structure. It is clear that the Rashba coupling constant $\alpha$ can be tuned to a very large extent in a carefully designed selected quantum well.

The status quo regarding the remaining two prerequisites for realizing the spin transistor – spin injection and spin-selective collection – will emerge through discussions in subsequent sections of this chapter, which review recent work. We shall provide an overall summary of the state-of-the-art in Sect. 2.7.

## 2.4 Initial Experiments on Spin Injection in Semiconductor Heterostructures

### 2.4.1 Motivation and Initial Data

We begin this section by applying the model of JS to estimate the performance of a spin device in which the paramagnetic metal is replaced by an InAs quantum well. The result we obtain is of historical significance because the profound increase in the spin transresistance that was indicated provided part of the original incentive to attempt experiments in high mobility two-dimensional electron gases. However this simple estimate is expressly carried out for a diffusive system; similar ideas for *ballistic* spin polarized electron transport in high mobility electron gases are explored in Sect. 2.6.

For a two-dimensional system such as an InAs quantum well, the Fermi energy is

$$E_\mathrm{F} = \frac{\hbar^2 k_\mathrm{F}^2}{2m^*} = \frac{\pi\hbar^2 n_\mathrm{s}}{m^*} \tag{2.8}$$

with $m^*$ the electron effective mass in InAs ($\sim 1/40$ of the free electron mass
[63]), $k_F$ the magnitude of the Fermi wavevector, and $n_s$ the sheet density
of electrons. Plugging this into (2.4) and substituting $w$ (the width of the
injection region) in this two-dimensional case for $A$ (the injection area, which
was relevant in 3D), we arrive at:

$$R_s = \frac{2\eta_1\eta_2}{3e^2} \frac{\pi\hbar^2\tau_{sf}}{m^*wd} . \tag{2.9}$$

The appropriate value of $\tau_{sf}$ that should be used to characterize the decay
of non-equilibrium magnetization decay in InAs devices is not obvious at the
outset. As a first approximation, we take the value $\tau_{sf} = 1.75$ ps, obtained
from weak antilocalization experiments [64]. This is likely to be a significant
underestimate of the spin lifetime, since weak antilocalization experiments
probe the electron's phase-breaking time, a much more sensitive quantity.
For suppression of the spin accumulation, the entire distribution must be
relaxed. On the other hand, Boggess et al. [65] determined a spin relaxation
time around 19 ps by subpicosecond pump-probe measurements in bulk InAs,
even at room temperature. But such experiment involve hot carriers, and
relaxation at the Fermi surface might be quite different.

Also unknown are the appropriate values of the spin transfer efficiencies,
$\eta_1$ and $\eta_2$. To be conservative, in the absence of detailed knowledge of interfacial spin transfer in these devices, we use a value of 0.05 for the parameters $\eta_1$
and $\eta_2$. This is in the middle of the range found in the original JS experiments.
Of course the ferromagnetic metal/InAs interface is completely different than
the metal/metal interfaces of the JS experiment, but these initial guesses provided a point of departure for subsequent, more enlightened estimates. For a
device with $w = d = 1$ μm, equation (2.9) gives us $R_s \sim 200$ mΩ, about *five
orders of magnitude* larger than in the all-metallic Au devices (the "Johnson
spin transistor"), and *eight orders* larger than in the original JS experiments
on Al foils. More importantly, a signal of this strength should be easily detectable, precluding the need for sensitive SQUID voltmeters. Moreover, if
realized, this would offer substantial device potential.

There are two significant problems with the estimate above: (1) Transport
is presumed diffusive, whereas 2DEG gas channels are known to be ballistic,
or quasi-ballistic, over typical microdevice dimensions. Hence different spin-
scattering mechanisms are operative [66]. (2) The physics of spin transfer
across the interface is lumped into the phenomenal constant $\eta$, which subsumes important concerns such as the nature of interfacial magnetism, interfacial spin relaxation, and band matching and electron transmission through
the interface, which may actually be spin dependent [67–69]. These are, in
general, not likely to be constant factors but instead may exhibit sensitive dependence upon the non-equilibrium spin population itself. In recent, relevant
experiments with optically excited (hot) electrons, the detailed and complex
physics of the spin transport across interfaces is beginning to emerge [see

Chap. 5]. Similar work near the Fermi surface is clearly necessary and important to understand spin dynamics in electrical devices.

Despite these uncertainties, the surprising magnitude of the estimate for $R_{\mathrm{spin}}$ motivated several groups to pursue spin injection into InAs 2DEGs [70–73]. Experimental results obtained by the Caltech group are reviewed first.

Monzon et al. carried out an extensive set of experiments using several different device geometries patterned by photolithography (PL) and, in later work, by electron beam lithography (EBL). Two representative PL geometries are shown in Fig. 2.7. Measurements were obtained by sending a low-frequency a.c. drive current through **F1** and detecting the voltage at **F2** by a lock-in amplifier while slowly sweeping an applied, in-plane, magnetic field. In the diffusive regime these polarize/analyze experiments should produce the characteristic signature of spin transport, namely a dip in detector voltage when **F1** and **F2** are anti-aligned. This group explored a very large number of devices at both small and large separations – and many showed such signals, suggestive of spin-coupled transport. However, as exemplified in Fig. 2.7d, other forms of hysteretic behavior was also observed. After much effort this group arrived at the conclusion that, despite their similarity with the expected polarize/analyze signals, the data obtained such as represented

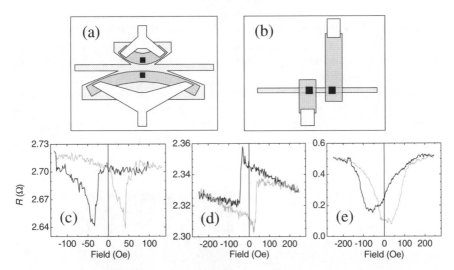

**Fig. 2.7.** Two geometries used in an InAs devices patterned by photolithography (**a** and **b**). *Black areas* denote contacts (NiFe on top of exposed InAs), *densely spotted areas* are NiFe, *sparsely spotted areas* are conducting mesas, and *white regions* show metal interconnections. Channel widths were either 3 or 6 μm and separations varied from 6 to 64 μm. In (**c**) and (**d**) data are shown for devices similar to that in (**a**), while in (**e**) shows data for a device that of (**b**). *Black traces* are for downfield sweeps, *grey traces* are for upfield.

by the traces of Figs. 2.7c,e are unlikely to be the result of spin injection and accumulation in the 2DEG. The origin of the signals that were seen could ultimately be traced to what is termed the "local Hall effect".

## 2.4.2   Local Hall Effect

Two-dimensional electron gases (2DEGs) have extremely large Hall resistances compared to metals. This is a consequence of their vastly smaller carrier densities, which makes them extremely sensitive to magnetic fields. And strong magnetic fringe fields near the edges of ferromagnets are, of course, the rule rather than the exception when working at micro- or nanometer scale dimensions. The Caltech group performed extensive experiments to understand the role that fringe magnetic fields might play in their devices [75], [74]. Fringe field close to the edge of NiFe ferromagnets can be of order 1 T, and it is reasonable to assume that such a field, though it dies off quickly away from the ferromagnet, can induce sizable "local Hall" voltages in the low density InAs. Special devices were developed which allowed separate evaluation of these local fields, in the absence of any spin injection phenomena. Figure 2.8 shows the device configuration utilized to characterize this local Hall effect. A small patterned ferromagnet film, $\mathbf{F}$, is deposited atop of a semiconducting cross-junction so that one edge of $\mathbf{F}$ is positioned directly over the center of the cross. A small AC bias current is applied to the cross junction while an external, in plane magnetic field is varied. Fringe fields from the edge of $\mathbf{F}$ induce an AC Hall voltage that is directly proportional to the magnitude and direction of the particle's magnetization. This Hall voltage is then detected with a lock-in amplifier.

Figure 2.9 shows data NiFe magnets deposited by electron beam evaporation onto Hall cross junctions patterned from 75nm thick n+ GaAs ($n \approx 10^{18}$ cm$^{-3}$). These data not only confirm the important role of these local Hall fields, but also demonstrate the utility of small, low electron density

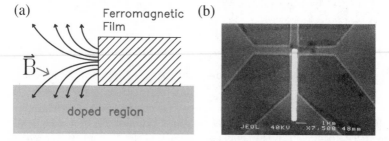

**Fig. 2.8.** (a) Schematic side view of a device showing the magnetic fringe field whose perpendicular component, $B_\perp$ induces Hall voltages in the conducting underlayer. The SEM micrograph in (b) shows a NiFe magnet of width 500 nm positioned over a GaAs cross-junction. Magnetization and current flow are directed vertically, while Hall voltage is read out across the horizontal legs.

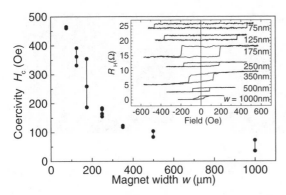

**Fig. 2.9.** $H_c$ versus $w$ for numerous magnets with nominally the same deposition conditions and with aspect ratios of 10. All data were taken at room temperature. *Vertical lines* connect data points from magnets of the same width. *Inset*: A series of hysteresis loops for magnets of aspect ratio 10, but with varying widths. Traces are offset vertically.

Hall devices for characterizing nanomagnets. Figure 2.9 displays results for a family of magnets of different widths, all with aspect ratio (length/width) ~10. The shape of the hysteresis loops, and the coercivity of the magnets, $H_c$, both vary greatly with aspect ratio and width. As seen, for the narrowest magnets the hysteresis loops become very square, indicative of quasi-single domain behavior. This work demonstrated that the switching characteristics for micro- and nanofabricated NiFe magnets were quite reproducible, indicating that the previous polarize/analyze experiments were not compromised by uncontrolled magnet variation.

These results confirmed the suspicion that much of what initially appeared to be polarize/analyze effects in spin injection was, instead, most likely the result of uncontrolled local Hall fields. The sizable Hall coefficients of semiconductors place a very high threshold for believability in such experiments; the experimenter must conclusively prove that local Hall fields play no role in their experiments. It no longer seems likely that that there is a substantial body of work following these initial efforts that lacks sufficient safeguards against the local Hall effect. One must conclude that these latter efforts fail to surmount this threshold. The Caltech group subsequently used these results to design refined configurations to suppress such phenomena in nanoscale semiconductor spin devices; this is described in the next section.

### 2.4.3   Results from Smaller, Optimized Devices

Figure 2.10 shows SEM micrographs of two devices fabricated by electron beam lithography. These devices differed in several significant ways from the previous optically-patterned devices investigated by the Caltech group. First, the magnets were completely planar. In the earlier devices the magnetic films

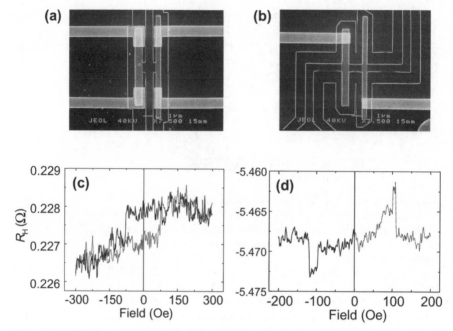

**Fig. 2.10.** SEM micrographs of InAs quantum wire devices patterned by electron beam lithography. The channels are formed by trench etching on either side of the narrow channels (**a,b**). These devices have **F1-F2** spacings of 1.5 μm. Magnet dimensions are 500nm ×10 μm and 750 nm ×7.5 μm. Data from similar devices are shown in (**c,d**).

were patterned over the edge of a mesa and the resulting height variations, albeit small, are likely to have given rise to local domain structure in the magnetic films. Second, the magnets were sufficiently narrow to be in the regime where square hysteresis loops, indicative of nearly single domain behavior, were found in the local Hall measurements. The ends of these nanomagnets, from which the fringe fields of these nanomagnets emanate, were situated far from the InAs conduction channels to minimize local Hall phenomena. Finally, each chip had a set of diagnostic devices including local Hall measurements on separate nanomagnets co-fabricated with those on the spin devices, and magnetotransport devices (without nanomagnets) allowing complete characterization of the InAs quantum wires beneath. The former indicated that magnets switched sharply (within a few Oe) and at well separated coercivities attained by patterning the magnets with different aspect ratios. The latter indicated that the carrier mobility decreased in the patterned channels decreased somewhat due to the fabrication processes, but the elastic mean free path always remained above 2 μm. Many dozens of spin devices patterned by electron beam lithography as exemplified by Fig. 2.10, were studied [70].

In contrast to the previous generations of optically patterned devices, hysteretic phenomena in this refined family of devices were extremely scarce, occurring on only a few devices. Figure 2.10c shows data from a device similar to that of Fig. 2.10a, with 750 nm **F1-F2** separation. The hysteresis loop has ~1 mΩ full-scale deflection assumed to be the result of local Hall fields. Noteworthy is the fact that the scale of these local Hall signals is hundreds of times smaller in these devices compared those observed in previous, unoptimized device configurations. This underscores the importance of controlling the role of local Hall fields in such work. Here, both moving the edges of the ferromagnets far from the conduction channel of the device and implementing quasi-single domain magnets profoundly reduced the role of magnetic fringe fields in nanoscale device operation. It was concluded that convincing spin transresistance signals were completely absent in these very large families of optimized devices. The observed magnitude of the residual, strongly suppressed local Hall signals ~1 mΩ, (Fig. 2.10c) were taken as an *upper* bound on the spin transresistance of these devices. This upper bound can, in turn, be used to establish an upper bound on the spin transfer efficiency operative at the interfaces. The previous, simplistic estimates indicated that a spin-coupled resistance of 175 mΩ should arise for a spin transfer efficiency across the ferromagnet/InAs interface of $\eta = 0.05$. As mentioned, however, in most of the devices measured by the Caltech group a signal of only a *few* mΩ would have been readily detected. Within this simple model, obtaining such small spin transresistance values would require $\eta \leq 0.005$, meaning that the spin polarization of the current injected into the InAs must be significantly less than 1%.

As a final, cautionary note, we describe data displayed in Fig. 2.10d from one particular device studied by the Caltech group. This structure was similar to the one shown in Fig. 2.10b with a 1.5 μm injector-to-collector separation. For this geometry one does not expect to see a hysteresis loop even if the LHE is significant. The trace, obtained from just this one device out of the very large family, looks tantalizingly like the type of spin injection data expected from a polarize/analyze experiment. But these data were inconsistent with expectations; instead of exhibiting two features with the same polarity, both a peak and a dip were observed. Furthermore, the widths of these features were not consistent with the measured coercive transitions in co-fabricated nanomagnets. The conclusion one must draw here is that the polarize/analyze experiment, while an excellent method of quickly determining whether magnetoelectronic phenomena are at work in a device, cannot alone provide conclusive proof of spin injection. The more rigorous demonstration provided by the Hanle effect is more convincing. Hanle effect experiments performed on the particular device of Fig. 2.10d failed to show the expected behavior. However even Hanle effect experiments are not a panacea. As shall be described in Sect. 2.6, in the ballistic or quasi-ballistic regime it may prove difficult to separate the spin injection phenomena from magnetoresistive backgrounds

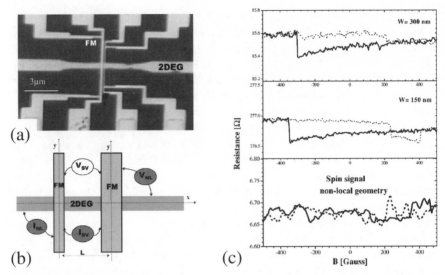

**Fig. 2.11.** (a) Scanning electron microscope micrograph of an InAs spin injection device fabricated by the Groningen group. The 1 μm wide 2DEG channel is horizontal, and two ferromagnetic electrodes are vertical. (b) Sketch of the two measurement configurations. The indices "SV" and "NL" refer to the classic spin-valve and the nonlocal geometry, respectively. (c) Non-local measurements for a Py/2DEG/Py device. Top two curves give the AMR traces for the two ferromagnetic electrodes, showing different coercive fields in one sweep direction. No spin signal is observed in any of the geometries. The dashed lines correspond to a sweep of the magnetic field towards positive fields.

arising from ballistic junction scattering phenomena. But without the clear demonstration of Hanle phenomena, any claim of spin-coupled transport must be treated with skepticism, because local Hall fields from micro- or nanomagnetic contacts are clearly unavoidable (and can play a role even in all-metal devices).

A second significant body of work on spin injection from ferromagnetic metals into small InAs 2DEG devices originated from the University of Groningen [73] employing materials developed at IMEC Leuven. An electron micrograph and schematic layout for a representative device is shown in Fig. 2.11a,b. Both non-local measurements, and standard spin-valve type measurement were made on these devices.

Fabrication involved patterning a two-dimensional electron gas (2DEG), existing within a 15 nm thick InAs layer, into 1 μm wide mesa structure. Before mesa etching and metal deposition, the whole cap layer was removed by selective wet etching. After this process the channel mobility decreased to $1.5 \text{ m}^2/\text{V} \cdot \text{s}$. Ferromagnetic contacts were then made across the 2DEG mesa, after removing the oxide by Kaufmann sputtering.

Various ferromagnetic metallic contacts (Py, Co, Ni) were investigated. These metallic ferromagnets were configured to incorporate multiple terminals on each, enabling on-chip characterization of their magnetic properties. In contrast to the characterization by Caltech group carried out on co-fabricated nanomagnets via the local Hall effect measurements, characterization here of the actual magnets used in the spin devices was possible by four-terminal anisotropic magnetoresistance (AMR) measurements. Different magnet widths were chosen to obtain different coercive fields for the injector and collector.

Figure 2.11c shows the results obtained at 4.2 K. By measuring the magnetoresistance of the ferromagnetic electrodes, as shown in the top two traces, it can be established that the two electrodes have different coercive fields. However, the non-local voltage measurements displayed in the bottom trace of Fig. 2.11c show that no spin signal was detected when the two ferromagnets switch from a parallel to an antiparallel configuration. More than 20 devices with different ferromagnetic materials were carefully characterized. None showed signals that could be attributed to spin injection, confirming the results of the Caltech group.

An number of groups have recently reported the observation of very weak spin injection signals among these are: Hu et al. [109], Gardelis et al. [71], Hammar et al. [106]. However, our view is that, to date, concerted effort failed to produce a *conclusive* demonstration of electrical spin injection phenomena in ohmic all-electrical ferromagnetic/semiconductor (2DEG) devices. Principal obstacles present in such structures appear to be the strong "local Hall effect", as described above, but perhaps more importantly, the effects of "conductivity mismatch", as described in the following section, which are most severe for *Ohmic* contacts between ferromagnetic metals and non-magnetic semiconductors. In this context, it is noteworthy that the all of the initial efforts centered upon *Ohmic* contacts to the electron gas. This was motivated by the desire to realize low impedance devices which, it was presumed, would be most amenable to large scale integration for memory or logic applications. In the intervening period, however, engineering for memory devices employing (by comparison) relatively high impedance magnetic tunnel junctions has been developed. So the presumed constraint to create Ohmic devices has been lifted. As we shall describe, this "evolved" perspective opens important new horizons for efficient spin injection into semiconductors

## 2.5   Spin Injection in Diffusive Systems

This section deals with the basic physics of spin injection, accumulation and detection in media where the transport is diffusive, which means that the electron mean free path $\ell_0$ is shorter than the dimensions of the system. For metallic systems this is usually the case. For semiconductors both diffusive and ballistic (e.g. in two-dimensional electron gases) transport regimes are encountered. A description of spin injection in the ballistic case is discussed in Sect. 2.6. Here we discuss the linear transport regime, where the measured voltages are linear functions of the applied currents. This should be contrasted with the non-linear transport regime, which is relevant for various semiconductor diode structures (see Chap. 5). For further reading, particularly concerning the role of spin accumulation for the giant magneto resistance (GMR) effect in multilayer structures we refer to reviews [76], [77].

### 2.5.1   Basic Model for Spin Transport in Diffusive Systems

We first give a review of the basic model for the spin transport (which we will call the "standard" model). The description of electrical transport in a ferromagnet in terms of a two-current (spin-up and spin-down) model dates back to Fert and Campbell [78]. van Son et al. [79] extended the model to describe transport through ferromagnet-normal metal interfaces. A firm theoretical underpinning, based on a Boltzmann transport description, has been given by Valet and Fert [80]. They applied the model to describe the effects of spin accumulation and spin dependent scattering on the giant magnetoresistance effect (GMR) in magnetic multilayers. This model allows for a detailed quantitative analysis of the experimental results. An alternative model, based on thermodynamic considerations, has been put forward and applied by Johnson [14]. In principle both models describe the same physics, and should therefore be equivalent. However the Johnson model has a drawback in that it does not allow a direct calculation of the spin polarization of the current ($\eta$ in Ref. [81]), whereas in the standard model all measurable quantities can be directly related to the parameters of the system.

In the standard model transport in the ferromagnet is described by spin dependent conductivities:

$$\sigma_\uparrow = N_\uparrow e^2 D_\uparrow \text{ with } D_\uparrow = 1/3 \, v_{F\uparrow} l_\uparrow \tag{2.10a}$$

$$\sigma_\downarrow = N_\downarrow e^2 D_\downarrow \text{ with } D_\downarrow = 1/3 \, v_{F\downarrow} l_\downarrow \tag{2.10b}$$

Here $N_{\uparrow,\downarrow}$ denotes the spin dependent density of states ($DOS$) at the Fermi energy, and the spin dependent diffusion constants $D_\uparrow$ and $D_\downarrow$, expressed in spin-dependent Fermi velocities $v_{F\uparrow,\downarrow}$ and electron mean free paths $l_{\uparrow,\downarrow}$. Note therefore that the spin dependence of the conductivities is determined by *both* density of states and diffusion constants. This should be contrasted with magnetic F/F or F/N tunnel junctions, where the spin polarization of the tunneling electrons is determined by the $DOS$ only. According to (2.10a), (2.10b), current is spin-polarized in a bulk ferromagnet, with a polarization is given by:

$$\alpha_F = (\sigma_\uparrow - \sigma_\downarrow)/(\sigma_\uparrow + \sigma_\downarrow) \tag{2.11}$$

The next step is the introduction of spin flip processes, which can be described by a spin flip time $\tau_{\uparrow\downarrow}$, the average time to flip an up-spin to a down-spin, and $\tau_{\downarrow\uparrow}$ for the reverse process. Note that the detailed balance principle imposes that $N_\uparrow/\tau_{\uparrow\downarrow} = N_\downarrow/\tau_{\downarrow\uparrow}$, so that in equilibrium no net spin scattering takes place. This implies that in a ferromagnet $\tau_{\uparrow\downarrow}$ and $\tau_{\downarrow\uparrow}$ are generally not the same. Usually these spin-flip times are larger than the momentum scattering time $\tau = l/v_F$. The transport can then be described in terms of parallel diffusion of the two spin species, where spin-flip processes couples the carrier densities in the two spin reservoirs. It should be noted, however, that in certain ferromagnets (e.g. various permalloy alloys [82]), the spin flip times can become comparable to the momentum scattering time. In this case an (additional) spin-mixing resistance arises [76], which will not be discussed further here.

The effect of spin-flip scattering can be described by the following equation (assuming diffusion in one dimension only):

$$D\frac{\partial^2(\mu_\uparrow - \mu_\downarrow)}{\partial x^2} = \frac{\mu_\uparrow - \mu_\downarrow}{\tau_{sf}} \tag{2.12}$$

Here $D = D_\uparrow D_\downarrow (N_\uparrow + N_\downarrow)/(N_\uparrow D_\uparrow + N_\downarrow D_\downarrow)$ is the spin-averaged diffusion constant, and the spin-relaxation time $\tau_{sf}$ is given by: $1/\tau_{sf} = 1/\tau_{\uparrow\downarrow} + 1/\tau_{\downarrow\uparrow}$. The general solution of (2.10) and (2.12) for a uniform one-dimensional ferromagnet or nonmagnetic wire is now given by:

$$\mu_\uparrow = a + bx + c/\sigma_\uparrow \exp\left(-\frac{x}{\lambda_{sf}}\right) + d/\sigma_\uparrow \exp\left(\frac{x}{\lambda_{sf}}\right) \tag{2.13a}$$

$$\mu_\downarrow = a + bx - c/\sigma_\downarrow \exp\left(-\frac{x}{\lambda_{sf}}\right) - d/\sigma_\downarrow \exp\left(\frac{x}{\lambda_{sf}}\right) \tag{2.13b}$$

Here, as before, the spin diffusion length is $\lambda_{sf} = \sqrt{D\tau_{sf}}$. The coefficients $a,b,c,$ and $d$ are determined by the boundary conditions imposed at the junctions where the conducting channel is coupled to other conducting channels, etc. In the absence of spin flip scattering at the interface, the boundary conditions are: 1) continuity of $\mu_\uparrow$, $\mu_\downarrow$ at the interface, and 2) conservation of spin-up and spin down currents $j_\uparrow, j_\downarrow$ across the interface.

It should be noted that physics of spin injection and detection with F/N tunnel junctions is quite different from that of clean, Ohmic F/N contacts. The most important difference is that for tunnel contacts the spin polarization of the current is determined by the composition and properties of the first few atomic layers of the ferromagnet next to the tunnel barrier. A non-magnetic ("magnetically dead") region of only a few atoms thick can completely suppress any polarization in the transmitted current. By contrast, as will be shown in the next section, for clean contacts the spin polarization is determined by a region of thickness $\lambda_F$ away from the interface. Since usually $\lambda_F > l$, this can be considered as a "bulk" property of the ferromagnet.

## 2.5.2   The F/N Interface

van Son *et al.* [79] applied (2.10a)–(2.13b) to describe the spin accumulation and spin-coupled resistance at a clean F/N interface. Figure 2.12 shows how the spin polarized current in the (bulk) ferromagnet is converted into a non-polarized current in the paramagnetic region (a non-magnetic metal or semiconductor) away from the interface. As can be seen, two phenomena occur. First a "spin-coupled" interface resistance arises given by:

$$R_{\mathrm{I}} = \frac{\Delta\mu}{eI} = \frac{\alpha_{\mathrm{F}}^2(\sigma_{\mathrm{N}}^{-1}\lambda_{\mathrm{N}})(\sigma_{\mathrm{F}}^{-1}\lambda_{\mathrm{F}})}{(\sigma_{\mathrm{F}}^{-1}\lambda_{\mathrm{F}}) + (1 - \alpha_{\mathrm{F}}^2)(\sigma_{\mathrm{N}}^{-1}\lambda_{\mathrm{N}})} \tag{2.14}$$

The second phenomenon is that at the interface the electrochemical potentials $\mu_\uparrow, \mu_\downarrow$ of spin-up and spin-down electrons diverge. This implies that spin accumulation occurs, which has the maximum value at the interface:

$$\mu_\downarrow - \mu_\uparrow = \frac{2\Delta\mu}{\alpha_{\mathrm{F}}} \tag{2.15}$$

with $\Delta\mu$ given by (2.14). In addition the expression for the spin polarization of the current at the interface is

$$P = \frac{I_\uparrow - I_\downarrow}{I_\uparrow + I_\downarrow} = \frac{\alpha_{\mathrm{F}}\sigma_{\mathrm{N}}\lambda_{\mathrm{F}}}{\sigma_{\mathrm{N}}\lambda_{\mathrm{F}} + (1 - \alpha_{\mathrm{F}}^2)\sigma_{\mathrm{F}}\lambda_{\mathrm{N}}} \tag{2.16}$$

For conventional ferromagnets $\alpha_{\mathrm{F}}$ is expected to be in the range $0.1 < \alpha_{\mathrm{F}} < 0.7$. Thus, the expressions above show that the magnitude of spin-coupled resistance, spin accumulation and polarization of the current is essentially limited by $\sigma_{\mathrm{N}}^{-1}\lambda_{\mathrm{N}}$ or $\sigma_{\mathrm{F}}^{-1}\lambda_{\mathrm{F}}$, whichever is largest. Since the condition $\lambda_{\mathrm{F}} \ll \lambda_{\mathrm{N}}$ holds in almost all cases for metallic systems, this implies that the spin-flip length in the ferromagnet is the limiting factor. This problem becomes

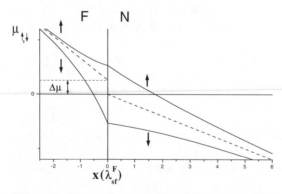

**Fig. 2.12.** Electrochemical potentials (or densities) of spin-up and spin-down electrons with a current $I$ flowing through an F/N interface. Both spin accumulation as well as spin coupled resistance can be observed (see text). The figure corresponds to $\lambda_{\mathrm{N}} = 5\lambda_{\mathrm{F}}$.

progressively worse, when (high conductivity) metallic ferromagnets are used to inject spin polarized electrons into (low conductivity) semiconductors. We will discuss this "conductivity mismatch" problem in Sect. 2.5.8.

### 2.5.3   Spin Accumulation in Multiterminal Spin Valve Structures

The fundamental way to study spin accumulation effects is to use spin-valves. These can be F/F, F/N/F, or the corresponding multilayer (...F/N/F/N/...) structures. Spin accumulation effects play an important role in these systems (for reviews see [76], [77]). However, the effects of spin accumulation are often difficult to separate from the magnetoresistance of the ferromagnets – effects such as anisotropic magnetoresistance (AMR), spin dependent interface scattering, (anomalous) Hall effects, etc. A solution is provided by using a multiterminal geometry, in which separate current and voltage contacts are and can be configured for a non-local measurement. This technique was employed by JS, in their spin injection on single crystal Al at low temperature (Fig. 2.2). However, due to the relatively large size of their samples and the extremely low resistances (in the $n\Omega$ range) that resulted, SQUID based measurements were required to measure the small signals. A second experimental system employing nonlocal measurements was the Johnson spin transistor (Sect. 2.2.4). This technique was also employed in the experiments of Jedema et al. [83], who fabricated and studied mesoscopic spin valve structures for the study of spin accumulation in metal systems. The device geometry is depicted in Fig. 2.13. Two 50 nm thick ferromagnetic permalloy (Py) strips, with composition of $Ni_{0.8}Fe_{0.2}$, are deposited on a substrate. Their aspect ratio (length/width) is chosen in order to obtain different coercive fields. In a second evaporation step, a Cu cross is deposited. Care is taken that the interface between the Py and the Cu is kept clean, by using Kaufman ion source to ion mill any oxide or contamination before deposition.

Referring to Fig. 2.13, in the "potentiometric" [14], or (equivalently) "nonlocal" [83] geometry, current is sent from 1 to 5 and the voltage is measured between 6 and 9. The voltage drop in the injector ferromagnet is not measured, whereas there is no net (charge) current, and therefore no voltage drop, within the ferromagnetic electrode that measures the voltage. Therefore all magnetoresistance effects emanating from the ferromagnets themselves are eliminated from these measurements.

By matching solutions (2.13a), (2.13b) at the F/N interfaces and the center of the Cu cross, one obtains the following expression for the measured resistance in parallel $R_{\uparrow\uparrow}$, and in anti-parallel configuration $R_{\uparrow\uparrow,\downarrow\uparrow} = \pm\frac{\Delta R}{2}$ [83]:

$$R_{\uparrow\uparrow,\downarrow\uparrow} = \pm\frac{1}{2}\frac{\alpha_F^2\lambda_N \exp(-L/2\lambda_N)}{\sigma_N A(M+1)[M\sinh(L/2\lambda_N) + \cosh(L/2\lambda_N)]} \qquad (2.17)$$

Here $A$ is the cross sectional area of the Cu wires, and $M = (1 - \alpha_F^2)(\sigma_F\lambda_N/\sigma_N\lambda_F)$. In typical experiments, one expects $\lambda_N \gg \lambda_F$. Then (2.17) has two

**a**

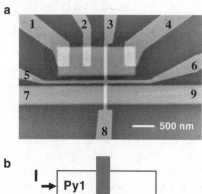

**b**

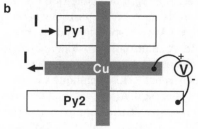

**Fig. 2.13.** Sample layout in the experiment of Jedema et al. (**a**) Scanning electron microscope image of the mesoscopic spin valve junction. The two wide strips are the ferromagnetic electrodes Py1 and Py2. The vertical arms of the Cu cross (with contacts 3 and 8) lay on top of the Py strips, the horizontal arms of the Cu cross form contacts 5 and 6. Contacts 1, 2, 4, 7, and 9 are attached to Py1 and Py2 to allow four-terminal AMR measurements of the Py electrodes. (**b**) Schematic representation of the non-local measurement geometry. Current is entering from contact 1 and extracted at contact 5. The voltage is measured between contacts 6 and 9.

limits: For $L \ll \lambda_N$ the spin signal scales as: $\Delta R \propto 1/L$, whereas for for $L \gg \lambda_N$ the signal is attenuated exponentially: $\Delta R \propto \exp(-L/\lambda_N)$. Also, depending on the ratio of spin flip lengths and conductivities in the ferromagnet and nonmagnetic regions, the signal can be attenuated, even in the absence of spin-flip scattering. We will discuss this "conductivity mismatch" in Sect. 2.5.8.

For comparison, in the "conventional" spin valve measurement current is send from electrode 1 to 7 and the voltage is measured between 4 and 9. For this case the following expression is obtained:

$$R_{\uparrow\uparrow,\uparrow\downarrow} = R_{\text{series}} \pm \Delta R \qquad (2.18)$$

with $\Delta R$ given by (2.17). The magnitude of the spin signal is twice that in the other geometry. However, as pointed out already, this geometry suffers from the fact that a resistance $R_{\text{series}}$ of the ferromagnet is measured. The AMR contribution is given by $\Delta R/R = C\cos^2(\varphi)$, where $\varphi$ is the angle between the magnetization direction and the current direction, and $C$ is of the order of 3%. This shows that in the case of imperfect switching, when the

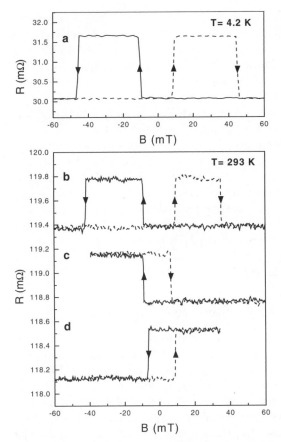

**Fig. 2.14.** Spin valve effect at T=4.2 K (**a**), and room temperature (**b**) in the non-local geometry for a sample with 250 nm Py electrode spacing. An increase in resistance is observed, when the magnetization configuration is changed from parallel to anti-parallel. The *solid (dashed) lines* correspond to the negative (positive) sweep direction (**c**),(**d**) illustrate the "memory effect". For clarity (**c**) and (**d**) are offset downwards. The sizes of the Py1 and Py2 electrodes are $2.0 \times 0.5 \ \mu m^2$ and $14 \times 0.1 \ \mu m^2$.

magnetization does not switch through 180° at the coercive field, a problem arises, this has been indeed observed by the authors of Ref. [83].

### 2.5.4  Observation of Spin-Injection and Spin-Accumulation in an All-Metal Spin Valve

Jedema et al. [83] succeeded in using the principles described above to demonstrate spin injection and detection in an all metal spin valve structure. Representative data from their measurements are shown in Fig. 2.14. When the applied magnetic field is swept from negative to positive, an increase in the resistance is obtained when the ferromagnetic strip with the smallest coercive field switches, and an anti-parallel magnetization is obtained. When the coercive field of the second strip is exceeded, a drop in the resistance is observed, corresponding to parallel orientation of the magnetization of the injector and collector. The hysteresis behavior observed in Figs. 2.14c,d corresponds to the situation where the direction of the magnetic field sweep is reversed, while the second strip has not switched yet. As can be seen in the data, the spin valve signal increases when the temperature is reduced to 4.2 K.

Equation (2.17) shows that there are three unknowns in the experiment: $\alpha_F$, $\lambda_F$, and $\lambda_N$. In order to determine the values of these, a series of measurements was performed with increasing electrode spacing $L$, ranging from 250 nm to 2 μm. By fitting the measured $\Delta R$ to (2.17), the spin flip length could be extracted. At room temperature it was deduced that $\lambda_N \approx 350$ nm, and $\lambda_N \approx 1000$ nm at 4.2K. It was also concluded that the fitting procedure was not sensitive enough to obtain $\alpha_F$ and $\lambda_F$ separately, however the product $\alpha_F \lambda_F$ was deduced to be ∼1 nm. Using an estimated value $\lambda_F = 5$ nm obtained from the literature [82] yields $\alpha_F \sim 0.22$.

### 2.5.5  Comparison with the Johnson "Spin Transistor"

The signals of Jedema et al. [83], when scaled to the cross sections utilized in the impure Au film devices of Ref. [17] (the "Johnson spin transistor"), are about $10^4$ times *smaller* than obtained in that previous work. However, in that earlier work it was necessary to invoke a spin-polarization exceeding 100% to explain the results in terms of spin accumulation [67], [84]. In contrast the recent results of Ref. [83] correspond to a spin polarization $P$ at the interface of about 1–2%. (Note that the amplitude of the spin signal scales with the square of the spin polarization). Given the unexplained discrepancies of the earlier work, and the more consistent values obtained in the recent work, it is our opinion that the results of [17] cannot be associated with spin injection and spin accumulation.

In this light, it should be noted that for the sandwich structure used in [17], the current not only flows perpendicular to the F/N/F layers, but also parallel to them. Therefore the magnetoresistance of these layers (in particular the AMR) can contribute to the measured signal, if the magnetic switching of the layers is not perfect. One expects this to be the case if a multidomain structure is formed before or during the switching. The data of [17] indeed show that the switching of the layers is not abrupt. Jedema et al. "conventional" spin valve measurements shows that a contribution of the AMR to the signal could also be observed. In particularly unfavorable cases, this could fully overwhelm spin accumulation signals.

## 2.5.6 Future Prospects for Spin Accumulation and Spin Transport in All Metal Devices

The average time for spin relaxation time found in the recent metal-based spin injection work at Groningen is about $10^3$ times the momentum relaxation time. Similar ratios have also been found from conduction electron spin resonance (CESR) experiments, which allow the determination of the spin flip time $T_2$ from the width of the observed resonance [85]. The spin-orbit interaction was found to be the dominant mechanism for spin flip in non-magnetic materials. It can be induced by phonon scattering, or by scattering from static imperfections, such as impurities, dislocations and surface scattering. This implies that the spin-diffusion length $\lambda_{sf} = \sqrt{D\tau_{sf}}$ scales in direct proportion to the electron mean free path. We believe that by reducing the scattering (in particular surface scattering which is the limiting factor for thin films), it should be possible to extend the spin-flip length to several micrometers. This will make it possible to study the effects of spin accumulation in more complicated devices, such as the "spin-flip" transistor proposed by Brataas et al. [86]

The second issue is that the observed spin signals in all-metal lateral spin valve devices are small, in the m$\Omega$ range. This is due to their high conductivity, but also, as pointed out above, is due to the fact that the polarization obtained so far in lithographically defined structures is only about 1–2%. As will be discussed in the next section, however, by using tunnel junctions as injectors and detectors it should be possible to increase the polarization to about 10% or more, and this should make it possible to increase the signals to about 1$\Omega$.

## 2.5.7 Spin Injection in a Diffusive Semiconductor

The discussion above shows that, since the time of the anomalous results of [17], significant additional understanding of spin injection and accumulation in thin film metallic systems has been attained. These recent experiments can be fully understood within the framework of the standard model, and the parameters extracted are in acceptable agreement with those obtained from CESR experiments and the analysis of magnetotransport in multilayers. This raises once again the question whether spin injection in semiconductors can also be achieved in similar fashion, i.e. by using direct, Ohmic injection from ferromagnetic electrodes. In the next section the diffusive model is applied to this situation.

## 2.5.8 Conductivity Mismatch

If we apply the standard model to deduce the expected (nonlocal) signal for the device geometry pictured in Fig. 2.15a we find that

$$R_{\uparrow\uparrow,\downarrow\uparrow} = \pm \frac{\alpha_F^2 \lambda_N}{\sigma_N A[(M^2+1)\sinh(L/\lambda_N) + 2M\cosh(L/\lambda_N)]} \tag{2.19}$$

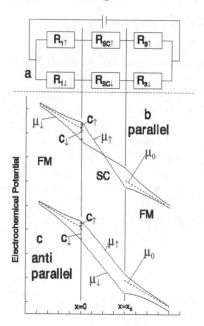

**Fig. 2.15.** (a) Simplified resistor model for a device consisting of a semiconductor (SC) with two ferromagnetic contacts (FM) 1 and 3. The two independent spin channels are represented by the resistors $R_{1\uparrow,\downarrow}$, $R_{SC\uparrow,\downarrow}$, and $R_{3\uparrow,\downarrow}$. (b) and (c) show the electrochemical potentials in the three different regions for parallel (b) and antiparallel (c) magnetization of the ferromagnets. The *solid lines* show the potentials for spin-up and spin-down electrons, the *dotted line* for $\mu_0$ (undisturbed case). For parallel magnetization (b), the slopes of the electrochemical potentials in the semiconductor are different for both spin orientations. They cross in the middle between the contacts. Because the conductivity of both spin channels is equal, this results in a (small) spin-polarization of the current in the semiconductor. In the antiparallel case (c), the slopes of the electrochemical potentials in the semiconductor are equal for both spin orientations, resulting in unpolarized current flow. (Note that the slope of $\mu$ in the metals is exaggerated). From [87].

with $M - 1 = (1 - \alpha_F^2)(\sigma_F \lambda_N / \sigma_N \lambda_F)$. The crucial point to be noted here is that, unlike the case for metal/metal contacts, for metal/semiconductor (2DEG) contacts the ratio of conductivities is about $10^3$, and therefore one finds $M \approx 10^3$. As seen from (2.19), this leads to a very large attenuation of the spin signal, even when $L \ll \lambda_N$, and no spin flip scattering takes place within the 2DEG. Although it cannot be excluded that there may be additional reasons for the suppression of the spin signal in metal/semiconductor devices (such as random spin precession induced by the spin-orbit interaction in InAs channel, or non-ideal interfacial transfer), it follows that the "conductivity mismatch" has very major consequences for this system. The

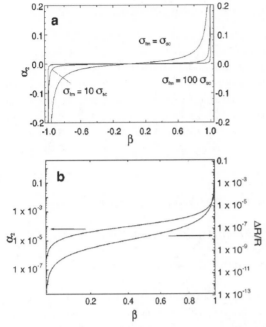

**Fig. 2.16.** Dependence of the spin polarization of the current in the semiconductor $\alpha_2$ and $\Delta R/R$ on the spin polarization $\beta$ in the bulk ferromagnet. In (**a**) $\alpha_2$ is plotted over $\beta$ for different ratios $\sigma_{\mathrm{fm}}/\sigma_{\mathrm{sc}}$. For a ratio of 100, 2 is well below 0.1% for $\beta < 99\%$. In (**b**), again $\alpha_2$ is plotted versus $\beta$ with $\sigma_{\mathrm{fm}}/\sigma_{\mathrm{sc}}=100$, with the corresponding values for $\Delta R/R$ on a logarithmic scale. For $\beta$ between 0 and 90%, $\Delta R/R$ is smaller than $10^{-7}$ and thus difficult to detect in the experiment. After Ref. [87].

conductivity mismatch problem has been described in detail by Schmidt et al. [87]

The fundamental reason for suppression due to conductivity mismatch is that the ratio of spin-up and spin-down currents is determined by the total resistance encountered in a region of the length $\lambda_{\mathrm{F}} + \lambda_{\mathrm{N}}$. In the above case the spin-independent resistance of the semiconductor dominates, leading to an equalization of both currents, see Fig. 2.15.

Figure 2.16 shows the calculated results from [87]. It is clear that, except for $\alpha_{\mathrm{F}}$ very close to one, the attenuation of the spin signal due to conductivity mismatch is substantial. In our opinion a direct ohmic contact between a high conductivity ferromagnetic injector/detector and a low conductivity non-magnetic medium will always yield a strong supression of spin injection phenomena.

### 2.5.9   Possible Solutions to Conductivity Mismatch

Two solutions offer themselves: The first is to use injectors which have $\alpha_F$ close to 1, e.g. using magnetic semiconductors (see Chap. 1). For the case where non fully spin polarized ferromagnets are used, a possible solution to the problem is the use of tunnel barriers, so that the spin dependent resistance of the tunnel barrier becomes comparable to the spin independent resistance of the normal metal [73], [88]. It should be noted however that the mechanism for spin selectivity changes, and this will impose more stringent conditions on the fabrication technology. Recent work on tunnel injection will be discussed further in Sect. 2.7.2.

## 2.6   Spin Transport in the Ballistic Regime

Pursuit of spin injection phenomena in two-dimensional electron gas systems has been motivated by the expectation spin transport phenomena should become greatly enhanced in these materials (see Sect. 2.4.1 in this chapter). This is especially true for devices with submicron dimensions, where mean free paths for momentum scattering can easily exceed device dimensions. Furthermore, sufficiently narrow conductors of such materials become quantum wires, in which current is carried by only a small number of transverse subbands [27]. If a quantum wire is short enough, intersubband scattering becomes inconsequential and quantization of electrical conductance is manifested [28], [29]. Most previous modeling of electrical spin injection and spin transport phenomena has been predicated on the assumption that electron transport is *diffusive*. In this section, we describe simple modeling allowing exploration of spin injection and precession phenomena when transport is *ballistic*. Although the approach taken is straightforward in concept, it provides a very important intuitive understanding about the unique aspects of ballistic spin transport, and underscores significant gaps in our current understanding of the problem's details. These models also provide important benchmarks for observational expectations in ballistic systems. The previous discussion makes if it clear that this is in a realm where the signatures of spin transport phenomena may be ambiguous or even obfuscated by other phenomena. As mentioned (Sect. 2.4.2), semiconducting systems can be profoundly sensitive to extraneous magnetic fields given their large Hall coefficients. The models described here culminate in predictions for observable precessional phenomena, analogous to the Hanle effect in the diffusive regime, which should constitute definitive experimental demonstrations of spin injection, transport, and detection in all-electrical semiconductor devices. To date these remain to be realized.

Ballistic transport in nanoelectronic devices is usually described within the Büttiker-Landauer model. Current flow is pictured as reservoir-to-reservoir transport – in contrast to the case for diffusive systems; here we do not usually define local electrochemical potential, which would give rise to a continuous voltage drop within the conducting channel. Instead, current flow is viewed as a ballistic, or, in certain situations, phase-coherent, flux of carriers that propagates without scattering through the conducting channel. Equilibrium occurs solely at the reservoirs formed by Ohmic contacts at the ends of the channels. Conductance is then described as the ratio of the electron flux (proportional to the current) to the steady state electrochemical potentials of these reservoirs.

For spin electronics, one needs to transform these ideas to the cases where transport is spin polarized. One is immediately faced with a complication: what is the underlying physics of spin-selective contacts? Büttiker's multi-terminal formalism is based on the ansatz that the reservoirs are *perfectly absorbing*. Specifically, he assumed that every electron impinging from the channel upon a "collecting" contact is admitted, and thereby contributes to the development of that reservoir's electrochemical potential in steady state. For the rather low density electron gases which form quantum -wires, -point contacts and -dot structures, this is physically justifiable. The density of states of the metallic contacts is immense compared to that of the channels themselves. It is therefore almost a certainty that a state can always found within the reservoir that is well-matched, both in energy and momentum, to the discrete "modes" of the electron gas channels.

It is not so clear, however, what the correct picture should be for *spin-selective* contacts. Is it possible that spin-polarized reservoirs can be created that will reflect impinging carriers in the "wrong" spin state? For the ferro-magnetic metal/semiconductor contacts, this seems highly unlikely. Except for the so-called "half metallic" ferromagnets, there is appreciable density of states in both (polarized) $d$-like and (unpolarized) $s$-like bands at the Fermi surface in metallic ferromagnets (Fig. 2.17). Any carrier of the wrong spin type that impinges upon a polarized $d$-band state is quite likely to find a match within the $s$-band. This brief discussion, rather than striving to be definitive, is meant to serve as an introduction to important issues regarding spin-polarized contacts in the ballistic regime.

The semiconducting spin injection transistor envisaged by Datta and Das [26] was predicated upon single-subband, ballistic transport from source to drain. In this first formulation of the problem, the specific details of how spin polarized current might be injected into the channel were side stepped – their focus was upon controllable, spin-selective *transport* within the channel itself. Below we shall explore spin transport and precession in ballistic regime in more detail, investigating the role that effects such as multisubband trans-port and scattering will play in real devices. We can also make some general statements about spin transmission at contacts. Concrete microscopic models for the contacts, however, are only possible for the case of specific, well-controlled experimental realizations of the interfacial transport involved. Recent advances in epitaxy are now permitting growth of heterostructures with nearly ideal interfaces between ferromagnetic and non-magnetic semi-

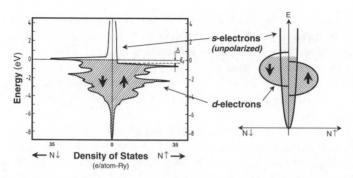

**Fig. 2.17.** Calculated DOS for nickel, [after J. Callaway and C.S. Wang: Phys. Rev. B **63**, 1096 (1973)], and a simple Stoner model for ferromagnetic electrodes.

conductors. This research is greatly accelerating our rate of understanding of interfacial spin transmission – and it should also begin to clarify what is physically possible regarding the realization of spin selective contacts.

A three- or four-terminal spin transistor device (cf. Figs. 2.1 and 2.2) is most clearly characterized by spin transresistance $R_s$, defined in Fig. 2.18a, which provides the most direct demonstration of spin transport. In the *diffusive* regime for the geometry shown, if the current contact **F1** is replaced with one that is unpolarized no voltage will appear between the analyzer contact **F2** and a suitably defined ground reference **R**. In this case these voltage contacts, being well outside the net current path, remain at the same equipotential. However, if the current from **F1** is spin polarized, the injected magnetization can lead to steady state *spin accumulation* that persists over the entire length of the channel when $\lambda_{sf} \geq L$. This spin accumulation induces disequilibrium between the electrochemical potentials at the collecting contacts on the far end of the device, **F2** and **R**, and thereby yields a finite value of $R_s$. As before $\lambda_{sf} = \sqrt{\ell_0 \ell_s / 2}$ is the spin diffusion length, $\ell_s = v_F \tau_s$ and $\ell_0 = v_F \tau_0$ are the spin and momentum mean free paths, $\tau_s$ and $\tau_0$ are the effective spin and momentum relaxation times.

### 2.6.1 Multiprobe Model for Ballistic Spin Polarized Transport

To describe spin transport in the *ballistic* regime, Büttiker's picture for mesoscopic transport within a multiprobe conductor is employed [90]. Here it is augmented with a simple, experimentally-motivated model that describes the essential physics of spin-selective contacts [89]. The procedure involves four principal steps: (a) a simple description of spin-resolved contacts is developed, based upon careful consideration of the ferromagnetic/semiconductor contacts in real devices. (b) An 8-reservoir model is constructed, in the spirit of Büttiker's formalism, to describe the spin injection experiment. (c) Careful consideration of the boundary conditions yields constraints upon the spin-resolved currents and chemical potentials. This process leads to an expression for $R_s$ in terms of reservoir-to-reservoir, *spin-resolved* transmission probabilities, $T_{ij}^{\alpha\beta}$, of the 2DEG that forms the device conduction channel. Here the indices $i$, $j$ specify the reservoirs themselves, and $\alpha, \beta$ their constituent spin bands. Finally, (d) the requisite $T_{ij}^{\alpha\beta}$ are calculated semiclassically, using a modified Monte Carlo numerical technique. This is carried out by following the electrons' ballistic trajectories *and* the phase of their spin wavefunctions as they pass through the device, while the phase of their spatial wavefunctions is ignored. To account for the properties of real devices, it is also necessary to include both junction scattering and scattering at the surfaces and in the interior of the channel. For unpolarized ballistic systems, this semiclassical approach has provided a theoretical description remarkably consistent with experimental data at T $\sim$ 4K, the regime where the electron phase coherence length is smaller than typical dimensions of nanoscale devices [91]. We expect that the semiclassical model is capable of providing equally valuable insights for spin polarized systems.

Figure 2.18b depicts a simple model for spin-selective contacts based on experimental considerations. It comprises two elements: **F2**, a fully spin-polarized reservoir which is in perfect contact with what we shall term, a disordered (i.e. momentum-randomizing) but spin-preserving region (hereafter denoted as DSPR). The DSPR consists of separate spin-up and spin-down bands that internally equilibrate, in the sense of momentum relaxation, while preserving the overall spin imbalance. The separate spin-resolved reservoirs comprising the DSPR (1↑, 1↓, 2↑, 2↓) model low mobility regions always present beneath unalloyed ferromagnetic metal contacts, e.g. in typical InAs devices (Sect. 2.4 of this chapter). Disorder within these special regions yields significant momentum randomization and, hence, a short $\ell_0$. However, in contrast to the usual picture describing unpolarized reservoirs [90], we assume these special contacts are small compared to $\lambda_{sf}$, thus any spin disequilibrium within them is preserved. This, in fact, is consistent completely general and with the more restrictive constraint, $\lambda_{sf} \geq L$, which is fundamental to *any* *spin injection experiment*. If significant spin relaxation occurs anywhere in a "spin transistor", including in the vicinity of the ferromagnetic contacts, spin-selective transport will be suppressed. For sake of clarity, we consider the most ideal situation, initially assuming that **F1** and **F2** are fully polarized at the Fermi surface (half-metals). This approximation serves to illustrate the most important aspects of the underlying physics. Of course, many complexities in real devices may diminish spin transport effects, such as partial spin polarization and additional unpolarized bands at $E_F$ in **F1** and **F2**, variations in band structure near the interfaces, magnetic disorder and spin scattering at the $F/S$ interface, momentum and spin scattering in the 2DEG, and thermal smearing. Here our aim is to clarify at the outset what may be expected in ballistic systems under the most favorable of conditions.

Measurement of $R_s$ involves four terminals (Fig. 2.18a), two that are spin-selective (**F1**, **F2**), and two that are conventional, i.e. momentum- and spin-

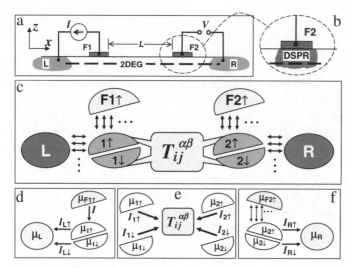

**Fig. 2.18.** Model for ballistic spin transport in a two-dimensional electron gas. (a) Measurement configuration: a current $I$ is injected through the 2DEG via a ferromagnetic contact **F1** and an Ohmic contact **L**. The spin transresistance $R_s = V/I$ arises from spin-polarized carriers traversing a distance $L$ from the net path of the current, which induce a nonlocal voltage, $V$, between a second, similar, pair of contacts **F2** and **R**. (b) The conductor beneath the ferromagnetic contacts (DSPR) is assumed to be a *disordered*, but *spin preserving* region. (c) The full-eight-reservoir model; *complete ellipses* represent spin-relaxing reservoirs, and *half ellipses* represent spin-resolved reservoirs. **F1** and **L** are current contacts, **F2** and **R** are voltage probes. $T_{ij}^{\alpha\beta}$ denotes the 2DEG device channel in which spin precession occurs. Other multimode leads are denoted by three *arrows* and *ellipsis*. Panels (**d**), (**e**), and (**f**) illustrate decomposition of the eight-reservoir model. [Panel (**e**) depicts the reduced four-reservoir problem.]

relaxing (**L**, **R**). Here we assume the spatial extent of these contacts is sufficient to allow both spin and momentum relaxation; they are "conventional" reservoirs as described by Büttiker [90]. As depicted in Figs. 2.18d–f, the full problem separates into three sub-components. Figure 2.18d represents the spin-up and spin-down currents ($I_{L\uparrow}$, $I_{L\downarrow}$) that flow between **F1**, 1↑, 1↓, and **L**. A Sharvin resistance [92], $R_{\rm sh} = (h/2e^2)(k_F w)/\pi = (h/2e^2)N_{\rm ch}$, arises between 1↑, 1↓ and the multichannel conductors connecting them to **L**. Under conditions of current flow this yield the spin-resolved electrochemical potential differences $\mu_{1\uparrow} - \mu_L = 2eR_{\rm sh}I_{L\uparrow}$ and $\mu_{1\downarrow} - \mu_L = 2eR_{\rm sh}I_{L\downarrow}$. Here, the factors of 2 arise because transport is spin resolved; $k_F$, $w$, and $N_{\rm ch}$ are the Fermi wave vector, channel width, and number of occupied modes within the 2DEG device channel, respectively. Similarly, at the rightmost side of Fig. 2.18f, current flow between the reservoirs 2↑, 2↓ and **R** establishes the electrochemical potential differences $\mu_{2\uparrow} - \mu_R = 2eR_{\rm sh}I_{R\uparrow}$ and

$\mu_{2\downarrow} - \mu_R = 2eR_{\rm sh}I_{R\downarrow}$. Also, $\mu_{F2\uparrow} = \mu_{2\uparrow}$ since no current flows between these reservoirs. Note that all $I$'s here represent net currents (forward minus reverse contributions). In our model, the following sum rules hold: $I = I_{L\uparrow} + I_{1\uparrow}$, $I = I_{L\uparrow} + I_{L\downarrow}$, $0 = I_{R\uparrow} + I_{R\downarrow}$, and $I_{1\uparrow} + I_{1\downarrow} = I_{2\uparrow} + I_{2\downarrow} = 0$. As the reservoirs in Fig. 2.18f are voltage contacts, net current is conserved separately for each spin band, $I_{R\uparrow} + I_{2\uparrow} = I_{R\downarrow} + I_{2\downarrow} = 0$. These expressions can be manipulated to yield

$$\begin{pmatrix} \mu_{1\uparrow} \\ \mu_{1\downarrow} \\ \mu_{2\uparrow} \\ \mu_{2\downarrow} \end{pmatrix} = \begin{pmatrix} \mu_L + 2eR_{\rm sh}(I - I_{1\uparrow}) \\ \mu_L + 2eR_{\rm sh}I_{1\uparrow} \\ \mu_R - 2eR_{\rm sh}I_{2\uparrow} \\ \mu_R + 2eR_{\rm sh}I_{2\uparrow} \end{pmatrix} \tag{2.20}$$

Given these relations, calculation of $R_s$ reduces to a four-terminal problem that solely involves the four spin-resolved reservoirs: $1\uparrow$, $1\downarrow$, $2\uparrow$, and $2\downarrow$ and the 2DEG device channel that connects them (Fig. 2.18e). Modifying Büttiker's formula to account for the spin-resolved transport, the four-terminal linear response at zero temperature becomes

$$I_{i,\alpha} = \frac{e}{h}\left[(N_{\rm ch} - R_{ii}^{\alpha\alpha})\mu_{i,\alpha} - T_{ij}^{\alpha\beta}\mu_{j,\beta}\right] \equiv \frac{e}{h}U_{ij}^{\alpha\beta}\mu_{j,\beta} \tag{2.21}$$

Transport within the ballistic multimode 2DEG conductor, is hence fully represented by the transmission coefficients $T_{ij}^{\alpha\beta}$, and the reflection coefficients $R_{ii}^{\alpha\alpha}$. These describe carriers incident from the lead $i$ with spin polarization $\alpha$ that are transmitted into lead $j$ with final spin state $\beta$ ; and carriers incident from $i$, $\alpha$ that are reflected back into same lead and spin channel. The coefficients $U_{ij}$ in (2.20) satisfy the sum rule $\sum_j U_{ij} = \sum_i U_{ij} = 0$ ensuring the current sum rules, and that all currents vanish when the $\mu_i$ are equal.

Simplification of (2.20) and (2.21) yields

$$\begin{pmatrix} I_{1\uparrow} \\ I_{1\downarrow} \\ I_{2\uparrow} \\ I_{2\downarrow} \end{pmatrix} = \boldsymbol{S} \begin{pmatrix} \tilde{\mu}_L + I \\ \tilde{\mu}_L \\ \tilde{\mu}_R \\ \tilde{\mu}_R \end{pmatrix} \tag{2.22}$$

where $\tilde{\mu}_L = \mu_L/2R_{\rm sh}e$ and $\boldsymbol{S} \equiv (1+\boldsymbol{U})^{-1}\boldsymbol{U}$. The elements of $\boldsymbol{S}$ satisfy the same sum rules that constrain $\boldsymbol{U}$ (for identical reasons). These steps lead to an explicit expression for the $R_s$. For parallel alignment of polarizer and analyzer (**F1,F2**), which we denote by the superscript ($\uparrow\uparrow$), these steps yields

$$R_s^{(\uparrow\uparrow)} = -2\frac{S_{31}S_{42} - S_{32}S_{41}}{S_{31} + S_{32} + S_{41} + S_{42}}R_{\rm sh} \tag{2.23}$$

For antiparallel alignment, only the sign changes: $R_s^{(\uparrow\downarrow)} = -R_s^{(\uparrow\uparrow)}$.

We obtain the requisite elements of $\boldsymbol{S}$ numerically, extending the semiclassical billiard model [91] to allow tracking of an electron's spin wavefunction along ballistic trajectories between the spin-resolved reservoirs ($1\uparrow,1\downarrow,2\uparrow,2\downarrow$)

at either end of the 2DEG device channel. We consider electrons confined within a hard-wall channel, of length $L$ and width $w$, at zero temperature. $T_{ij}^{\alpha\beta}$ are calculated by injecting and following a large number of electron trajectories (typically $>10^4$), made up of straight line or circular arc segments that specularly reflect from the walls until they are collected into one of the reservoirs. We implicitly assume that, compared to the Fermi energy, the spin splitting is small (and, hence, that the $T_{ij}^{\alpha\beta}$ are essentially energy-independent on this scale). This is valid for the narrow gap, low density semiconductor systems that are the focus of our study. For a uniform external field, $B_{\text{ext}}$, we decompose the spin wavefunctions into linear combinations of spin eigenstates. If, say, $B_{\text{ext}}$ is applied along $z$ (normal to the 2DEG plane, Fig. 2.18), then the $\pm z$-polarized states (hereafter denoted as $|+\rangle, |-\rangle$) form the basis, and we then represent $\pm x$ and $\pm y$ polarized states as $|\pm x\rangle = (|+\rangle \pm |-\rangle)/\sqrt{2}$ and $|\pm y\rangle = (|+\rangle \pm i |-\rangle)/\sqrt{2}$, respectively. For this field orientation, the phase of an electron's spin wavefunction evolves as it traverses the 2DEG device channel. For example, if at time zero an electron subject to $B_{\text{ext}} \parallel z$ is characterized by $|\sigma\rangle = C_+ |+\rangle + C_- |-\rangle$, then at a later time, $t$, $|\sigma, t\rangle = C_+ \exp(-i\omega_L t/2) |+\rangle + C_- \exp(i\omega_L t/2) |-\rangle$. Here $\omega_L = g^* eB/2m$ is the Larmor frequency for that particular path segment, $g^*$ the effective electron g-factor, and $m$ the free electron mass. (The case involving a Rashba field, described below, is more complex.) Total precession is then accumulated for each complete trajectory as part of our Monte Carlo procedure.

### 2.6.2   Results of Spin Resolved 4-Probe Model

In Fig. 2.19a we display $R_s^{(\uparrow\uparrow)}$ as a function of perpendicular magnetic field strength. The prominent and striking new feature is that $R_S$ is *oscillatory*, a ballistic phenomenon not found in the diffusive regime. In Figs. 2.19b and 2.19c, we display $R_s^{(\uparrow\uparrow)}$ calculated for three orientations of the external field – two that are in-plane and the perpendicular case, displayed again for comparison. In all three cases the **F1**, **F2** magnetizations are parallel and $y$-oriented.

When the external field is along $y$, the injected carriers remain in spin eigenstates and do not precess. In this situation $R_s^{(\uparrow\uparrow)}$ is a positive constant. However, with an $x$-oriented field precession is maximal, and $R_s$ oscillates. Since orbital effects are absent for an in-plane field, the oscillations in this case arise purely from spin precession and the oscillation *period*, $\Delta B$, is determined by the condition $2n\pi = \omega_L t_{\text{TR}}$, i.e., $\Delta B = h/(g^* \mu_B t_{\text{TR}})$. Here $t_{\text{TR}} = S/v_F$ is a typical transit time from $1 \rightarrow 2$, and $\mu_B$ is the electronic Bohr magneton. $\Delta B$ is thus inversely proportional to $S$, a typical path length averaged over the injection distribution function. The *decay* of $R_s$ occurs on a field scale where $\omega_L \delta t_{\text{TR}} \sim \pi$; i.e. for $B = \hbar\pi/(g^* \mu_B \delta t_{\text{TR}})$, beyond which precession among the different contributing trajectories tends to get out of step. Here $\delta t_{\text{TR}} =$

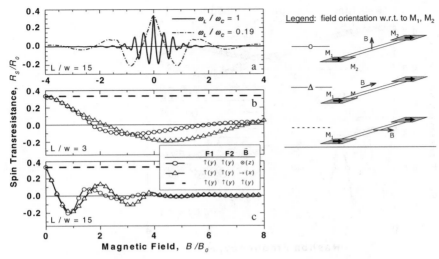

**Fig. 2.19.** Ballistic spin transresistance in an external field normalized to $B_0 = p_F/eW$. (**a**) For a channel with $L/w = 15$ in a perpendicular field, we plot two traces representing $\omega_L/\omega_c = 1$ and 0.19, appropriate for a typical metal and for InAs, respectively. (**b**) and (**c**) Spin transresistance for three different configurations and two channel lengths $L/w = 3$ and 15. Here, $\omega_L/\omega_c = 0.19$ (InAs).

$\sqrt{t_{\mathrm{TR}}^2 - \langle t_{\mathrm{TR}}\rangle^2}$ is the variance in path lengths traversed while propagating from 1→2.

The perpendicular field ($B_{\mathrm{ext}} \parallel \hat{z}$) is special – it induces both spin and *orbital* effects. (The characteristic field scale for the latter is $B_0 = p_F/ew$, at which the cyclotron diameter, $d_c = 2v_F/\omega_c$, equals the channel width, $w$.) The frequency ratio $\omega_L/\omega_c = (g^*/2)(m^*/m)$ describes the relative importance of orbital and spin transport phenomena. Here, $p_F$ is the Fermi momentum, $\omega_c = eB/m^*$ the cyclotron frequency, and $m^*$ the effective mass. For InAs ($m^*=0.025$, $g^*=15$) this ratio is $\sim0.19$, for InGaAs $\sim0.1$, whereas it is roughly 1.0 for most metals. In the latter case spin and orbital effects will have similar periodicity so disentangling them is difficult.

As mentioned, electrons confined within an InAs heterostructure are subject to an internal Rashba field that is present even for zero applied magnetic field. This can be modeled by a term $H_R = \alpha[\boldsymbol{\sigma} \times \boldsymbol{k}] \cdot \hat{z}$. Comparing $H_R$ to the Zeeman term we write the effective Rashba field as

$$\boldsymbol{B}_R = \frac{2\alpha\,\boldsymbol{k} \times \hat{z}}{g^*\mu_B} \qquad (2.24)$$

Using data from Heida et al. [38], we estimate the typical strength of this internal field to be about 5T for an InAs 2DEG. The spin precession induced by $\boldsymbol{B}_R$ dramatically alters the $T_{ij}^{\alpha\beta}$. For each straight-($B_{\mathrm{ext}} = 0$) or arc-($B_{\mathrm{ext}} \neq 0$) segment of the electron's trajectory (traversed between reflections

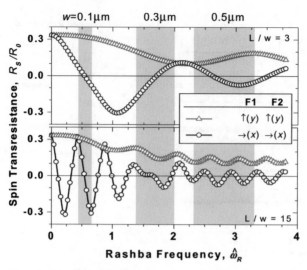74     H.X. Tang et al.

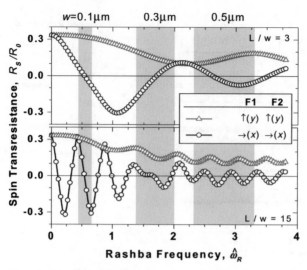

**Fig. 2.20.** Spin transresistance *vs.* reduced Rashba frequency, $\hat{\omega}_R = 2m^*\alpha w/\hbar^2$, at zero applied field, for two different device channel lengths, $L/w = 3$ and 15. The Rashba field strength, characterized by $\hat{\omega}_R$, can be tuned by an external gate voltage. *Shaded regions* delineate the range of tunability expected for InGaAs devices[38] of three widths, 0.1, 0.3, and 0.5 μm.

from the boundaries) its spin precession can be calculated analytically and incorporated into the Monte Carlo procedure.

Figure 2.20 displays how Rashba-induced spin precession is manifested in $R_s$ for zero external field. We represent the effective Rashba field strength by the dimensionless frequency $\hat{\omega}_R = 2\alpha m^* w/\hbar^2$ ; at $\hat{\omega}_R = 1$ an electron precesses one radian after traversing a distance $w$. As shown, the oscillations initially decay quickly but exceedingly slowly thereafter. No spin precession occurs for $\hat{\omega}_R = 0$, hence $t = 1$ yielding $R_s^{(\uparrow\uparrow)} = R_s^{(\rightarrow\rightarrow)} = R_{sh}/3$. For finite $\hat{\omega}_R$, $R_s$ displays strong dependence upon the orientation of the magnetizations, $\boldsymbol{M}$, (of **F1**,**F2**; assumed parallel) in relation to device channel's principal axis ( $\hat{x}$). For $\vec{M} \parallel \hat{x}$ (parallel to the channel), precessional effects are maximal. With increasing Rashba field, the variance in contributing path lengths causes the oscillations in $R_s$ to decay, as was described previously for the case of finite external field. Here, however, contributions from short paths (direct propagation between the DSPR's involving few or no boundary reflections) continue to add coherently for large $\hat{\omega}_R$, resulting in very slow decay. For $\vec{M} \parallel \hat{y}$, most of the injected carriers experience a Rashba field nearly aligned with their spin. At intermediate Rashba field these yield small oscillations that center about a *finite* value of $R_s$. The other carriers make a contribution to at small $\hat{\omega}_R$, but this becomes incoherent and thus quickly decays for large $\hat{\omega}_R$.

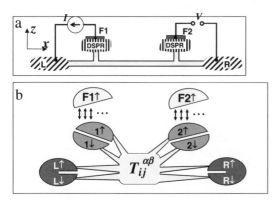

**Fig. 2.21.** Full eight probe model for ballistic spin injection in two dimensions which includes ballistic junction scattering effects. (**a**) The measurement configuration: a current, $I$, is injected through the 2DEG device channel via an a ferromagnetic contact (**F1**) and ohmic contact (L). The spin transresistance, $R_S = V/I$, arises from spin polarized carriers a distance $L$ from the net current path, which induce a nonlocal voltage, $V$, between a second, similar, pair of contacts (**F2**, **R**). (**b**) The full 8-reservoir model; *complete ellipses* represent spin-relaxing reservoirs, *half ellipses* represent separate spin-resolved reservoirs. **F1** and **L** are current contacts, **F2** and **R** are voltage probes. $T_{ij}^{\alpha\beta}$ denotes the 2DEG device channel in which spin precession occurs.

### 2.6.3   8-Probe Model: Junction, Bulk, and Boundary Scattering

In general, transport under the ferromagnetic contact will not be completely diffusive. In this situation one must consider the effects of junction scattering in this region [27]. Figure 2.21 represents the simplest idealized case that captures the essential physics of junction scattering in a ballistic spin transport device. The polarized contacts are now modeled as a DSPR offset from the main channel by a ballistic "t"-junction. To analyze this situation, one must be explicitly consider 8 reservoirs; i.e. two spin reservoirs for each of the four physical contacts. The junction scattering matrix couples *all* of these ballistic channels. The requisite analysis is conceptually depicted by Fig. 2.21b. For simplicity of the analysis, the external field is assumed to be applied only within the ballistic channel itself. In this configuration, the boundary constraints on currents and potentials can be succinctly represented: $\mu_{F2\uparrow} = \mu_{2\uparrow}$, $\mu_{L\uparrow} = \mu_{L\downarrow}$, $\mu_{R\uparrow} = \mu_{R\downarrow}$, $I_{1\uparrow} = I$, $I_{1\downarrow} = 0$, $-I = I_{L\uparrow} + I_{L\downarrow}$, $0 = I_{R\uparrow} + I_{R\downarrow}$, and $I_{2\uparrow} = I_{2\downarrow} = 0$. As a result of the increased number of trajectories able to carry the current in this model, oscillations in the spin transresistance damp more quickly than for the previous four-probe model. Figure 2.22 illustrates that only one oscillation peak is observable for this case.

In real devices we must consider the role played by scattering. Spin flip scattering will serve to equalize the spin populations and uniformly suppress the spin transresistance. The effects of momentum scattering processes on $R_s$

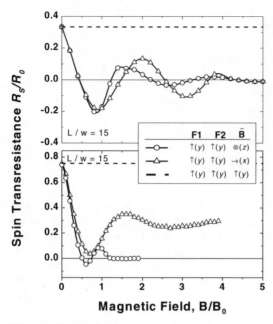

**Fig. 2.22.** Spin transresistance as a function of applied magnetic field for three field configurations (indicated) and two measurement geometries. We assume a frequency ratio, $\omega_L/\omega_c = 0.19$, appropriate for InAs.

will also be important – but are more subtle. There are two kinds of mechanisms of random scattering that must be considered: boundary scattering and bulk elastic scattering. In the simplest relaxation time treatment, they can be characterized by a mean free path $\ell_0$, and by a boundary specularity parameter $p$ respectively. For the latter, $p = 1$ implies completely specular reflection, while $p = 0$ represents completely diffuse scattering. Operationally, bulk scattering is modeled by imposing a probability $P(\ell) = \exp(-\ell/\ell_0)$ on all ballistic trajectories, which is essentially the approach of Chambers [93]. It represents a given carrier's likelihood of survival (i.e. remaining in a specific momentum state) on a given trajectory after traversing a distance $\ell$. Similarly, as we follow an electron's successive reflections at the channel boundaries (modeled as hard walls) the probability of a momentum-randomizing diffuse boundary collision is $1 - p$ per interaction. The evolution of spin transresistance in the presence of two type of diffusive scattering is depicted in Fig. 2.23. As expected, the scattered trajectories act to suppress the "coherence" of spin transport in the junction. The oscillation peaks are notably suppressed when the mean free paths is about an order of magnitude larger than the channel width, or when the boundary specularity is less than about 90%. (The effects of these phenomena are coupled). In the completely diffuse limit, negative transresistance disappears and we get a monotonically-decreasing Hanle

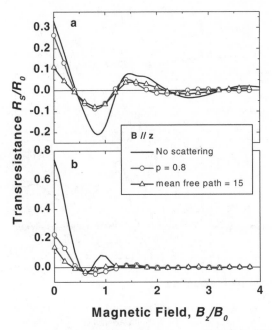

**Fig. 2.23.** Spin transresistance as a function of perpendicular magnetic field in the presence of diffusive scattering. The clean limit is plotted as *black curves* for comparison. (**a**) Results for the four-probe model. (**b**) Results for the eight-probe model that includes junction scattering. Here the mean free path is normalized to the channel width; the specularity parameter $p$, is defined in the text.

effect, such as observed in the first spin injection experiments (Fig. 2.2), emerges [94].

The dependence of the spin transresistance upon the spin-orbit coupling strength for finite bulk and boundary scattering is shown in Fig. 2.24. We find that both tend to wash out spin related phenomena, i.e. both the asymptotic effects and the high field oscillations.

### 2.6.4  The Spin Transistor: A Closer Look

The spin transistor was originally envisaged as allowing external control of the spin precession rate, via a gate potential acting in concert with the intrinsic confinement potential. In Fig. 2.20 we show how a typical range of tunability, here assumed to be of order 30% (Sect. 2.3.6), translates into direct modulation of $R_s$, for three device widths. Our calculations clearly illustrate that the "conventional" spin transistor configuration, $M \parallel \hat{y}$, (which is most easily fabricated) is, in fact, *not* optimal – *even for a very short channel* $(L \sim \ell_s)$. We find that tunability is maximized for $M \parallel \hat{x}$.

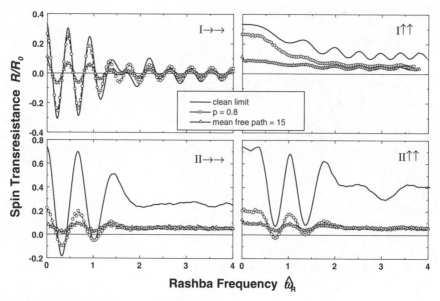

**Fig. 2.24.** The influence of random scattering on spin transresistance as a function of the strength of the Rashba effect. Three scattering mechanisms are considered: junction scattering, impurity scattering, diffuse boundary scattering. The upper two panels are results from the four-probe model. The lower two panels are from eight-probe model.

As originally conceived, the spin transistor was pictured as a one-dimensional device, with only a single transverse subband populated. Realizable devices in the near term will more likely employ two-dimensional or, perhaps, quasi-one dimensional channels. Their increased phase space for scattering can lead to quick suppression of $R_S$, especially in the presence of moderate levels of diffuse boundary or bulk scattering within the channel. These results confirm that an extremely narrow channel is a basic requirement for a spin transistor.

### 2.6.5  Other Theoretical Treatments

Ballistic spin polarized transport in the presence of Rashba precession in semiconductor nanowires has also been recently addressed by Mireles et al. [97] and by Seba et al. [95]. The first authors include interfacial (Rashba) spin-orbit scattering, and extend the spin-independent Ando method [98] to generate a spin-dependent transfer matrix. By applying spin selective contacts, an oscillatory conductance is calculated. The results they obtain are very similar in form to those generated previously by the simple approach described here, as displayed in Figs. 2.20 and 2.22.

In the second recent approach to this problem, carried out by Seba et al. [95], treats a situation in which solely spin-orbit scattering considered. A wave function similar to that originally employed by Elliott [96] is introduced to describe the spin flip process. They derive the rather surprising result that the spin dependent transmission probabilities oscillate with increasing device length and may even reverse sign after traveling a certain distance through a region of disordered semiconductor containing a fixed concentration of scatterers.

In both of these recent theoretical models, the two-terminal conductance from a ferromagnetic source to a ferromagnetic drain is calculated. Their motivation is to interpret the recent experiment by Hu et al. [109]. We point out that hysteretic features in two-terminal resistance observed by Hu et al. are not a definitive demonstration of successful spin injection, since local Hall phenomena can not be precluded and a Hanle effect was not observed.

## 2.7 Projections and Conclusions

### 2.7.1 Retrospective: The Spin Transistor

The concept of a spin transistor is founded upon the assumption that spin polarized electrons can survive traversal over an entire device structure. This involves three distinct and important pieces of physics. First, the transfer of spins across the *injection* interface, i.e. between a magnetic and a non-magnetic material, must be possible without wholesale loss of polarization. We have seen that conductivity mismatch can seriously impair this process for Ohmic contacts between metals and semiconductors. Second, once the "source" interface is traversed, spin polarization must be largely preserved while the spin current undergoes *propagation* through the device. This means that there must be minimal spin relaxation within the paramagnetic channel, i.e. the initially unpolarized electron gas in which the electron propagates from "source" to "drain". Third, the final prerequisite is implementation of a spin analyzer at the far end of the device, which is capable of *collection* – i.e. resolving the chemical potential imbalance between spin bands arising from spin injection. It is clear that understanding the detailed physics of these three elements of device operation – injection, propagation, and collection – and using this information to engineer materials and device geometries for next-generation structures, will be the focus of much spin electronics research in the next decade.

Realizing the spin transistor is one of the longest standing hopes of spin-based electronics [122]. Although the initial motivation for work on this has diminished somewhat, advances on many fronts of spin electronics (briefly outlined in Sect. 2.3.1) have heightened interest in achieving robust and controllable spin-polarized currents in small devices. Despite the widespread interest generated by the proposal of Datta and Das, and the past decade of effort that ensued, the "spin transistor" has not yet been achieved. It is thus fair to say that it remains a singular challenge for the community of researchers interested in these issues. In fact, since the initial demonstration in 1985 of spin injection in metal-based devices, the recent work of Jedema et al. [83] represents the only unambiguous work confirming that all-electrical spin injection devices are possible. To be fair, most effort has been placed on semi-conducting systems with *Ohmic contacts* – apparently owing to the prevailing assumptions that they offer both the greatest device potential and the highest likelihood for adoption by the mainstream microelectronics community. We shall examine these assumptions below. Thus far, however, most putative claims of all-electrical, spin injection devices in semiconducting materials have been engulfed in a shroud of controversy. The disappointingly small signals observed have failed to convince the greater magnetoelectronics community that robust spin injection effects have yet been attained. Nonetheless, if we take the most positive of interpretations, data from these experiments allows an upper bound to be placed on the magnitude of the current polarization induced – and this appears extremely small, of order 1% at best [72], [73], [75], [123], [104], [109], [124]. As mentioned, what is most troublesome about signals at this level is that conventional magnetoresistance measurements are plagued with a host of comparable, obfuscating, non-spin-injection related magnetoelectronic phenomena, such as AMR and the local Hall effect (Sect. 2.4.2 and 2.5.5, respectively). Accordingly, obtaining the correct interpretation of the signals obtained in such devices is very problematic.

But the situation is actually not as grave as this discourse might initially lead one to believe. Recently, very significant advances within three separate areas of spin electronics have emerged, and these provide growing optimism that useful all-electrical spin injection transistors can be realized. We list these advances here, and then separately review each of them in more detail below. First, robust electrical spin transfer across interfaces has been demonstrated in spin polarized light emitting diodes (LEDs). The emission of circularly-polarized light from these devices provides rather compelling evidence that electrical injection of hot, spin polarized carriers can indeed be achieved. What is interesting about these structures is that they involve electrical spin injection across *epitaxial* interfaces. Such interfaces are far more ideal than those employed in efforts with all-electrical, Ohmic spin injection devices to date. Second, there is also a growing body of work confirming the first reports of successful spin injection, carried out almost a decade ago, across a tunnel barrier. These two sources of evidence soundly demonstrate that spin polarization, in amounts varying from very small to substantial, can be preserved in injected currents if they traverse an *optimal* contact interface. Third, long lifetimes for spin polarized carriers have been observed in optically excited systems [32], [125]. This demonstrates that carriers can propagate within a semiconducting electron gas without significant loss of spin polarization. Collectively, this recent work – and the body of spin injection research that has preceded it – serves to demonstrate that new materials, ideal interfaces, and optimal device geometry are crucial for realizing new classes of magnetoelectronic devices.

### 2.7.2   Recent Advances in Spin Transport Across Interfaces

**(a) Optical Methods.** Optical methods have recently provided compelling evidence that, with optimal conditions, spins can be *efficiently* transferred through epitaxial semiconductor heterointerfaces. This has been most clearly demonstrated by Malajovich et al. [126] in experiments where optically generated, spin polarized carriers were allowed to diffuse across an epitaxial interface between ZnSe and GaAs. Loss of polarization was surprisingly minimal. This group subsequently demonstrated that spin-transfer efficiency could be increased by five to forty times, when electric fields are used to 'drag' spins across a semiconductor interface [127]. They applied laser pulses, tuned to create a long-lived 'reservoir' of spin-polarized carriers in the sourcing epilayer and then, in two separate experiments, showed that both externally-applied and "built-in" internal electric fields could be highly effective at transferring spins from this reservoir, across an epitaxial interface, to another semiconductor. In the latter case, "built-in" fields were created during growth, through controlled doping of semiconductor layers, to employ the natural potential arising at $p/n$ junction.

**(b) Ohmic Contacts.** As originally envisaged, the spin-FET involved ferromagnetic source/drain electrodes and a semiconducting electron gas "channel" that were electrically connected by *Ohmic* contacts. With Ohmic contacts the two-terminal (drain/source) device impedance is kept quite low, and this circumvents problems arising in high impedance devices such as limited fan-out (due to source loading), slow dynamical response, and large power dissipation at steady bias. These were the principal reasons that motivated efforts on the spin transistor over the past decade that were focused upon Ohmic structures.

Efficient spin injection is possible, even for diffusive transport in a "half metallic" ferromagnet where the conduction electrons are nearly 100% spin-polarized. There are a number of materials that have this nearly complete polarization [128], [129] however these have not yet been applied to semiconducting spintronics, presumably due to the significant difficulties associated with integrating them into advanced growth and fabrication processes for high quality semiconductors.

Recently Hammar et al. [106] reported a novel, indirect means of detecting spin injection across a single interface between a metallic ferromagnet and a heavily doped InAs 2DEG. The observations, reported to persist even above liquid nitrogen temperature, were interpreted as the consequence of a spin dependent resistance that the authors believe is intrinsic to such an interface. As in other spin injection experiments involving Ohmic contacts between metallic ferromagnets and semiconductors, however, only a very minute change in device resistance could be correlated with changes in ferromagnet orientation. Given the host of other phenomena of comparable magnitude known to be operative at ferromagnetic/semiconductor interfaces, compounded with the unknown quality of the interfaces studied, these results have not met with widespread acceptance to date. Alternative explanations, including ones based upon the local Hall effect, have been advanced [107], [108], [130]. The underlying physical mechanism proposed for this effect involves a non-equilibrium spin polarization induced by current flow in a spin-orbit-split conduction band [131]. Recently theoretical models for a spin-dependent interfacial resistance involving spin-orbit coupling have been formulated in both the ballistic [132], [133] and diffusive [134] regimes. However very fundamental questions have been raised regarding correctness of such interpretations [135]. To date, the feasibility and effectiveness of this current-induced spin polarization for spin injection appears to be unsettled.

Convincing evidence of spin injection has been realized in efforts involving polarized light emission induced by spin polarized carriers injected across a $p/n$ junction. Among such experiments, the spin-LED recently reported by Ohno et al. [102] employs spin injection from a $p$-type ferromagnetic semiconductor (GaMnAs) into an intrinsic GaAs region. The spin polarized current subsequently flows through the intrinsic region into an $n$-doped InGaAs quantum well. Hence one may consider this device as first involving spin injection across an Ohmic, unipolar junction, and then involving injection across a subsequent barrier.

Finally, we note that recent experiments with nanoscale point contacts formed between ferromagnetic and non-ferromagnetic metals have demonstrated that ballistic injection of highly ($> 40\%$) spin-polarized currents into the non-ferromagnet is possible [140], [141].

**c) Tunnel Contacts.** We have mentioned the underlying reasons why Ohmic contacts have, been pursued most vigorously to date. Given the significant difficulties mentioned above, however, it now seems worthwhile to explore the possibility of electrical spin injection by alternate methods – i.e. by tunnel or Schottky contacts despite the fact that the resulting two-terminal device impedances may be larger. Especially noteworthy in this context is the recent adoption of magnetic tunnel junction structures for memory applications by the commercial sector. This demonstrates that large two-terminal impedances need not preclude utility for commercial electronics. A tunnel contact has the additional advantage that it can provide a diffusion barrier between ferromagnetic metals and semiconductor materials. This is important given the extensive past work (described below) demonstrating the ease with which these systems can interdiffuse, even at relatively low growth or processing temperatures.

Of course the most ideal barrier is vacuum itself. A decade ago, Alvarado and Renaud [136] showed that spins can be injected into a semiconductor by vacuum tunneling. They demonstrated this using a scanning tunneling microscope outfitted with a ferromagnetic tip. Recently, La Bella and coworkers have extended this work to image the orientation-dependent spin injection probability and the spin lifetime for spin polarized carriers injected into a pristine surface of GaAs [137].

Recent theoretical work by Rashba [88], Smith [138], Heersche et al. [139] and Flatte et al. [Chap. 4], has explored the effectiveness of spin injection via tunneling contacts. Collectively, what is emerging from these investigations is that, for the case of small transmission, the spin-dependent density of states for the two electrodes determines transport across a tunnel barrier. In this limit of regime of linear response the electrodes remain in equilibrium and their respective spin-dependent conductivities play a minimal role in determining transport across the interface. Hence, either a FM (ferromagnet metal)-semiconductor Schottky diode or a FM-insulator-semiconductor tunnel diode may be effective for injecting spins into a semiconductor system.

**d) Schottky Contacts.** There has been growing interest in the study of epitaxial ferromagnetic metal / semiconductor heterostructures. Such efforts have a long history; more than twenty years ago Prinz showed that single crystal Fe films could been stabilized on GaAs through molecular beam epitaxy [142]. The lattice constant of GaAs ($a_0 = 5.654$ Å) is almost exactly twice that of Fe ($a_0 = 2.866$Å). Much recent effort has continued the focus upon GaAs materials; among systems explored are Fe/GaAs [111], [112], [113], [143], Py(permalloy)/GaAs [144], and Fe-Co/GaAs [145]. It was realized very early in the course of these studies that, under non-optimal growth conditions, a magnetically dead layer can exist at the interface between these materials. This has been shown to arise from intermixing, yielding a magnetically depleted phase [111] or half-magnetized [113] phases such as $Fe_3Ga_{2-x}As{x}$. However it has also been found that, under optimal conditions, growth of only a few monolayers is sufficient for the ferromagnetic film to recover the full magnetic properties of the bulk [147]. Additional work has been carried out on the Fe/InAs [114], [115], [116] system. The crystal structure of InAs is very similar to that of GaAs, but the lattice mismatch of Fe and InAs ($a_0 = 6.058$Å) is 5.4%, poorer than that of Fe/GaAs. It has proven possible to grow single crystal bcc Fe on InAs (100) surface, however Ohmic rather than Schottky contacts are achieved [115]. In all these grown materials, clear *in situ* MOKE (magneto-optic Kerr effect) hysteresis loops have been demonstrated after only 5-8 monolayers of Fe is deposited.

For spin injection devices, Schottky barriers represent another alternative to Ohmic contacts. The differences between the spin-split conduction bands of the ferromagnetic metal and the spin-degenerate conduction band of a semiconductor yield a spin-dependent interfacial transmissivity. The picture is conceptually clear for ballistic electrons; for a high quality epitaxial interface it is generally assumed that the transverse momentum of incident electrons is conserved after transport through the barrier. In this case a straightforward description emerges for the spin-dependent *ballistic* transmission and reflection probabilities of the interface [148], [149]. Such simple pictures are almost certainly *not* expected to hold for ordinary interfaces, however ballistic electron emission microscopy (BEEM) studies demonstrate that, for "generic" surfaces, diffuse scattering is predominant [150].

Spin injection across a Schottky barrier between a ferromagnetic metal and a semiconductor appears to be very promising. Zhu et al. [146] have recently reported a room temperature spin-injection efficiency of 2% with a Fe-GaAs Schottky contact. In their experiments the injected spin polarization was detected by polarization-resolved electroluminescence. This has been confirmed by recent experiments of Hanbicki et al. [151] who studied spin injection from an epitaxial Fe layer grown on AlGaAs/GaAs quantum well LED structures. Also using polarization-resolved electroluminescence, they deduced a 30% spin injection efficiency for such structures at 4.5K, which shrank by a factor of about three (to $\sim$9%) at room temperature.

An interesting variant on spin injection involves the use of spin-polarized "hot" electrons with energies much greater than $E_F$ [152]. Filipe et al. [99] demonstrated a novel spin filter effect involving spin injection across a Schottky barrier. Quasi-monoenergetic, spin-polarized electrons are tunneled across a vacuum gap into a thin ferromagnetic metal layer deposited upon an n-doped GaAs substrate. Spin-dependent transmission probabilities are deduced by measuring the current transmitted through this Schottky barrier. Before the deposition of ferromagnet, the n+ GaAs surface was oxidized by an ozone system to form a 2-nm-thick oxide layer. This oxide layer proves to be very efficient in preserving spin polarization. A spin filter efficiency up to about 25% has been achieved.

In recent experiments [153], hot electrons have been tunnel injected from a STM tip across a barrier into a ferromagnetic layer, in a technique termed BEMM (ballistic electron magnetic microscopy). Since the majority-spin and minority-spin electrons have much different inelastic mean free paths, passage of hot electrons through, for example, a 3 nm Co layer has proven sufficient to result in a ballistic electron current with > 90% spin polarization [146]. This highly polarized hot-electron current can then be employed to interrogate, for example, underlying metal-semiconductor interfaces. In this situation the spin-dependent transmission probability of the barrier can be probed. If spin-flip scattering at the interface is minimal, then the ballistic electron current entering the semiconductor will be highly polarized (> 90%). Furthermore, the energy of the injected electrons, relative to the bottom of the conduction band in the semiconductor, is tunable via the bias voltage across the tunnel barrier. However, the disadvantage of hot electron injection is that the overall efficiency (yield) is low.

### 2.7.3   Recent Advances in Spin Injection
### Via Semimagnetic Semiconductors

Oestreich et al. have explored use of a semimagnetic Mn-based diluted magnetic semiconductor as the spin-injecting contact. In the presence of a large applied magnetic field, electron spins in this large-$g$ material will be strongly polarized [100]. They used time- and polarization-resolved photoluminescence to demonstrate that optically excited carriers became spin aligned in a $Cd_{1-x}Mn_xSe$ layer, and that spin-polarized electrons were transferred into an adjacent CdTe layer with little loss in spin polarization.

Fiederling et al. [101] have demonstrated efficient electrical injection of spin-polarized carriers from the epitaxially grown semimagnetic semiconductor, ZnMnSe, into a GaAs semiconductor heterostructure. A light-emitting diode structure (spin-LED) was constructed which enabled direct proof, by spin polarized emission measurements, that the current was spin polarized. As in other spin LED structures, the diode is always biased at more than 1 V to achieve appreciable emission. This voltage scale is greater than the band gaps of the quantum wells employed. The device incorporates an intrinsic spacer layer (GaAs), in which the electrons have a very high energy relative to the Fermi level. Recent efforts in this area have been reported by Jonker et al. [103].

Given the ability to grow ZnMnSe epitaxially upon GaAs, semimagnetic contacts appear to be a very interesting alternative for spin injection devices, at least in initial research efforts [154]. There are several notable drawbacks that must be considered, however. Large magnetic field operation precludes the commercial viability of this approach. Perhaps most important is the fact that the applied (large) magnetic field obviously determines the direction of polarization. For fundamental experiments there is no clear way to perform polarize/analyze or spin precession (Hanle effect) experiments – such as can be done with *ferromagnets* given their spontaneous, remnant polarization, which enables both configurations with antiparallel magnetic contacts, and the application of magnetic fields non-collinear with respect to the injected polarization.

### 2.7.4   Recent Advances in Spin Propagation in Semiconductors

In the past four years, it has been proven experimentally that spin coherence can persist in doped semiconductors on very long temporal scale, exceeding $> 100$ns at low temperatures. Local optically-excited, spin polarized "puddles" of carriers have been transported laterally through semiconducting materials for distances over 100 μm by applied electric fields –while maintaining spin coherence [32], [33], [34], [35], [36]. For a detailed introduction and discussion, interested readers are referred to Chap. 5 of this book. As mentioned, the spin polarized holes are found to relax quickly, but the spin polarized electrons persist for long times. At present, the extent to which the spin dynamics of these hot electrons differ from those of spin-polarized electrons at Fermi surface is unclear. Since spin scattering is closely related to momentum scattering, one might expect the spin relaxation rate at the Fermi surface to be *even slower* than that observed for hot carriers. But, to date, difficulties in manipulating near-equilibrium electrons, have precluded measurement of spin lifetime at the Fermi surface. But the picture that emerges is extremely promising for electrical spin injection devices.

### 2.7.5   Detection of Nonequilibrium Spin Polarization

We have described the use of polarization-resolved electroluminescence for measurement of spin injection across Schottky barriers. However, more general approaches are desired, especially for developing completely electrical devices, e.g., those within the general class of spin transistors, amenable to large scale integration. The most obvious approach is to employ the process inverse to that of injection, namely *electrical detection* of spin populations in semiconductors, via the spin-dependent transport properties of semiconductor/FM interfaces. Most efforts to date based upon this scheme have employed Ohmic contacts. But, as for the case of injection, similar difficulties will apply to spin collection. By analogy, Schottky or tunneling contacts may be helpful, however their efficiency is quite low. Hence, for a pair of such contacts employed in series to realize a spin transistor, the effect on the spin transresistance will be compounded (i.e. "squared"). With highly transmissive tunnel barriers, it may be possible to achieve the favorable situation where spin-dependent tunneling into the ferromagnetic electrode is more probable than spin relaxation within the semiconductor [88].

Alternative means for detecting spin polarized currents in semiconductors are on the horizon that have promise for exceptional sensitivity [155]. Quantum point contacts (QPC) or quantum dots (QD) can be employed to resolve the spin polarization attained after electric injection. In a QPC at zero magnetic field, conductance steps appear at quantized values proportional to $2g_0 = 2e^2/h$. It is well known that in a sufficiently large applied field, the quasi-1D modes within the constriction become spin-split [27]. Hence, as the applied magnetic field is increased, the fundamental step height evolves to half its zero field value, $g_0 = e^2/h$. In effect, spin splitting induces a new set of plateaus that are not observed at zero field. The same phenomenon should be manifested when the spin bands become unequally populated within the constriction through *spin injection*. Estimates indicate that a polarization sensitivity of 0.01% may be achievable with ideal point contacts at low temperatures [155]. Recently clear $\sim 0.5(2e^2/h)$ conductance steps have been observed [121] in a quantum point contact with a ferromagnetic Ni dot embedded at its center. Although how this Ni dot makes contact to 2DEG, and how it affects the electron gas are both unknown – this observation appears to confirm the alternate scheme for detecting spin polarization mentioned above.

Another potentially important form of detection may be achievable through the emerging capabilities of magnetic resonance force microscopy (MRFM) [156–158]. The magnetic resonance force microscope is a novel scanned probe instrument that combines the three-dimensional capabilities of magnetic resonance imaging with the high sensitivity and resolution of atomic force microscopy. In the present context, it may allow non-destructive, scanned probe imaging of spin polarized carriers within miniature spintronic devices [159]. Here, estimates are that the polarization sensitivity of, again, $\sim 0.01\%$ may be achievable – but in this case it appears achievable within a scanned volume of $< 1 \, \mu m^3$. If successful, this tool may provide unprecedented insight into the detailed physics of spin injection.

# References

1. S. S. P. Parkin, IBM J. Res. Develop. **42**, 3 (1998)
2. J. S. Moodera, G. Mathon, J. Magn. Magn. Mater. **200**, 248 (1999)
3. P. M. Tedrow, R. Meservey, Phys. Rev. Lett. **26**, 192 (1971); R. Meservey, D. Paraskevopoulos, P. M. Tedrow, Phys. Rev. Lett. **37**, 858 (1976)
4. P. M. Tedrow, Physics Reports **238**, 173 (1994)
5. B. D. Cullity, *Introduction to Mangetic Materials* (Addison-Wesley, Reading MA, 1972) pg. 617, 526
6. M. B. Stearns, J. Magn. Magn. Mater. **5**, 167 (1977)
7. J. C. Sloncezwski, Phys. Rev. B **39**, 6995 (1989)
8. S. Zhang, P. M. Levy, Phys. Rev. Lett. **81**, 5660 (1998)
9. J. M. Maclaren, W. H. Butler, X-G. Zhang, J. Appl. Phys. **83**, 6521 (1998); J. M. Maclaren, X.-G. Zhang, W. H. Butler, X. Wang, Phys. Rev. B **59**, 5470 (1999)
10. E. Y. Tsymbal, D. G. Pettifor, J. Phys., Condens. Matter **9**, L411 (1997); Phys. Rev. B **58**, 432 (1998)
11. M. Julliere, Phys. Lett. **54A**, 225 (1975)
12. A. G. Aronov, Pis'ma Zh. Eksp. Teor. Fiz. **24**, 37 (1976) [JETP Lett. 24, **32** (1976)]
13. M. Johnson, R.H. Silsbee, Phys. Rev. Lett. **55**, 1790 (1985)
14. M. Johnson, R. H. Silsbee, Phys. Rev. B **37**, 5712 (1988); M. Johnson, R.H. Silsbee, Phys. Rev. B **37**, 5326 (1988)
15. D. Lubzens, S. Schultz, Phys. Rev. Lett. **36**, 1104 (1976)
16. P. C. van Son, H. van Kempen, P. Wyder, Phys. Rev. Lett. **58**, 2271 (1987); **60**, 378 (1988); M. Johnson, R. H. Silsbee, Phys. Rev. Lett. **60**, 377 (1988)
17. M. Johnson, Science **260**, 320 (1993)
18. E. I. Rashba, Fiz. Fiz. Tverd. Tela (Leningrad) **2**, 1224 (1960) [Sov. Phys. Solid State **2**, 1109 (1960)]
19. W. J. Broad, *New York Times*, 6 July 1993, p. C1.; J.J. Keller, *The Wall Street Journal*, 19 April 1993, p. C21; M. Johnson, Nanotechnology **7**, 390 (1996)
20. G.A. Prinz, Science **282**, 1660 (1998)
21. J. A. Katine, F. J. Albert, R. A. Buhrman, E. B. Myers, D. C. Ralph and references therein, Phys. Rev. Lett. **84**, 3149 (2000)

22. R. K. Kawakami, Y. Kato, M. Hanson, I. Malajovich, ·J. M. Stephens, E. Johnston-Halperin, G. Salis, A. C. Gossard, D. D. Awschalom, Science **294**, 131 (2001)
23. I. Zutic, J. Fabian, S. Das Sarma, Phys. Rev. B **64**, 121201 (2001)
24. I. Zutic, J. Fabian, S. Das Sarma, Appl. Phys. Lett. **79**, 1558 (2001)
25. M.E. Flatté, G. Vignale, Appl. Phys. Lett. **78**, 1273 (2001)
26. S. Datta, B. Das, Appl. Phys. Lett. **56**, 665 (1990)
27. C. W. J. Beenakker, H. van Houten, *Solid State Physics*, **44**, (Academic Press, 1991) pp 1-170
28. B. J. van Wees, H. van Houten, C. W. J. Beenakker, J. G. Williamson, L. P. Kouwenhoven, D. van der Marel, C. T. Foxon, Phys. Rev. Lett. **60**, 848 (1988)
29. D. A. Wharam, T. J. Thornton, R. Newbury, M. Pepper, H. Ahmed, J. E. F. Frost, D. G. Hasko, D. C. Peacock, D. A. Ritchie, G. A. C. Jones, J. Phys. C **21**, L209 (1988)
30. A. Yacoby, H. L. Stormer, N. S. Wingreen, L. N. Pfeiffer, K. W. Baldwin, K. W. West, Phys. Rev. Lett. **77**, 4612 (1996)
31. S. Frank, S. P. Poncharal, Z. L. Wang, W. A. de Heer, Science **280**, 1744 (1998)
32. J. M. Kikkawa, I. P. Smorchkova, N. Samarth, D. D. Awschalom, Science **277**, 1284 (1997)
33. D. Hägele, M. Oestreich, W. W. Rühle, N. Nestle, K. Eberl, Appl. Phys. Lett. 73, 1580 (1998)
34. J. M. Kikkawa, D. D. Awschalom, Phys. Rev. Lett. **80**, 4313 (1998)
35. J. M. Kikkawa, D. D. Awschalom, Nature **397**, 139 (1999)
36. M. Potemski, E. Pérez, D. Martin, L. Viña, L. Gravier, A. Fisher, K. Ploog, Sold. State Commun. **110**, 163 (1999)
37. J. Nitta, T. Akazaki, H. Takayanagi, Phys. Rev. Lett. **78**, 1335 (1997). C. M. Hu, J. Nitta, T. Akazaki, H. Takayanagi, Phys. Rev. B **60**, 7736 (1999)
38. J. P. Heida, B. J. van Wees, J. J. Kuipers, T. M. Klapwijk, Phys. Rev. B **57**, 11911 (1998)
39. J. P. Lu, J. B. Yau, S. P. Shukla, M. Shayegan, L. Wissinger, U. Rössler, R. Winkler, Phys. Rev. Lett. **81**, 1282 (1998)
40. G. Engels, J. Lange, Th. Schäpers, H. Lüth, Phys. Rev. B **55**, 1958 (1997)
41. J. Nitta, T. Akazaki, H. Takayanagi, T. Enoki, Physica E **2**, 527 (1998)
42. D. Grundler, Phys. Rev. Lett, **84**, 6074 (2000)
43. T. Matsuyama, R. Kürsten, C. Meißner, and U. Merkt,, Phys. Rev. B **61**, 15588 (2000)
44. S. J. Papadakis, E. P. De Poortere, H. C. Manoharan, M. Shayegan, R. Winkler, Science, **283**, 2056 (1999)
45. U. Rössler, Solid State Commun. **49**, 943 (1984)
46. M. Cardona, N. e. Christensen, G. Fasol, Phys. Rev. B **38**, 1806 (1988)
47. G. Lommer, F. Malcher, U. Rössler, Phys. Rev. Lett. **60**, 728 (1988)
48. H. Mayer and U. Rössler, Phys. Rev. B **44**, 9049 (1991)
49. For a review, see *Optical Orientation*, edited by F. Maier and B. Zakharchenya (North-Holland, Amsterdam, 1984)
50. H. Riechert, H. J. Drouhin, C. Hermann, Phys. Rev. B **38**, 4146 (1988)
51. P. D. Dresselhaus, C. M. A. Papavassiliou, R. G. Wheeler, R. N. Sacks, Phys. Rev. Let. **68**, 106 (1992)

52. B. Jusserand, D. Richards, H. Peric, B. Etienne, Phys. Rev. Lett. **69**, 848 (1992)
53. D. Stein, K. v. Klitzing, G. Weimann, Phys. Rev. Lett. **51**, 130 (1983).
54. G. Lommer, F. Malcher, U. Rössler, Phys. Rev. B **32**, 6965 (1985)
55. P. Ramvall, B. Kowalski, P. Omling, Phys. Rev. B **55**, 7160 (1997)
56. H. L. Stormer, Z. Schlesinger, A. Chang, D. C. Tsui, A. C. Gossard, W. Wiegmann, Phys. Rev. Lett. **51**, 126 (1983)
57. J. P. Eisensten, H. L. Störmer, V. Narayanamurti, A. C. Gossard, W. Wiegmann, Phys. Rev. Lett. **53**, 2579 (1984)
58. B. Das, D. C. Miller, S. Datta, R. Reifenberger, W. P. Hong, P. K. Bhattacharya, J. Singh, M. Jaffe, Phys. Rev. B **39**, 1411 (1989)
59. S. Yamada, Y. Sato, S. Gozu, T. Kikutani, Physica E **7**, 992 (2000)
60. J. Luo, H. Munekata, F. F. Fang, P. J. Stiles, Phys. Rev. B **38**, 10142 (1988)
61. S. Brosig, K. Ensslin, R. J. Warburton, C. Nguyen, B. Brar, M. Thomas, H. Kroemer, Phys. Rev. B **60**, 13989 (1999)
62. S. Sasa, K. Anjiki, T. Yamaguchi, M. Inoue, Physica B **272**, 149 (1999)
63. S. J. Koester, C. R. Bolognesi, M. Thomas, E. L. Hu, H. Kroemer, M. J. Rooks, Phys. Rev. B **50**, 5710 (1994)
64. G. L. Chen, J. Han, T. T. Huang, S. Datta, D. B. Janes, Phys. Rev. B **47**, 4084 (1993)
65. T. F. Boggess, J. T. Olesberg, C. Yu, M. E. Flatté, W. H. Lau, Appl. Phys. Lett. **77**, 1333 (2000)
66. M. I. D'yakanov, V. I. Perel', Zh. Eskp. Teor. Fiz. **60**, 1954 (1971) [Sov. Phys. JETP **33**, 1053 (1971)]
67. A. Fert and S. -F. Lee, Phys. Rev. B **53**, 6554 (1996)
68. M. W. J. Prins, H. van Kempen, H. van Leuken, R. A. Degroot, W. van Roy, J. Deboeck, J. Phys. Cond. Matt. **7**, 9447 (1995)
69. M. W. J. Prins, R. Jansen, H. van Kempen, Phys. Rev. B **53**, 8105 (1996)
70. F. G. Monzon, M. L. Roukes, J. Magn. Magn. Mater **195**, 19 (1999)
71. S.Gardelis, C.G.Smith, C.H.W.Barnes, E.H.Linÿeld, D.A.Ritchie, Phys. Rev. B **60**, 7764 (1999)
72. W.Y.Lee, S.Gardelis, B.-C.Choi, Y.B.Xu, C.G.Smith, C.H.W.Barnes, D.A.Ritchie, E.H.Linÿeld, J.A.C.Bland, J. Appl. Phys. **85**, 6682 (1999)
73. A.T. Filip, B. H. Hoving, F.J. Jedema, B. J. van Wees, B. Dutta, S. Borghs, Phys. Rev. B **62**, 9996 (2000)
74. F.G. Monzon, M. Johnson, M.L. Roukes, Appl. Phys. Lett. **71**, 3087 (1997)
75. F.G. Monzon, D.S. Patterson, M.L. Roukes, J. Magn. Magn. Mater. **198**, 632 (1999)
76. M.A.M. Gijs, G.E.W. Bauer, Advances in Physics **46**, 285 (1997)
77. J-Ph. Ansermet, J. Phys. Condens. Matter **10**, 6027 (1998)
78. A. Fert, I.A. Campbell, J. de Physique, Colloque C! **32**, C1-46 (1971)
79. P.C. van Son, H. van Kempen, P. Wyder, Phys. Rev. Lett. **58**, 2271 (1987)
80. T. Valet, A. Fert, Phys. Rev. B **48**, 7099 (1993)
81. M. Johnson, R. H. Silsbee, Phys. Rev. B **35**, 4959 (1987)
82. S. Dubois, L. Piraux, J. M. George, K. Ounadjela, J. L. Duvail, A. Fert, Phys. Rev. B **60**, 477-484 (1999)
83. F.J. Jedema, A.T. Filip, B.J. van Wees, Nature **410**, 345 (2001)
84. S. Herschfield, H.L. Zhao, Phys. Rev. B **56**, 3296 (1997)
85. F. Beuneu, P. Monod, Physica B **86-88**, 265 (1977)

86. A. Brataas, Yu.V. Nazarov, G.E.W. Bauer, Phys. Rev. Lett. **84**, 2481(2000)
87. G. Schmidt, D. Ferrand, L. Molenkamp, A.T. Filip, B.J. van Wees, Phys. Rev. B **62**, R4790 (2000)
88. E.I. Rashba, Phys. Rev. B **62**, R16267 (2000)
89. H. X. Tang, F. G. Monzon, R. Lifshitz, M. C. Cross, M. L. Roukes, Phys. Rev. B **61**, 4437 (2000)
90. M. Büttiker, Phys. Rev. Lett. **57**, 1761 (1986)
91. C.W.J. Beenakker, H. van Houten, Phys. Rev. Lett. **63**, 1857 (1989)
92. Yu. V. Sharvin, Zh. Eksp. Teor. Fiz. **48**, 984 (1965) [Sov. Phys. JETP **21**, 655 (1965)]
93. R.G. Chambers, Proc. Phys. Soc. (London) **81**, 877 (1963). See also, N. W. Ashcroft, N. D. Mermin, Solid State Physics, p247, Harcourt Brace College Publishers, 1976.
94. M. Johnson, R.H. Silsbee, Phys.Rev. B **35**, 4959 (1987)
95. P. Seba, P. Exner, K. N. Pichugin, A. Vyhnal, P. Streda, Phys. Rev. Lett. 86, 1598 (2001)
96. R. J. Elliott, Phys. Rev. **96**, 266 (1954)
97. F. Mireles, G. Kirczenow, Phys. Rev. B **64**, 024426 (2001)
98. T. Ando, Phys. Rev. B **44**, 8017 (1991)
99. A. Filipe, H. -J. Drouhin, G. Lampel, Y. Lassailly, J. Nagle, J. Peretti, V. I. Safarov, A. Schuhl, Phys. Rev. Lett. **80**, 2425 (1998)
100. M. Oestreich, J. Hübner, D. Hägele, P. J. Klar, W. Heimbrodt, W. W. Rühle, d. E. Ashenford, B. Lunn, Appl. Phys. Lett. **74**, 1251 (1999)
101. R. Fiederling, M. Keim, G. Reuscher, W. Ossau, G. Schmidt, A. Waag, L. W. Molenkamp, Nature **402**, 787 (1999)
102. Y. Ohno, D. K. Young, B. Beschoten, F. Matsukura, H. Ohno, D. D. Awschalom, Nature **402**, 790 (1999)
103. B. T. Jonker, Y. D. Park, B. R. Bennett, H. D. Cheong, G. Kioseoglou, A. Petrou, Phys. Rev. B **62**, 8180 (2000)
104. Y. Q. Jia, R. C. Shi, S. Y. Chou, IEEE Trans. Magn. **32**, 4707 (1996)
105. F.G. Monzon, M.L Roukes, J. Magn. Magn. Mater. **199**, 632 (1999); F.G. Monzon, Caltech thesis, unpublished
106. P.R. Hammar, B. R. Bennett, M.J. Yang, M. Johnson, Phys. Rev. Lett. **83**, 203 (1999)
107. F. G. Monzon, H. X. Tang, M. L. Roukes, Phys. Rev. Lett. **84**, 5022 (2000)
108. B. J. van Wees, Phys. Rev. Lett. **84**, 5023 (2000)
109. C. M. Hu, J. Nitta, A. Jensen, J. B. Hansen, H. Takayanagi, Phys. Rev. B **63**, 125333 (2001)
110. M. Tanaka, H. Shimizu, T. Hayashi, H. Shimada, K. Ando, J. Vac. Sci. Technol. A **18**, 1247 (2000)
111. J. J. Krebs, B. T. Jonker, G. A. Prinz, J. Appl. Phys. **61**, 2596 (1987)
112. C. Daboo, R. J. Hicken, E. Gu, M. Gester, S. J. Gray, D. E. P. Eley, E. Ahmad, J. A. C. Bland, Phys. Rev. B **51**, 15964 (1995)
113. A. Filipe, A. Schuhl, P. Galtier, Appl. Phys. Lett. **70**, 129 (1997)
114. Y. B. Xu, E. T. M. Kernohan, M. Tselepi, J. A. C. Bland, S. Holmes, Appl. Phys. Lett. **73**, 399 (1998)
115. Y. B. Xu, D. J. Freeland, E. T. M. Kernohan, W. Y. Lee, M. Tselepi, C. M. Guertler, C. A. F. Vax, J. A. C. Bland, S. N. Holmes, N. K. Patel, D. A. Richie, J. Appl. Phys. **85**, 5369 (1999)

116. W. Y. Lee, Y. B. Xu, S. M. Gardiner, J. A. C. Bland, B. C. Choi, J. Appl. Phys. **87**, 5926 (2000)
117. K. Tsukagoshi, B.W. Alphenaar, H. Ago, Nature **401**, 572 (1999)
118. A. Bachtold, C. Strunk, J. P. Salvetat, J. M. Bonard, L. Forro, T. Nussbaumer, C. Schonenberger, Nature **397**, 673 (1999)
119. K. J. Thomas, J. T. Nicholls, M. Y. Simmons, M. Pepper, D. R. Mace, D. A. Ritchie, Phys. Rev. Lett **77**, 135 (1996)
120. J. A. Folk, S. R. Patel, K. M. Birnbaum, C. M. Marcus, C. I. Duruoz, J. S. Harris, Phys. Rev. Lett. **86**, 2102 (2001)
121. T. Kikutani, N. Aoki, S. Yamada, Phys. Rev. B **61**, 9956 (2000)
122. See, e.g., G. A. Prinz, Physics Today **48**, 58 (1995)
123. J. Nitta et al., Private Communication
124. B. W. Alphenaar, K. Tsukagoshi, M. Wagner, J. Appl. Phys. **89**, 6863 (2001)
125. D. D. Awschalom, J. M. Kikkawa, Physics Today **52**, 33 (1999)
126. I. Malajovich, J. M. Kikkawa, D. D. Awschalom, J. J. Berry, N. Samarth, Phys. Rev. Lett. **84**, 1015 (2000)
127. I. Malajovich, J. J. Berry, N. Samarth, D. D. Awschalom, Nature **411**, 770 (2001)
128. K. P. Kamper, W. Schmitt, G. Guntherodt, R. J. Gambino, R. Ruf, Phys. Rev. Lett. **59**, 2788 (1987)
129. R. A. de Groot, F. M. Mueller, P. G. van Engen, K. H. J. Buschow, Phys. Rev. Lett. **50**, 2024 (1983)
130. P. R. Hammar, B. R. Bennett, M. J. Yang, M. Johnson, Phys. Rev. Lett. **84**, 5024 (2000)
131. M. Johnson, Phys. Rev. B **58**, 9635 (1998)
132. D. Grundler, Phys. Rev. Lett. **86**, 1058 (2001)
133. C. M. Hu, T. Matsuyama, Phys. Rev. Lett. **87**, 6803 (2001)
134. R. H. Silsbee, Phys. Rev. B **63**, 155303 (2001)
135. L. W. Molenkamp, G. Schmidt, G. E. W. Bauer, Phys. Rev. B **64**, 121202 (2001)
136. S. F. Alvarado, Ph. Renaud, Phys. Rev. Lett. **68**, 1387 (1992)
137. V. P. LaBella, D. W. Bullock, Z. Ding, C. Emery, A. Venkatesan, W. F. Oliver, G. J. Salamo, P. M. Thibado, M. Mortazavi, Science **292**, 1518 (2001)
138. D. L. Smith, R. N. Silver, Phys. Rev. B **64**, 045323 (2001)
139. H. B. Heersche, Th. Schäpers, J. Nitta, H. Takayanagi, Phys. Rev. B **64**, 161307 (2001)
140. S. K. Upadhyay, A. Palanisami A, R. N. Louie RN, R. A. Buhrman RA, Phys. Rev. Lett. **81**, 3247 (1998)
141. S. K. Upadhyay, R. N. Louie RN, R. A. Buhrman, Appl. Phys. Lett. **74**, 3881 (1999)
142. See, for example, G. A. Prinz, G. T. Rado, J. J. Krebs, J. Appl. Phys. **53**, 2087 (1982)
143. T. L. Monchesky, R. Urban, B. Heinrich, M. Klaua, J. Kirschner, J. Appl. Phys. **87**, 5167 (2000)
144. J. A. C. Bland, A. Hirohata, C. M. Guertler, Y. B. Xu, M. Tselepi, J. Appl. Phys. **89**, 6740 (2001).
145. L. C. Chen, J. W. Dong, B. D. Schultz, C. J. Palmstrom, J. Berezovsky, A. Isakovic, P. A. Crowell, N. Tabat, J. Vac. Sci. Tech. B **18**, 2057 (2000)
146. H. J. Zhu, M. Ramsteiner, H. Kostial, M. Wassermeier, H. P. Schönherr, K. H. Ploog, Phys. Rev. Lett. **87**, 016601 (2001)

147. F. Bensch, G. Garreau, R. Moosbühler, G. Bayreuther, E. Beaurepaire, J. Appl. Phys. **89**, 7133 (2001)
148. G. Kirczenow, Phys. Rev. B **63**, 054422 (2001)
149. D. Grundler, Phys. Rev. B **63**, R161307 (2001)
150. D. L. Smith, M. Kozhevnikov, E. Y. Lee, V. Narayanamurti, Phys. Rev. B **61**, 13914 (2000)
151. A. T. Hanbicki, B. T. Jonker, G. Itskos, M. Furis, G. Kioseoglou, A. Petrou, *preprint* (2001)
152. D. Monsma, J. C. Lodder, Th. J. A. Popma, B. Dieny, Phys. Rev. Lett. **74**, 5260 (1995); R. Jansen, O. M. J. van 't Erve, S. D. Kim, R. Vlutters, P. S. Anil Kumar, J. C. Lodder, J. Appl. Phys. **89**, 7431 (2001)
153. W. H. Rippard, R. A. Buhrmann, Phys. Rev. Lett. **84**, 971 (2000)
154. G. Schmidt, L. W. Molenkamp, Physica E **10**, 484 (2001)
155. H. X. Tang, M. L. Roukes, to be published.
156. J. A. Sidles, Appl. Phys. Lett. **58**, 2854 (1991); J. A. Sidles, Phys. Rev. Lett. **68**, 1124 (1992)
157. D. Rugar, C.S. Yannoni, J.A. Sidles, Nature **360**, 563 (1992)
158. P. C. Hammel, Z. Zhang, G. J. Moore, M. L. Roukes,r J. Low Temp. Phys. **101**, 59 (1995)
159. D. Orgassa, H. X. Tang, M. L. Roukes, to be published.

# 3 Electrical Spin Injection: Spin-Polarized Transport from Magnetic into Non-Magnetic Semiconductors

Georg Schmidt and Laurens W. Molenkamp

## 3.1 Introduction

Spin polarized transport in semiconductors presently attracts a great deal of interest. For the past five years or so, several research groups have tried to achieve efficient injection of spin polarized electrons into non-magnetic semiconductors using ferromagnetic metals as spin injecting contacts. All these experiments have failed to produce conclusive results; the reasons for this lack of success were outlined in Chap. 2 and in [1]. The main outcome of the theory for spin injection given in Chap. 2 is that the spin polarization $\alpha$ in the semiconductor is simply not proportional to the bulk spin polarization $\beta$ in the ferromagnet. This is mainly due to the different conductivities of semiconductor ($\sigma_{sc}$) and ferromagnet ($\sigma_{fm}$) and to the short spin scattering length $\lambda_{fm}$ in the ferromagnet. For a one dimensional device with a two dimensional electron gas extending from 0 to $x_0$ contacted by two magnetic contacts with a thickness well beyond the spin flip length $\lambda_{fm}$, $\alpha$ in the semiconductor is given by

$$\alpha = \beta \, \frac{\lambda_{fm}}{\sigma_{fm}} \, \frac{\sigma_{sc}}{x_0} \, \frac{2}{\left(2\frac{\lambda_{fm}\sigma_{sc}}{x_0\sigma_{fm}} + 1\right) - \beta^2}, \tag{3.1}$$

if the spin flip length in the semiconductor is much longer than $x_0$ and the magnets are aligned in parallel ($\beta$ is the bulk spin polarization of the magnets). From (3.1) we can see that the expression for $\alpha$ indeed contains a factor $\beta$, reduced by a factor of $(\lambda_{fm}/\sigma_{fm})/(x_0/\sigma_{sc})$, times a term non-linear in $\beta$. Even when it is possible to realize ferromagnets and semiconductors with comparable conductivities, the short spin flip length in the ferromagnet will always present a major obstacle to spin injection. On one hand, a short spin flip length is necessary to guarantee for an efficient spin alignment in the ferromagnet, on the other hand an $x_0$ of at least several hundreds of nanometers is necessary to enable the fabrication of transport devices for spin manipulation [2]. However, (3.1) also shows that, because of the $1/(1 - \beta^2)$ term, in the limit of $\beta \to 1$, also $\alpha \to 1$. In simple words: If a spin injector with a sufficiently high spin polarization is used, the conductivities of the materials and the spin scattering length loose their influence on the spin injection efficiency. In Fig. 3.1 the spin polarization at the interface is plotted over $\beta$ and a typical set of parameters.

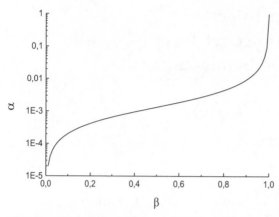

**Fig. 3.1.** Spin polarization $\alpha$ plotted over the bulk spin polarization $\beta$ in a one dimensional device with a contact spacing of 100 nm, a spin flip length in the ferromagnet of $\lambda_{fm} = 10$ nm and a ratio of respective conductivities of $\sigma_{fm} = 100\sigma_{sc}$

## 3.2   Electrical Spin Injection

It is thus desirable to use a spin injector which exhibits a spin polarization close to one. We have already learned that typical ferromagnetic metals have a spin polarization which is merely in the range of 70% and thus not suitable for spin injection. However, possible candidates for a close to 100% spin polarization are the diluted magnetic semiconductors.

### 3.2.1   Diluted Magnetic Semiconductors

DMS are semiconductors that contain a certain percentage of magnetic ions, usually Mn [3]. They exist both in the II-VI and the III-V semiconductor families. In the following, we will describe in detail the properties of II-VI DMS materials (III-V materials are described in Chap. 1). II-VI DMS are usually not ferromagnetic. Although heavily p-doped II-VI DMS like (Be,Mn)Te and (Zn,Mn)Te show ferromagnetism at very low temperatures, the n-doped materials described in this chapter are only remarkable paramagnets. In the absence of a magnetic field the spins of the manganese ions are not aligned in parallel. In this state, the net magnetization of the material is zero, the conduction and valence band states are degenerate for the two spin directions and the charge carriers are hence unpolarized. In II-VI DMS, Mn is isoelectronic. So n and p-type doping are possible. This is an advantage as the spin-precession (and thus loss of spin information in a 3D or 2D system) is much stronger for holes than for electrons.

When an external magnetic field is applied and the temperature is low enough (typically less than 10K), the situation changes dramatically. The g-factor of a DMS is typically of the order of 100 and the conduction band

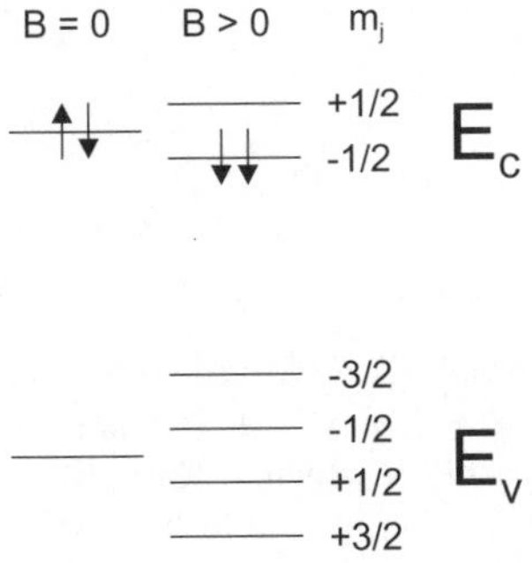

B = 0      B > 0      $m_j$

————  +1/2   $E_c$
————  -1/2


————  -3/2
————  -1/2   $E_v$
————  +1/2
————  +3/2

**Fig. 3.2.** Zeeman splitting in a typical magnetic semiconductor

as well as the valence band exhibit a so called giant Zeeman-splitting. In $Zn_{0.9}Mn_{0.1}Se$ for example, this splitting is 100 meV in the valence band and 20 meV in the conduction band at 4.2 K and 2 T [4]. At low temperatures, where kT is in the range of 0.1 meV, the splitting in the conduction band guarantees a spin polarization of the electrons (for n-doped material) of almost 100% because the strong spin scattering of the Mn ions quickly establishes the equilibrium which is close to a Boltzmann distribution. The II-VI dilute magnetic semiconductors should thus be ideal candidates for efficient spin injection [5,6].

### 3.2.2   The Optical Detection of Spin Injection

In Chap. 2, we have already learned that besides the difficulty of injecting spin polarized electrons into a semiconductor, a second problem exists in the detection of the spin polarization. The more or less standard detection method is the use of two ferromagnetic contacts on a semiconductor and the determination of the difference in overall device resistance for parallel and antiparallel magnetic alignment of the contacts [7]. This experiment is equivalent to the spin valve effect known from all-metal GMR. However, at least for a small $\alpha$, the magnitude of the effect is roughly proportional to the square of the obtainable spin polarization $\alpha$. For small values of $\alpha$ it will thus be difficult to detect and it seems wise to look for a detection mechanism which gives a signal which is linear in $\alpha$. In zincblende type semiconductors like GaAs, the selection rules for the radiative recombination of electrons and holes are spin selective. Four different recombination processes are allowed, two of them involving a heavy hole, the others involving light hole states. The optical polarization of the photons which result from the recombination depend on the spin polarization of the electrons. An electron with spin +1/2 can only recombine with a light hole with spin -1/2 or a heavy hole with spin +3/2, giving a photon with a circular polarization of $\sigma^-$ for the light hole and $\sigma^+$ for the heavy hole (Fig. 3.3). For the spin down electron (-1/2) all signs and the handedness of the polarizations are reversed. It is important to note that the probability for the heavy hole process is three times as large as that for the light hole process.

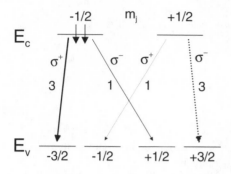

Fig. 3.3. Allowed radiative transitions in GaAs. The heavy hole transitions have a probability which is three times as large as for the light hole transition.

Because of this, the luminescence emitted by spin polarized electrons recombining with unpolarized holes will be circularly polarized [8], if the spin polarization is collinear with the direction of the light emission. If the electrons in the conduction band are partially spin polarized with the spin-polarization defined as

$$\alpha = \frac{n^\uparrow - n^\downarrow}{n^\uparrow + n^\downarrow} \tag{3.2}$$

and light and heavy hole states are degenerate, recombination processes with unpolarized holes will lead to the following circular polarization of the emitted radiation:

$$P_{circ}^{opt} = \frac{\sigma^+ - \sigma^-}{\sigma^+ + \sigma^-} = \frac{(3n^\uparrow + n^\downarrow) - (3n^\downarrow + n^\uparrow)}{(3n^\uparrow + n^\downarrow) + (3n^\downarrow + n^\uparrow)} = \frac{1}{2} \frac{n^\uparrow - n^\downarrow}{n^\uparrow + n^\downarrow} = \frac{1}{2}\alpha \tag{3.3}$$

For the case of non degenerate light and heavy holes, the degree of optical polarization and the spin polarization will be equal. If we fit a GaAs light emitting p/n-diode with a non-magnetic contact on the p-side and a DMS contact on the n-side, the diode will be constantly fed with unpolarized holes on the p-contact. In the case of spin injection from the DMS n-contact the light which is emitted by the diode will show a degree of optical circular polarization which is proportional to the degree of spin polarization of the injected electrons. Thus, it presents a much more sensitive detector for small spin polarizations than the GMR-like device mentioned in the first paragraph.

### 3.2.3   The Spin Aligner LED

We have fabricated a device which makes use of the mechanisms described above [5]. A GaAs/(Al,Ga)As pin-diode was grown on a highly p-doped GaAs (100)-substrate. The diode consists of a p-type GaAs buffer layer (300 nm thickness) on which the lower (Al,Ga)As layer with a thickness of 500 nm and heavy doping was grown. Then the active GaAs layer (15 nm) is deposited followed by the upper (Al,Ga)As barrier with a thickness of 100 nm and an n-type doping of approx. $10^{16}$ cm$^{-3}$. Thickness and doping of the

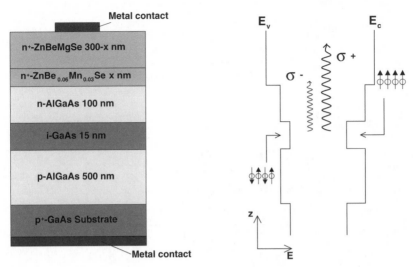

**Fig. 3.4.** Schematic drawing (*left*) of the GaAs/(Al,Ga)As light emitting diode with the spin aligning II-VI-DMS contact. On the right the corresponding bandstructure is sketched.

upper (Al,Ga)As layer are chosen in a way to guarantee a low spin scattering probability during the transition of the electrons.

The sample was then transferred into another growth chamber without breaking the UHV. In that chamber, a highly ($3 \times 10^{18}$ cm$^{-3}$) n-doped (Zn,Be,Mn)Se layer was deposited on top of the diode, followed by a similarly doped (Zn,Be)Se layer. The thickness of the non-magnetic (Zn,Be)Se layer was chosen in such a way that together with the DMS layer the total thickness of the II-VI layers is 300 nm. This was done in order to achieve identical current spreading conditions around the contact independent from the DMS thickness. Still in UHV, an Al layer was deposited top of the structure. Using optical lithography and wet etching, mesa-like diodes with a center top contact were fabricated (3.4). The structures were characterized in an optical cryostat equipped with a superconducting split coil magnet.

### 3.2.4   Experimental Results

All LEDs with spin aligner contacts showed a distinct circular polarization of the electroluminescence signal (Fig. 3.5). For a spin aligner thickness $d_{sa}$ of 300 nm, the degree of circular optical polarization was as high as 45%.

This same degree of polarization was maintained for the $d_{sa}$=100 nm sample. Even a spin aligner of 3 nm gave a degree of optical polarization of 10% and only the sample with no spin aligner gave a signal which was, to within the detection limit, unpolarized.

The efficiency of the spin injection indicates that the DMS is completely spin polarized. On first sight, this seems in contradiction with the high dop-

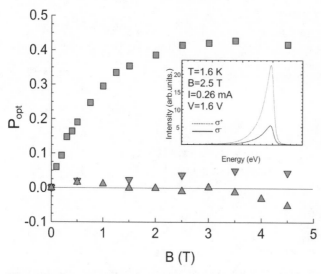

**Fig. 3.5.** Circular optical polarization of the light emitted by the LED. The squares and the *downward triangles* indicate the polarization of the electroluminescence signal for a diode with 300 nm spin aligner and without spin aligner, respectively. The *upward triangles* show the polarization of the photoluminescence signal of the sample with spin aligner emitted under unpolarized optical excitation. The *inset* shows a typical electroluminescence spectrum of a diode with a spin aligning contact. The main emission peak shows a FWHM of less than 2 meV and the light is strongly circularly polarized

ing in the DMS. A doping level of $3 \times 10^{18}/\text{cm}^3$ in a calculation which takes only into account a parabolic band structure would yield a Fermi Energy of more than 20 meV, which is the expected Zeeman splitting in the DMS at high fields. If the Fermi energy indeed exceeded the conduction band Zeeman splitting, the spin polarization would not exceed a few tens of percent and could not explain the experimental results. However, we know from additional experiments on samples with various doping densities, and at various temperatures, that the spin polarization of our injector is indeed close to 100%. At present, we believe that this contradiction can be solved by assuming the presence of an impurity band. At the high doping levels involved here, the DMS is far beyond the Mott criterion. The wavefunctions of the donors overlap and the donor levels form an impurity band with a density of states much higher than in the conduction band. The Fermi energy is drastically reduced and it is possible to achieve complete spin polarization of the electrons even at relatively low Zeeman splitting. This explanation is supported by additional measurements which show that conductivity and carrier concentration in the highly doped DMS layers are temperature independent.

### 3.2.5 Exclusion of Side Effects

As mentioned in Chap. 2, electrical spin injection experiments which use the change in device resistance to evidence the presence of a spin-polarized current, may suffer from side effects like stray hall voltages. Optical experiments also may be tainted by artifacts, which have to be carefully excluded. For instance, a polarization dependent absorption of light in a magnetic semiconductor – known as magnetic circular dichroism (MCD) (Fig. 3.6) – may influence the polarization of the light which is emitted by the diode [9].

As the selection rules explained above are only valid for emission which is perpendicular to the plane of a quantum well or for bulk material, the experiment has to be carried out in a perpendicular emission geometry. If the spin aligner is on top of the diode, the light typically passes through the magnetic layer and it is possible that one polarization of the electroluminescence is absorbed, while the other one is transmitted. Actually, because of the device layout, such an artifact is not to be expected in our experiment, because the bandgap of the spin aligner material used is almost twice the gap of GaAs, making reabsorption effects unlikely.

However, control experiments were carried out. Besides the electroluminescence experiment, we have also the possibility to excite carriers inside the quantum well by laser light. If the exciting light is unpolarized, the carriers will not be spin polarized and the photoluminescence signal should also be unpolarized. The light path for the emitted photons is the same for electroluminescence and for photoluminescence. If MCD had a critical influence on the polarization, this would be visible in both PL and EL experiments. In the case of the DMS spin aligner, the effect is not detectable in the photolu-

B = 0 ⇒ no absorption in DMS

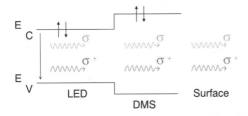

B ≠ 0 ⇒ spin dependent absorption in DMS

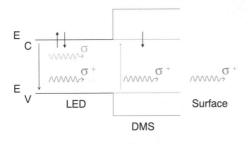

**Fig. 3.6.** Light coming from the diode may be unaffected when going through the magnetic semiconductor at zero field (*top*). Nevertheless, a polarization dependent absorption can occur in the presence of the giant Zeeman splitting in a magnetic field (*bottom*).

minescence experiment (Fig. 3.5, upward triangles). On the contrary, the PL measurements show a slight polarization with opposite handedness, which can be attributed to the Zeeman splitting in the GaAs which has a negative g-factor.

### 3.2.6  Hole Injection

A similar experiment was carried out by the groups of D. D. Awschalom and H. Ohno (see Chap. 1) [10]. In that case the light emitting diode consisted of an (In,Ga)As well embedded in between GaAs layers. Instead of a dilute

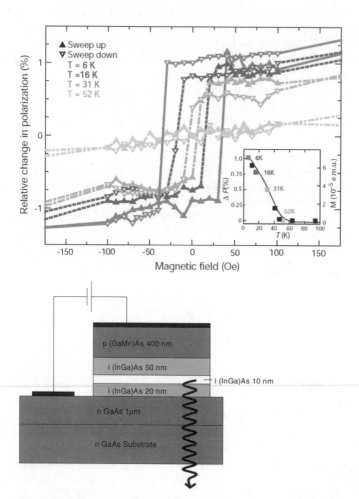

**Fig. 3.7.** Polarization curve (*top*) for a laterally emitting (Ga,Mn)As/InGaAs LED [10] and schematic drawing of a new vertically emitting LED structure (*bottom*). Emission is done through the substrate in order to avoid MCD.

magnetic semiconductor, the ferromagnetic semiconductor (Ga,Mn)As was used as a spin aligner. Because Mn is an acceptor in III-V-semiconductors, (Ga,Mn)As is strongly p-type (Chap. 1), so that spin polarized holes instead of electrons are injected. Spin polarized holes are more subject to spin precession and spin flip by spin orbit interaction in the non-magnetic semiconductor. Moreover, the spin polarization of the holes is in the range of 90% and thus smaller than in the dilute magnetic semiconductors, so a smaller spin injection efficiency can be expected. The light emitted by the diode showed a degree of circular polarization of 2%. This value is in good agreement with theory [1], if one assumes equal conductivities for the (Ga,Mn)As and the non-magnetic semiconductor. In a magnetic field sweep, the degree of optical circular polarization showed a distinct hysteresis corresponding to the switching behavior of the ferromagnetic (Ga,Mn)As layer. As the bandgap of (Ga,Mn)As is close to the gap of (In,Ga)As a different experimental setup than in the II-VI-DMS experiment had to be chosen. The light is either emitted at the side of the device [10] or in a perpendicular geometry through the substrate. In this way, the passage of the light through the magnetic semiconductor and a possible influence of MCD can be avoided.

## 3.3   A Novel Magnetoresistance Effect

Although the experiments described above demonstrate spin injection, different effects than optical polarization would be desirable for technical applications. For electronic devices, the spin polarization should be directly convertible into a voltage, without using light emitting devices. This can be accomplished if a spin dependent magnetoresistance like GMR is realized in semiconductors.

### 3.3.1   Theoretical Prediction

In a GMR geometry one measures the difference in resistance between parallel and antiparallel alignment of two ferromagnetic contacts. As dilute magnetic semiconductors are only magnetic in an external field, it is difficult to realize two contacts with antiparallel magnetization. However, if one contemplates the modeling described in [1] and in Chap. 2 in more detail, one may see that the dilute magnetic semiconductors allow for the observation of a different magnetoresistance effect. One assumption of the theory is that in a non-magnetic semiconductor the current is carried by electrons of both spin directions to equal parts. We have thus two spin channels which are independent over the range of the spin flip length and which have equal conductivities. If a spin polarized current is injected into the semiconductor, one of the two channels carries the current, while the other one is suppressed. However, the conductivity of the spin channels does not change and the total conductance of the semiconductor should decrease, in the ideal case - no spin flip and 100% of spin polarization - by a factor of 2. If one includes spin flip in the non-magnetic semiconductor, a calculation of $(R_{\text{magnetized}} - R_{\text{nonmagnetized}})/R_{\text{semiconductor}}$ gives the following result.

$$\frac{\Delta R}{R_{sc}} = \beta^2 \frac{\lambda_{dms}}{\sigma_{dms}} \frac{\sigma_{sc}}{x_0} \frac{2}{\frac{\lambda_{dms}}{\sigma_{dms}} \frac{\sigma_{sc}}{\lambda_{sc}} \left(1 + e^{-\frac{x_0}{\lambda_{sc}}}\right) + 2\frac{\lambda_{dms}}{\sigma_{dms}} \frac{\sigma_{sc}}{x_0} e^{-\frac{x_0}{\lambda_{sc}}} + 1 - \beta^2} \quad (3.4)$$

In comparison to GMR the effect shows one peculiarity. GMR only works if the spin polarized electrons coming from one contact can still reach the second contact without changing their spin. The GMR signal rapidly decays, when the spacing is larger than the spin flip length. The novel effect behaves differently. Even for a contact spacing which is larger than the spin flip length, the effect persists, as it only depends on the spin polarization of the injected current and the conductivity of the non-magnetic semiconductor. As soon as the contact spacing is much longer than the spin flip length, the resistance does even not depend on the relative magnetization of the contacts. The effect may even be demonstrated using only one spin aligning contact, but for a better comparison to GMR the calculations were performed for a two contact device.

### 3.3.2   Device Layout

In the following, a structure is presented, which allows for the demonstration of the effect [11]. It consists of a mesa of the non-magnetic II-VI-semiconductor (Zn,Be)Se which is fitted with two (Zn,Be,Mn)Se DMS contacts. A 500 nm thick $Zn_{0.97}Be_{0.03}Se$ layer with a doping of $9 \times 10^{18} \text{cm}^{-3}$ was grown on a semi-insulating GaAs substrate using MBE. On top of this layer a thin layer of $Zn_{0.89}Be_{0.05}Mn_{0.06}Se$ (doping $3 \times 10^{18}\text{cm}^{-3}$) was deposited. The thickness of this layer was varied between 0 and 200 nm for different samples. Finally, a highly n-doped ($1.5 \times 10^{19}\text{cm}^{-3}$) 10 nm thick ZnSe layer was deposited in order to achieve good contact properties. After transfer in UHV to another growth chamber, a 100 nm layer of Al was evaporated as a contact metal.

Using optical lithography and wet chemical etching, contact pads of $200 \times 250$ μm were defined in the aluminum layer. These contact pads were used as an etch mask for wet chemical etching through the ZnSe, the (Zn,Be,Mn)Se and some tens of nanometers into the non-magnetic (Zn,Be)Se. Subsequently, a mesa structure was defined by photolithography and etching through the (Zn,Be)Se down to the substrate, resulting in a device geometry which is sketched in Fig. 3.8.

The devices thus fabricated had spacings between the contact pads that varied from 10 to 30 μm. Note that, for all cases, the spacings are much larger than the spin flip length in highly doped (Zn,Be)Se ($9 \times 10^{18}$ cm$^{-3}$) which we estimate to be a few hundred nanometers. The measurements were carried

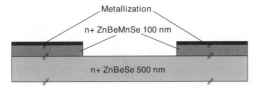

**Fig. 3.8.** Device for the demonstration of the novel magnetoresistance effect. The contact pads are $200 \times 200$ μm and the spacing between the contacts varies between 10 and 30 μm.

out in a $He^4$ bath cryostat equipped with a superconducting magnet. Much in contrast to the optical experiments described above, where a voltage in the range of 2 V was applied and the devices operated far from equilibrium, the resistance measurements were carried out using an excitation voltage of 100 μV guaranteeing an experiment in the range of linear response.

### 3.3.3  Results and Interpretation

All devices containing a magnetic spin aligner layer exhibit a positive magnetoresistance (Fig. 3.9). The resistance increases with the magnetic field and saturates at a field of about 2 T.

The total effect is approx. $1.2\Omega$ giving a $\Delta R/R$ of more than 8%. However this is only a lower limit, as the device resistance includes also the contact resistance. The magnitude of the effect is independent from the orientation of the magnetic field, so Hall effect can be excluded. In addition magnetoresistance measurements of the semiconductor materials used in the experiment were performed. Both materials, the non-magnetic (Zn,Be)Se and the DMS

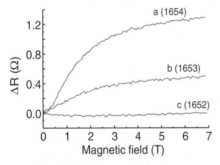

**Fig. 3.9.** Total device resistance plotted over the magnetic field for two DMS/NMS hybrid structures and one NMS control sample. 1653 (**b**) and 1654 (**a**) include a DMS contact with a thickness of 100 and 200 nm, respectively. The zero field resistances is 15 $\Omega$ (1653) and 20.5 $\Omega$ (1654). 1652 (**c**) is a reference device without DMS ($R_0=14$ $\Omega$). The magnetoresistance is less than 0.02 $\Omega$.

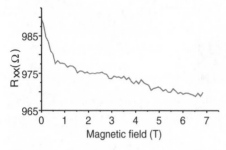

**Fig. 3.10.** Longitudinal resistance of a Hall bar of the DMS used in the magnetoresistance device. The sample shows a negative magnetoresistance of approx $-2\%$.

(Zn,Be,Mn)Se showed either no or a slightly negative magnetoresistance, respectively (Fig. 3.10). Only in combination of both materials, the positive magnetoresistance occurs, indicating that indeed the suppression of one spin channel leads to the effect.

On first sight, the effect seems to be relatively large for a device whose dimensions are far beyond the spin scattering length. However, the magnitude of the effect can be explained when the current path in the device is taken into account. We have performed two dimensional simulations based on the theoretical model using a finite difference poisson solver. The results indicate that the slightly higher resistivity of the DMS compared to the (Zn,Be)Se leads to a current flow which is more or less perpendicular to the DMS/NMS interface. The main voltage drop occurs underneath the contacts and not in-between. This also explains the magnitude of the effect, as the spin polarization is highest close to the interface between DMS and NMS. Furthermore the calculations show that, due to the low current density in the DMS, the DMS layer contributes very little to the total device resistance, making the device behavior insensitive to a possible magnetoresistance in the DMS. More experiments with devices with the same geometry but reduced doping have been performed showing a higher total device resistance but also an increased positive magnetoresistance of $\Delta R/R = 25\%$. This increase in magnitude with reduced doping can be attributed to the increasing spin flip length in the non magnetic material.

## 3.4   Outlook

The spin injection experiments with DMS spin aligners have shown that high efficiency spin injection is possible and that spin polarization levels of up to 90% can be achieved. The magnetoresistance effect demonstrated at low excitation voltages indicate clearly that spin injection is a single particle phenomenon. All results are in well agreement with theory, showing that efficient spin injection is only possible when the spin aligner exhibits an electron spin polarization very close to 1. Although the dilute magnetic semiconductors are only suitable spin aligners at low temperatures, they can already be used to study the physics of spin polarized transport in semiconductors and of possible spintronic devices. However, for applications, new materials have to be developed which are ferromagnetic and fully spin polarized at room temperature, in order to combine the non volatile memory function of the ferromagnet with spintronic applications. Possible candidates for this issue can be the Heusler compounds like NiMnSb, double perovskites, or ferromagnetic semiconductors like (Ga,Mn)N, which is predicted [12] to show ferromagnetism at room temperature.

**Acknowledgments**

This work was supported by the German Bundesministerium für Bildung und Forschung, the European Commission, and DARPA.

# References

1. G. Schmidt, D. Ferrand, L. W. Molenkamp, A. T. Filip, B. J. vanWees, Phys. Rev. B **62**, R4790 (2000).
2. S. Datta, B. Das, Appl. Phys. Lett. **56** (7), 665 (1990).
3. J. K. Furdyna, J. Appl. Phys. **64**, R29 (1988).
4. J. A. Gaj, R. Planel, G. Fishman, Solid State Comm. **29**, 435 (1979).
5. R. Fiederling, M. Keim, G. Reuscher, W. Ossau, G. Schmidt, A. Waag, L. W. Molenkamp, Nature (London) **402**, 787 (1999).
6. B.T Jonker, Y.D. Park, B.R. Bennett, H.D. Cheong, G. Kioseoglou, A. Petrou, Phys. Rev. B **62** (12), 8180 (2000).
7. F. G. Monzon, M. L. Roukes, J. Magn. Magn. Mater. **198-199**, 632 (1999).
8. A. G. Aronov, G. E. Pikus, Sov. Phys. Semicond. **10** (6), 698 (1976). **15** (3), 1215, (1997)
9. B. Beschoten, P.A. Crowell, I. Malajovich, D.D. Awschalom, F. Matsukura, A.Shen, H. Ohno, Phys. Rev. Lett. **83** (15), 3073 (1999).
10. Y. Ohno, D. K. Young, B. Beschoten, F. Matsukura, H. Ohno, D. D. Awschalom, Nature (London) **402**, 790 (1999).
11. G. Schmidt, G. Richter, P. Grabs, D. Ferrand, L. W. Molenkamp, Phys. Rev. Lett. **87**, 227203 (2001).
12. T. Dietl, H. Ohno, F. Matsukura, J. Cibert, D. Ferrand, Science **287**, 1019 (2000).

# 4 Spin Dynamics in Semiconductors

Michael E. Flatté, Jeff M. Byers, and Wayne H. Lau

## 4.1 Introduction

Recent advances in probing and manipulating spin dynamics in semiconductors suggest a new semiconducting electronics technology based on spin [1–3]. These advances include ultrafast all-optical manipulation of the spins of conduction electrons [4–9], core electrons of magnetic impurities [5], and nuclei [10–12], as well as all-electrical generation of optical orientation [13–15] and the development of a new class of III-V ferromagnetic semiconductors [16,17] (See Chaps. 1, 2, 3 and 5.) Thus the *material properties* of semiconductors essential to spintronic devices appear to be taking shape.

The detailed form of spin-based semiconducting electronics remains to be determined; suggested applications include non-volatile memory, reprogrammable logic, and even quantum computation [18–21]. To realize these goals requires a deep understanding of the physics determining both the coherence times and transport properties of inhomogeneous spin distributions in semiconductors. A full understanding of the relationship between the physical effects driving semiconductor spin electronics and those driving the mature area of metallic spin electronics [22] would help to determine whether semiconductor devices will compete for the market in magnetic read heads and magnetic non-volatile memory, and would assist in the development of hybrid semiconductor-metallic devices. Looking even farther ahead, deepening our understanding of semiconductor spin coherence may clarify if semiconductor spin electronics will play a significant role in the quest for a physical realization of a quantum computer.

In traditional semiconductor electronics the elements of device performance, such as rectification in a $p$–$n$ diode or gain in a transistor, are achieved through control of the spatial motion of carriers through that motion's sensitive dependence on potential energy variations. These potential energy variations come from applied electric fields or quasi-electric fields (such as heterostructure band offsets). Spin introduces a third element into this mix which couples sensitively to the other two. Starting from current semiconductor device structures, and anticipating the integration of semiconductor spin electronics into current architectures, one critical question is: how can the presence and motion of spin coherence be manipulated by electric or quasielectric fields and in turn influence the charge current and voltage?

The central device questions of quantum computer construction differ, for currently there is no experimentally-implemented scalable architecture. Here the principal concerns are the persistence and control of quantum coherence itself. Quantum coherence of a system is diminished by strong coupling to other systems, however, coupling to other systems is necessary for manipulation. Thus the proposed qubit, a spin-1/2 particle, must be sufficiently decoupled to remain coherent, but be sufficiently coupled to allow manipulation. Hence spin decoherence is an inevitable result of the ability to manipulate the spins – it can be controlled but not eliminated.

The outline of this chapter will be as follows. We begin with an analysis of the elements of mobile electron spin coherence in semiconductors, contrasting with other canonical coherence situations. We proceed to describe spin coherence near room temperature in bulk and nanostructure semiconductor systems, illustrating the different phenomenology as the dimensionality is reduced. We note that the number of possible qualitative regimes multiplies dramatically when $k_B T$ is much smaller than many other energy scales, so we will focus on the near-room-temperature regime. We then describe the role of the electronic charge in the motion of localized spin disturbances, or "spin packets", in structurally homogeneous semiconducting materials. Finally we briefly analyze a case which requires elements from all of these: spin injection across the ferromagnetic metal-semiconductor interface.

## 4.2    Fundamentals of Semiconductor Spin Coherence

The novel spin-dependent effects described in this book are characterized by the spin coherence times $T_1$ and $T_2$, representing the decay of longitudinal and transverse spin order, respectively. This nomenclature is common in the study of nuclear spins where the carriers of spin, the nuclei, are highly localized and interact only weakly with their environment. Usually fundamental material properties determine the coupling of nuclear spins to the perturbations responsible for spin relaxation. For example, coupling to phonons (inelastic spin-lattice relaxation) often determines $T_1$, whereas $T_2$ is set by direct interactions of magnetic dipoles with each other [23]. Electron spin relaxation can also be usefully described using $T_1$ and $T_2$ but with significant practical differences. Unlike nuclear spin, the spin of electrons can be intertwined with motion in space over distances of many micrometers. The resulting spin relaxation mechanisms of electrons are also more varied and larger in magnitude, suggesting an enhanced ability to manipulate these properties via external perturbation and materials engineering.

The simplest non-trivial spin system is the two-state system, appropriate to spin-1/2 nuclei or electrons. The dynamics of two-state systems have been explored in detail, particularly in the context of nuclear magnetic resonance (NMR) and electron spin resonance (ESR). A nonequilibrium perturbation to a two-state system can have two general elements: a change in the relative occupation of the two states, and a change in the coherent (phase) relationship between the two states. The evolution of these two components of the perturbation is described by the times $T_1$ and $T_2$. For a single two-state system the $T_1$ time can be described as a lifetime (or relaxation time), and the $T_2$ time as a coherence time. When the two-state system is a spin system and is placed in an external magnetic field, the $T_1$ time corresponds to the longitudinal magnetization decay time and the $T_2$ time corresponds to the transverse magnetization decay time.

### 4.2.1  Coherent Ensembles of Spins

In general the time evolution of a nonequilibrium perturbation to a *population* of two-state systems, within which interactions between pairs of states are permitted, is more subtle. Times characteristic of the evolution of the magnetization of the population ($T_1$ for the longitudinal magnetization and $T_2$ for the transverse magnetization), can still be defined without ambiguity, however the direct connection to microscopic characteristics such as single-spin lifetimes or coherence times is sometimes lost. The discussion below will emphasize those properties which are essential to the time evolution of spin perturbations in the conduction band of electron-doped semiconductor systems, particularly at temperatures above $\sim 50$ K. We will not discuss mechanisms of spin relaxation via electron-hole spin exchange (Bir-Aronov-Pikus mechanism [24]), which can dominate at low-temperature in the presence of both electrons and holes, or spin relaxation via hyperfine interaction with nuclear moments, which plays a key role at low temperatures in electron-doped GaAs. For a brief physical introduction to these two mechanisms we direct the reader to the presentation of Fishman and Lampel [25]. These mechanisms are treated in detail in *Optical Orientation* [26].

**Energy Bottleneck.** In a population of two-state systems there is a qualitative difference between $T_1$ and $T_2$ when the energies of the two states are different (such as for spins in the presence of a magnetic field). This can be seen in Fig. 4.1 below. The relaxation of a state occupation imbalance (Fig. 4.1a) to equilibrium requires energy transfer between the spin system and another system, whereas the decoherence of a phase relationship between the two states (Fig. 4.1b) does not. For example, phase decoherence between nuclear spins can occur via the elastic magnetic dipolar interaction between nuclei, whereas occupation relaxation in a magnetic field greater than the depolarization field (which sets the linewidth of the individual spin states) requires coupling to an energy reservoir (through inelastic spin-lattice coupling). Dipolar interactions between spins conserve the total angular momentum of the spin population, so *elastic* spin-lattice coupling is still required to transfer angular momentum into or out of the spin population at zero field. The dipolar interaction is usually much stronger than the inelastic spin-lattice

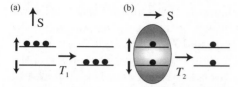

**Fig. 4.1.** (a) The decay of a nonequilibrium occupation (spin relaxation), characterized by the time $T_1$. Spin up and down eigenstates are shown split by a magnetic field. (b) Phase coherence between two states is indicated by the shaded circle. The decay of phase coherence, characterized by $T_2$, does not involve a change in occupation or a change in the energy of the population.

coupling, thus commonly $T_1 \gg T_2$ for fields greater than the depolarization field. This phenomenon is referred to as the "energy bottleneck", and is not restricted to nuclear spin systems. Core electrons in magnetic ions, for example, may also experience energy bottleneck.

### 4.2.2 Mobile Electron Decoherence Via the Spin-Orbit Interaction

In other electronic systems, however, principally those in which the electrons are mobile, the energy bottleneck is not present and $T_1 \sim T_2$. Shown in Fig. 4.2 is an example of an electronic spin system where the electrons are mobile and there is no energy bottleneck: electrons in a centrosymmetric semiconductor. In the absence of an applied magnetic field (Fig. 4.2a), all states in the electronic spectrum of a centrosymmetric crystal appear in (at least) doubly degenerate pairs. This characteristic emerges from the dual requirements of time-reversal (Kramers) symmetry and inversion symmetry:

$$E_n(\mathbf{k}) = E_{-n}(-\mathbf{k}), \qquad \text{(Time reversal symmetry)} \qquad (4.1)$$

$$E_n(\mathbf{k}) = E_n(-\mathbf{k}), \qquad \text{(Inversion symmetry)} \qquad (4.2)$$

imply

$$E_n(\mathbf{k}) = E_{-n}(\mathbf{k}). \qquad \text{(Pseudospin degeneracy)} \qquad (4.3)$$

Here $n$ refers to quantum numbers other than $\mathbf{k}$, including the spin of the electron. We assume for simplicity that there is no additional (band) degeneracy – this is the case for conduction electrons in group-IV semiconductors such as silicon and germanium. The presence of the low-energy valleys at $X$ and $L$ in group-IV semiconductors does not alter this argument – the double degeneracy exists throughout the Brillouin zone.

It is traditional to label the split eigenstates as "up",↑ and "down",↓. However, due to the spin-orbit coupling in the material, the eigenstates are not spin eigenstates – they should be more properly labelled as pseudospin eigenstates. This pseudospin nature emerges due to the inability to disentangle the spin component of the wavefunction from the orbital component; the

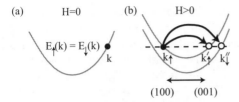

**Fig. 4.2.** Spectra for electrons in a centrosymmetric semiconductor. The pseudospin eigenstates along (100) are coupled to eigenstates along (001) by orbital (spin-independent) scattering.

wavefunction does not factorize into the product of a purely spin component and a purely orbital component. Lack of factorization in the wavefunction precludes constructing a spin eigenstate along an arbitrary direction from a linear combination of the two degenerate states. As a result one cannot find a *global* spin quantization axis good for all $\mathbf{k}$, and so $T_1$ cannot be rigorously associated only with changes in state occupation.

**Scattering-Driven Decoherence.** One of the consequences of this is that the spin quantum state of the electrons is not preserved if elastic scattering occurs, even via spin-independent scattering such as scattering from potential fluctuations. In general an eigenstate at momentum $\mathbf{k}$ and spin ↑ has a nonzero amplitude of scattering via *spin-independent processes* into each of two eigenstates, $\mathbf{k}'$ ↑ and $\mathbf{k}''$ ↓. These processes are shown for $H \neq 0$ in Fig. 4.2b for clarity (for $H \neq 0$ the spin up and spin down eigenstates are split), however these processes occur for $H = 0$ as well. The spin decoherence can be understood as an effective $\mathbf{k}'$-dependent rotation of the pseudospin resulting from the scattering event. The absence of the energy bottleneck shown in the schematic process of Fig. 4.2b permits this to occur, even for elastic scattering.

This mechanism of spin decoherence is often referred to as the Elliott-Yafet mechanism, and dominates in centrosymmetric crystals such as copper [27]. Each scattering event causes an effective spin rotation which contributes to the spin decoherence rate. Thus the more orbital scattering present in the material the more rapid the spin decoherence. Elliott-Yafet decoherence is characterized by spin coherence times decreasing as the material becomes dirtier.

**Precessional Dephasing and Decoherence.** Precessional dephasing in a population of two-state systems is distinct from the scattering-driven decoherence above. To illustrate this effect imagine a magnetic field pointing in the **x** direction, and an initial population of spins pointing along the **z** direction. The homogeneous magnetic field causes the spins to precess in the $yz$ plane, and the magnitude of the magnetization remains constant. If the magnetic field were inhomogeneous in magnitude, different spins would precess at different rates and the macroscopic magnetization would decay with time. This dephasing effect can be distinguished from decoherence, however, for with dephasing it remains possible to recover the initial macroscopic magnetization (an example being the "spin echoes" seen in NMR [23]). This recovery is possible so long as the carrier's spin orientation remains correlated with the carrier position. When this correlation is lost, then the initial magnetization has decohered, and cannot be recovered.

Loss of correlation between the spin orientation and the position can occur via a number of processes – when the spin is a part of a molecule in a liquid, the loss of correlation occurs by the molecule moving through a series of environments with different magnetic fields. This can also occur for electrons moving through a solid with a spatially inhomogeneous magnetic field or $g$-factor. The timescale marking the loss of correlation between spin orientation and position is the average time $\tau$ for carriers to move from their current location to another location where the inhomogeneous field points in the opposite direction. This time can be referred to as the field correlation time.

Two distinct regimes of precessional decoherence can be identified, depending on the rate of carrier transport. If the precessional rates are faster than $\tau$, then the distribution of phase angles spans 0 to $2\pi$ before spin orientation and position become uncorrelated, as shown in Fig. 4.3a. Once transport removes the correlation between spin orientation and position, the decoherence is complete. This regime is characterized by a static inhomogeneous field leading to dephasing, followed by carrier transport which produces complete decoherence. The decoherence time is thus $\tau$.

The other common physical regime, shown in Fig. 4.3b,c, is motional narrowing. Here the carrier transport is rapid enough that the effect of the inhomogeneous field can be modeled as the sum of a static homogeneous magnetic field $\mathbf{H}_0$ and an effective *time-dependent,* randomly oriented magnetic field. This field changes direction with the time $\tau$.

Shown in Fig. 4.3b is the precession of a single carrier in a spatially-inhomogeneous magnetic field $\mathbf{H}(\mathbf{r})$. The average orientation of the population precesses around $\mathbf{H}_0$. We now focus on the spatially inhomogeneous part $\delta\mathbf{H}(\mathbf{r}) = \mathbf{H}(\mathbf{r}) - \mathbf{H}_0$. After a time $\tau$ a spin has a component parallel to the macroscopic spin polarization of $S\cos(\Omega_\perp\tau)$ and a transverse component $S\sin(\Omega_\perp\tau)$. Here $\Omega_\perp$ refers to the precession frequency of the inhomogeneous field component perpendicular to the spin direction, $\Omega_\perp = g\mu\delta H_\perp/\hbar$, where $g$ is the $g$-factor and $\mu$ is the magnetic moment of the carrier. Between field correlation times the spin undergoes free precession around a definite axis – essentially every $\tau$ the spin precesses around a new effective field $\mathbf{H}(\mathbf{r}')$, which contains a randomly oriented $\delta\mathbf{H}(\mathbf{r}')$. Thus the spin undergoes a random walk around the population's steady precession under the influence of $\mathbf{H}_0$.

(a) fast dephasing - carrier population picture

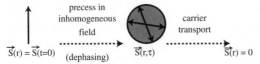

(b) motional narrowing - single carrier picture

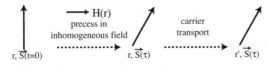

(c) motional narrowing - carrier population picture

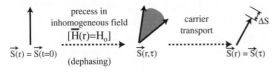

**Fig. 4.3. (a)** evolution of a population of carriers in the presence of a spatially-inhomogeneous magnetic field $H(\mathbf{r})$ and carrier transport. Here precession is so rapid that decoherence occurs as soon as carrier transport destroys the correlation between the carrier position and the spin orientation. **(b)** Motional narrowing regime. Here carrier *transport* is so rapid that the spin executes a random walk in orientation, and decoheres much more slowly than in **(a)**. **(c)** view of a population of carriers under the same conditions as **(b)**. Different precession rates lead to different spin directions for carriers in different positions. Carrier transport effectively rapidly averages the spin orientation, yielding a spin polarization reduced from the initial value, but reduced much less than in **(a)**.

If the free precession rate is independent of $\tau$, then the steps of the random walk have an average length $[\Omega_\perp^2(\mathbf{r})]^{1/2}\tau$, and in a given time $T$ there are $T/\tau$ steps, so the average precession angle away from the macroscopic magnetization per unit time is

$$\phi \sim [\overline{\Omega_\perp^2(\mathbf{r})}\tau]^{1/2}. \tag{4.4}$$

As the orientation of the precessional axis is arbitrary, only the spin component parallel to the macroscopic magnetization remains.

$$\frac{\Delta S}{S} = 1 - \cos\phi = [\overline{\Omega_\perp^2(\mathbf{r})}\tau]/2. \tag{4.5}$$

Fitting this with an exponential yields the well-known formulae [27]

$$T_1^{-1} = (\Omega_\perp^2)\tau, \tag{4.6}$$
$$T_2^{-1} = ([\Omega_\perp^2]/2 + \Omega_\parallel^2)\tau, \tag{4.7}$$

where the transverse ($\Omega_\perp$) and longitudinal ($\Omega_\parallel$) components of the precession vector from the random field are defined relative to the orientation of the static field $H_0$. Note that the principal difference in form between (4.6), (4.7), and (4.5) occurs because for (4.5) we have defined $H_\perp$ relative to the spin polarization, whereas in (4.6) and (4.7) the orientation is defined relative to an applied magnetic field.

Shown in Fig. 4.3c is the complementary view of precessional decoherence in the motional narrowing regime, viewed as the evolution of a population of spins rather than the random walk of a single spin. Between field correlation times the spin of each carrier in the population evolves according to free precession, and the spatially-varying component of the field produces dephasing. After every $\tau$ the spin orientation and carrier position become uncorrelated (producing *partial* spin decoherence), and the average spin polarization of the population is equally distributed among all the constituent carriers. Quantitative results for $T_1$ and $T_2$ from this picture are identical to those from the single-carrier random walk picture. Note that more rapid field correlation times $\tau$ mean dephasing has much less time to build up before it is turned into decoherence, and thus the decoherence rate is smaller.

**Precessional Decoherence in Noncentrosymmetric Crystals.** Evolution of carriers in inhomogeneous magnetic fields can occur in ways other than carrier transport through a spatially-varying magnetic field. A key example is the effective inhomogeneous **k**-dependent magnetic field in noncentrosymmetric crystals. In this category of crystals, which includes zincblende semiconductors [28], even in the absence of an applied magnetic field the pseudospin states are not degenerate. Pseudospin states of the electron are split at finite **k** by the relativistic transformation of internal electric crystal fields into magnetic fields in the rest frame of the moving electron. The splitting is described by the Hamiltonian

$$H = g\mu\mathbf{B(k)} \cdot \mathbf{S} = \hbar\mathbf{\Omega(k)} \cdot \mathbf{S}, \tag{4.8}$$

where $\mathbf{B(k)}$ is the effective crystal magnetic field and $\mathbf{\Omega(k)}$ is the resulting Larmor precession vector. In order to clearly distinguish this field from the applied field, $\mathbf{B}$ will refer to the effective crystal magnetic field, and $\mathbf{H}$ to the applied magnetic field. For direct-gap semiconductors the effective crystal magnetic field vanishes at the conduction minimum ($\mathbf{k} = 0$), because Kramers degeneracy (4.1) requires that

$$\mathbf{B(k)} = -\mathbf{B(-k)}. \tag{4.9}$$

Despite this, within 100 meV of the band edge in GaAs this crystal field approaches 1000 Tesla, which far exceeds the ability of an applied magnetic field (from a typical laboratory magnet) to split the electronic spin states near the band edges.

From (4.9) it is apparent that the effective magnetic field is inhomogeneous (as in our case of dephasing above), although the inhomogeneity now is in momentum space. For the purposes of further discussion we will neglect

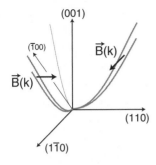

**Fig. 4.4.** Spectrum of a noncentrosymmetric direct-gap semiconductor such as GaAs. The spin splitting is finite along (110) and (1$\bar{1}$0), but vanishes along (100). The direction of the effective crystal magnetic field **B**(**k**) for **k** along (110) is perpendicular to its direction for **k** along (1$\bar{1}$0).

the pseudospin nature of these pairs of states (important for scattering-driven decoherence). As a spin only precesses in a transverse field, if **B**(**k**) were parallel to a fixed direction for all **k**, a macroscopic magnetization oriented along that axis would have an infinite lifetime from this mechanism (again, ignoring the pseudospin nature of the states). For a zincblende crystal, however, it is impossible to choose a global quantization axis because the *direction* of the crystal magnetic field varies with **k**.

As shown in Fig. 4.4, the effective field for **k** ∥ (110) is perpendicular to the field for **k** ∥ (1$\bar{1}$0), so no global quantization axis exists. Thus it is not possible to associate $T_1$ solely with processes which change occupation, implying that here both $T_1$ and $T_2$ should be referred to as a spin coherence times. An unusual special case exists to this rule, for there is (approximately) a global quantization axis for (110)-grown zincblende quantum wells as described below in Sect. 4.2.3.

The presence of this effective internal magnetic field implies that the spin polarization of a population of carriers will dephase due to the variability of the internal field with momentum. This effect is shown in Fig. 4.5 for the motional narrowing regime. D'yakonov and Perel' [29] have developed a theory for $T_1$ in bulk zincblende semiconductors based on precessional dephasing by the effective crystal magnetic field [**B**(**k**) in (4.8)], by identifying the effective crystal magnetic field with **H** in (4.6–4.7). A derivation based on the density matrix formalism is viewable in *Optical Orientation* [26]. In order to describe the origin of **B**(**k**) for a semiconducting material, and therefore that material's spin coherence times, a brief description follows of the origins of inversion asymmetry in semiconductor nanostructures.

### 4.2.3   Sources of Inversion Asymmetry

The three principal sources of inversion asymmetry in clean semiconductor nanostructures are shown in Fig. 4.6 below:

(1) bulk inversion asymmetry (BIA) – this is inversion asymmetry intrinsic to the crystal structure of bulk constituents of nanostructures, such as the asymmetry intrinsic to the zincblende crystal structure (Fig. 4.6a)

(a) single carrier

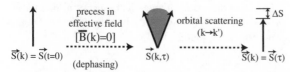

k, $\vec{S}(t=0)$ → precess in effective field → k, $\vec{S}(\tau)$ → orbital scattering (k→k') → k', $\vec{S}(\tau)$

(b) population of carriers

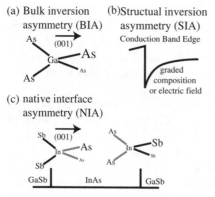

**Fig. 4.5.** (**a**) evolution of a single carrier in the presence of the effective magnetic field $B(\mathbf{k})$ and orbital scattering. Both the spin orientation and the momentum of the carrier are altered. (**b**) evolution of a population of carriers under the same conditions. Different precession rates and orientations of carriers with different momenta lead to different spin directions for different carriers. Orbital decoherence averages the spin orientation, yielding a spin polarization reduced from the initial value.

(a) Bulk inversion asymmetry (BIA)

(b) Structual inversion asymmetry (SIA)

(c) native interface asymmetry (NIA)

**Fig. 4.6.** Structural origins of inversion asymmetry in clean semiconductor heterostructures. (**a**) bulk inversion asymmetry associated with the zincblende crystal structure. The view looking down is slightly off the axis of (110). The size of the atomic label indicates its relative height along the vertical axis. (**b**) structural inversion asymmetry associated with position-dependent band edges due to electric fields and heterojunction band offsets, and (**c**) native interface asymmetry between non-common-atom materials. Shown is a (001) interface between InAs and GaSb. The InSb-like bonds at the InAs-on-GaSb interface are rotated 90 degrees from those at the GaSb-on-InAs interface.

(2) structural inversion asymmetry (SIA) – this refers to arrangements of semiconducting materials or electric fields which are not inversion symmetric, such as single-interface heterojunctions (Fig. 4.6b)

(3) native interface asymmetry (NIA) – this refers to configurations of bonding at interfaces which are not inversion symmetric. They are inevitable at certain interfaces between materials with no common atoms, and whose crystal structures have BIA. (Fig. 4.6c)

**Bulk Inversion Asymmetry (BIA).** For our purposes we will focus on the symmetry of zincblende crystals. The point group symmetry [31] of bulk zincblende is $T_d$, the same as that of a methane molecule (see Fig. 4.6a). The crystal magnetic field originates in spin-orbit coupling terms in the Hamiltonian. To lowest order in $p/mc$, spin-orbit coupling for the conduction electron has the form:

$$H_{so} = \frac{\hbar}{4m^2c^2} \nabla V(\mathbf{r}) \times \mathbf{p} \cdot \sigma = \hbar \mathbf{\Omega} \cdot \sigma/2, \tag{4.10}$$

where $V(\mathbf{r})$ is the crystal potential, $\boldsymbol{p}$ is the momentum operator and $\sigma$ are the Pauli spin matrices. $-\nabla V(\mathbf{r})$ is an effective electric field.

The splitting vanishes for the high-symmetry direction $\mathbf{k} \parallel (100)$. The (100) axis is a two-fold rotational symmetry axis for the crystal potential $V(\mathbf{r})$, so the components of $-\nabla V(\mathbf{r})$ which are perpendicular to $\mathbf{k}$ (and thus contribute to the cross product) are antisymmetric under this 180° rotation. The conduction band eigenstates along this high-symmetry direction are symmetric under this rotation. Thus the contributions to the splitting from positions related by this rotation cancel, yielding a vanishing spin splitting.

The general form of the spin splitting as a function of $\mathbf{k}$ can be outlined from general group-theoretical arguments. Both $-\nabla V(\mathbf{r})$ and $\mathbf{p}$ transform like vectors (representation $\Gamma_{15}$) in $T_d$. Thus the cross product of the two will transform like the representation $\Gamma_{25}$. Note that we will use molecular notation for most irreducible representations, but Bouckaert-Smoluchowski-Wigner (BSW) notation for these three-dimensional representations (whose molecular labels are $T_1$ and $T_2$!) [32] The lowest-order basis functions for $\Gamma_{25}$ are

$$x(y^2 - z^2), \qquad y(z^2 - x^2), \qquad z(x^2 - y^2). \tag{4.11}$$

Thus in perturbation theory the lowest-order form for $\mathbf{\Omega}(\mathbf{k})$ will be [28]

$$\mathbf{\Omega}(\mathbf{k}) \propto \left[ k_x(k_y^2 - k_z^2)\hat{\mathbf{x}} + k_y(k_z^2 - k_x^2)\hat{\mathbf{y}} + k_z(k_x^2 - k_y^2)\hat{\mathbf{z}} \right]. \tag{4.12}$$

Note that this expression indicates that $\mathbf{\Omega}(\mathbf{k}) = 0$ for $\mathbf{k} \parallel (111)$, which is true to lowest order in perturbation theory, but is not true to all orders. The microscopic model developed in Sect. 4.3.1 will allow the evaluation of $\mathbf{\Omega}(\mathbf{k})$ in general. Of course, because the crystal is cubic, in general the magnitudes of the fluctuating field along each of the three cubic axes are equivalent. Thus $B_x^2 = B_y^2 = B_z^2$, so $B_\perp^2 = 2B_\parallel^2$ and [see (4.6), (4.7)] $T_2 = T_1$.

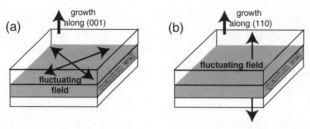

**Fig. 4.7.** orientation of the fluctuating field due to BIA (4.12) for (**a**) quantum wells grown along (001) and (**b**) quantum wells grown along (110).

We now turn to to the effects of BIA in quantum wells grown in the (001) and (110) directions. The symmetries of the crystal magnetic fields for these two cases are very different. The symmetry of (001)-grown quantum wells is $D_{2d}$. Using the compatibility relation

$$\Gamma_{25} = E \oplus B_1, \tag{4.13}$$

the in-plane magnetic field follows the two-dimensional $E$ representation and the growth-direction magnetic field follows the one-dimensional $B_1$ representation.

Using the lowest order perturbation theory expression of (4.12), and evaluating the fluctuating field for the lowest subband in the quantum well, the growth-direction magnetic field vanishes (Fig. 4.7a). Thus the relationship between $T_1$ and $T_2$ differs from that of a cubic crystal,

$$
\begin{aligned}
T_1^{-1}(\alpha) &= T_1^{-1}(\alpha = 0)(1 + \cos^2 \alpha)/2, \\
T_2^{-1}(\alpha) &= T_1^{-1}(\alpha = 0)(2 + \sin^2 \alpha)/4,
\end{aligned}
\tag{4.14}
$$

where $\alpha$ is the direction between a static uniform applied magnetic field $\mathbf{H}_0$ and the growth direction. $T_2$ ranges from $2T_1/3$ to $2T_1$ depending on $\alpha$. For $\alpha = 0$, $T_2 = 2T_1$. The value of this ratio deviates from 2 if $\Omega_z$ is non-zero. Non-zero values of $\Omega_z$ come from higher-order terms than those in (4.12).

The symmetry of (110)-grown quantum wells is $C_{2v}$, where the symmetry axis lies in the plane of the well [along (001)], rather than along the growth direction. $C_{2v}$ has only one-dimensional representations, so the compatibility relations now are

$$\Gamma_{25} = A_2 + B_1 + B_2, \tag{4.15}$$

the in-plane magnetic field along (001) follows $A_2$, along $(1\bar{1}0)$ follows $B_1$ and the growth-direction magnetic field follows $B_2$. In contrast to the case for (001), using the lowest order perturbation theory for (110) grown quantum wells the effective crystal magnetic field is entirely along the growth direction (see Fig. 4.7b), and

$$
\begin{aligned}
T_1^{-1}(\alpha) &= T_2^{-1}(\alpha = 0) \sin^2 \alpha, \\
T_2^{-1}(\alpha) &= T_2^{-1}(\alpha = 0)(1 + \cos^2 \alpha)/2.
\end{aligned}
\tag{4.16}
\tag{4.17}
$$

For $\alpha = 0$ the spin polarization is parallel to the fluctuating field, so $T_1^{-1}$ ($\alpha = 0$) vanishes. Physically, $T_1$ depends on the fluctuating field along two axes, whereas $T_2$ depends on all three components. Thus although $T_1^{-1}(\alpha = 0)$ vanishes, the same is not the case for $T_2^{-1}$ for any $\alpha$. Higher order terms in the spin splitting than (4.12) make $T_1^{-1}(\alpha = 0)$ nonzero as well.

**Structural Inversion Asymmetry (SIA).** Applying a uniform electric field to a centrosymmetric crystal will immediately break the inversion symmetry of the potential felt by the electrons. The resulting electrostatic potential is linear with position for a uniform electric field. This is a special case of inversion-asymmetric potentials which occur on length scales much greater than the bulk unit cell of the constituents. For example, in a nanostructure (even one constructed from centrosymmetric bulk constituents) the crystal potential is modified by the particular configuration of materials and their relative orientation. A structural configuration may break inversion symmetry, such as a linear grading of the alloy composition of a constituent. We choose this example because the effect on the electrons (linear crystal potential with position) is very similar to that of an applied electric field. Thus the structural fields may be referred to as quasi-electric fields. Both of these inversion-asymmetric phenomena: the effect of quasi-electric (heterostructure) fields and of ordinary electric fields, are grouped in the category of structural inversion asymmetry (SIA) [33].

The spin splitting arising from SIA is referred to as the Rashba effect, and takes the form

$$H_{so} = \alpha(\mathbf{r}) \times \mathbf{p} \cdot \sigma, \tag{4.18}$$

where $\alpha(\mathbf{r})$ is oriented along the direction along which inversion symmetry is broken by the structure (typically the growth direction for layered structures). For example, with an electric field applied parallel to the growth direction, the effective magnetic field is perpendicular to both the electric field and the electron momentum. As the carrier's crystal momentum is always perpendicular to the growth direction for a quantum well or single-interface heterostructure, the effective magnetic field from the Rashba effect for a growth-direction electric field is finite for all $\mathbf{k} \neq 0$, lies in the plane of the well, and is perpendicular to the carrier's crystal momentum. In the absence of BIA the symmetry is $C_{4v}$, and the relevant representation for the in-plane magnetic field is $E$. Thus in the absence of BIA the effect of SIA on spin decoherence is very similar to the effect of BIA alone on spin coherence in (001) quantum wells, yielding (4.14) for the relationship between $T_1$ and $T_2$.

The relative magnitude of BIA and SIA depends on the structure considered. In perfect quantum wells symmetrical along the growth direction, SIA will vanish. Other structures, including modulation-doped single-interface heterostructures (with similar band edges to Fig. 4.6b), have significant growth-direction asymmetry. Here under many circumstances SIA dominates over BIA [34].

SIA and BIA can even interfere with each other in unusual ways in (001) quantum wells. Once there is a growth-direction electric field the symmetry of the well drops from $D_{2d}$ to $C_{2v}$. In contrast to the (110) growth-direction quantum well, the symmetry axis is parallel to growth rather than in-plane. The growth-direction fluctuating field, which transformed like $B_1$ in $D_{2d}$, now transforms like $A_2$, whereas the representation of the in-plane field splits

$$E = B_1 \oplus B_2. \tag{4.19}$$

Thus the transformation properties of the effective fluctuating magnetic field oriented along (110) and along ($1\bar{1}0$) can differ [35]. This can cause anomalous beating in magneto-oscillation effects (de Haas-van Alphen and Shubnikov-de Haas) [36]. It has been pointed out that, within a simple approximation, the magnetic field component along one of these directions can vanish near zone center [37]. Quantitative calculations show that growth-direction electric fields in (110) quantum wells can dramatically change the $T_1$ time for the magnetic field parallel to (110), whereas the $T_2$ time in (110) quantum wells and both $T_1$ and $T_2$ in (100) quantum wells are much less sensitive [38].

**Native Interface Asymmetry (NIA).** An unusual form of inversion asymmetry arises in quantum wells in which the well and barrier have a different composition of both anions and cations [39–43]. This does not arise for GaAs/AlGaAs quantum wells, for the anion As is the same in both the well and barrier. A no-common-atom system is shown in Fig. 4.6c, an InAs/GaSb quantum well. There are two types of possible interface bonds – GaAs and InSb – neither of which appear in the well or barrier proper. If the bonding types at the two interfaces differ, then the system clearly is inversion asymmetric. Even if the bonding types are the same, say InSb-like, there is an inherent, or native, interface asymmetry. Moving from left to right in Fig. 4.6, which corresponds to moving along the (001) axis, the bonds from anion to cation are rotated 90° from those bonds from cation to anion. Thus the InSb bonds of the InAs-on-GaSb interface are rotated 90° from those of the GaSb-on-InAs interface. As a result these interfaces will contribute to inversion asymmetry and thus to spin decoherence. As NIA is associated with interfaces, its contribution to spin decoherence relative to BIA increases as the well thickness is decreased.

NIA depends strongly on the growth direction of the structure. Quantum wells grown along (001) have alternating planes of cations and anions. Thus interface bonds are aligned and oriented. Quantum wells grown along (110) have planes of mixed anions and cations. Thus NIA vanishes for (110) quantum wells. Even in systems where NIA can arise, imperfect interfaces which mix the bonding type may reduce the NIA. On the theoretical side the microscopic parameters necessary to calculate NIA are much less known than those characterizing the bulk behavior of zincblende semiconductors (and thus BIA). The importance of this effect is uncertain at present. Experiments must be done which correlate interface structure with spin decoherence rates in order to determine the true importance of this effect. One measurement indicates that in a 7 monolayer (ML) InAs/12 ML GaSb superlattice with InSb-like bonds the NIA contribution to spin decoherence is over an order of magnitude greater than the BIA contribution [44].

### 4.2.4   Comparison with Ultrafast Probes of *Orbital* Coherence

Earlier we described the relationship between nuclear (or core electron) spin decoherence and mobile electron spin decoherence. A key difference arises between the two because of the energy bottleneck present in nuclear spin decoherence and absent in mobile electron spin decoherence. In nuclear spin decoherence $T_1$ is determined by very different physics (spin-lattice relaxation) than $T_2$, but $T_1 \sim T_2$ for delocalized electrons.

Another situation where coherence plays a major role in ultrafast optics is in the coherent *orbital* excitation of a semiconductor. Here we can distinguish two general regimes, one of which corresponds to the behavior of nuclear spin coherence and the other to the behavior of electron spin coherence in semiconductors.

Table 4.1 below shows the correspondence between regimes of orbital coherence and spin coherence. The first orbital case is shown below in Fig. 4.8a. Optical excitation creates a coherent orbital state involving the conduction

**Table 4.1.** Correspondence between orbital and spin coherence in optical excitation

|  | interband orbital | intersubband orbital | spin |
|---|---|---|---|
| two levels | conduction state and valence state | two conduction subband states | spin up state and spin down state |
| coherent quantity | **P** (dipole polarization field) | **P** (dipole polarization field) | **S** (magnetization) |
| decoherence time ($T_2$) | decay of **P** | decay of **P** | decay of transverse magnetization |
| $T_1$ | carrier recombination time | intersubband relaxation time | longitudinal magnetization relaxation time |
| regime | $T_1 \gg T_2$ | $T_1 \sim T_2$ | $T_1 \sim T_2$ |

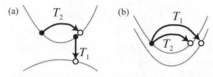

**Fig. 4.8.** Spectra for (a) interband ($T_1 \gg T_2$) and (b) intersubband ($T_1 \sim T_2$) optical transitions. Also shown are schematic processes for occupation relaxation ($T_1$) and phase decoherence ($T_2$).

and valence band of a semiconductor. The orbitally coherent state has been described in detail in the context of the dynamical (AC) Stark effect [45,46]. The macroscopic signature of orbital coherence in the AC Stark effect is a dipole polarization field $P$. In the absence of the driving field the interband orbital coherence (and thus $P$) decays rapidly – typically within a few hundred femtoseconds. The decay of the occupancy of the conduction band subsequent to optical excitation, however, is governed by the carrier recombination time (often of order nanoseconds).

The reason for the great differences in timescales is the same as that in nuclear spin systems – energy bottleneck. A transition of an electron from the conduction band to the valence band requires transferring energy to another excitation. For clean semiconductors at low excitation levels, the dominant process is the emission of a photon (radiative recombination). In contrast, loss of phase coherence within the conduction band (or within the valence band) does not require energy transfer, and can occur quasi-elastically. In fact typical $T_2$'s for orbital decoherence are on the order of the orbital scattering times.

Precessional decoherence is present in this system. Dephasing occurs because the difference in energy between the conduction band and the valence band is a function of the momentum **k**. The resulting time evolution of the phase relationship between conduction and valence states is **k**-dependent. The dephasing rate for this system, however, is typically much too rapid for motional narrowing. Therefore usually a single scattering event is enough to completely decohere the orbital phase.

Shown in Fig. 4.8b is another regime, where the relationship between $T_1$ and $T_2$ is more analogous to our case of electron spin decoherence in semiconductors. This is coherent optical excitation of an intersubband transition in a quantum well. Two processes are shown: one a scattering process within the upper subband, the second a scattering process from the upper subband to the lower subband. The first is analogous to $T_2$, the second to $T_1$. In contrast to the interband case, here both are elastic processes – no extra excitation is necessary to absorb energy for the relaxation of carriers from the upper to the lower subband. Thus in this case $T_1 \sim T_2$, again analogous to spin decoherence of mobile electrons.

### 4.2.5   Concluding Remarks

The phenomenology of spin coherence in semiconductor materials has been described in a variety of situations. Scattering-driven and precessional decoherence mechanisms have been introduced and contrasted. Origins of the inversion asymmetry which drives precessional decoherence have been identified in the structure of semiconducting materials. The importance of energy bottleneck has been emphasized in distinguishing the qualitative behavior of $T_1$ and $T_2$ in mobile electron spin decoherence from NMR and ultrafast optical coherence. Energy bottleneck is absent from mobile electron spin decoherence in both the bulk and the quantum well systems considered here. In principle it could be established in zero-dimensional semiconducting systems such as quantum dots. Whether or not that is feasible, and can lead to very long $T_1$ times (as predicted in Ref. [47]) is not yet known. As described above, though, energy bottleneck is not sufficient to make $T_2$ very long, and so it is unclear what advantage will ensue for quantum computation specifically from energy bottleneck in quantum dots.

In the subsequent section methods of calculating quantitatively accurate spin coherence times in bulk and quantum well systems via precessional decoherence will be described. This mechanism typically dominates near room temperature.

## 4.3   Precessional Spin Coherence Times in Bulk and Nanostructure Semiconductors

To guide further efforts in the controllable manipulation of room-temperature spin coherence it is essential to have a *quantitative* theory of spin decoherence both in bulk and in nanostructured environments. Spin decoherence near room temperature in direct-gap semiconductors is dominated by precessional decoherence (described in Sect. 4.2.3). Accurate calculations of precessional decoherence times, which will be the subject of this section, depend on accurate values for the effective time-dependent flucuating field experienced by a population of electrons. The elements of the effective field are

(1) The magnitude and direction of the momentum-dependent effective crystal magnetic field. The crystal magnetic field is determined by the composition and geometry of the semiconducting material. In this section we will limit our focus to direct-gap semiconductors in bulk or in symmetric (001)-grown quantum well geometries.

(2) The timescale of the fluctuating magnetic field (determined by orbital scattering times). The dependence of spin decoherence on orbital scattering times suggests that the maximal spin coherence times should be near the metal-insulator transition [7,48]. In the metallic regime, which we consider in this section, the relevant orbital scattering times depend on the scattering mechanism and the energy $E$ of the electron above the

conduction minimum. Scattering times appropriate for quasi-elastic scattering of electrons from ionized impurities (II), optical phonons (OP), as well as from point-like neutral impurities (NI) will be considered.

(3) The occupation of momentum states for a given population. Carrier occupation depends on the the details of the nonequilibrium condition of the semiconductor. Carrier pulses moving ballistically would have a much different occupation profile $f(\mathbf{k})$ than that of carrier pulses moving diffusively. Here we limit consideration to near-thermal occupation appropriate for diffusive transport.

To see the relationship between these three, consider a decomposition of the $\alpha$ component of the effective magnetic field into spherical harmonics (in 3D) or trigonometric functions (in 2D):

$$\Omega_\alpha(\mathbf{k}) = \sum_{\ell m} \Omega_{\alpha,\ell m}(E) Y_{\ell m}(\theta, \phi), \qquad (3D) \tag{4.20}$$

$$\Omega_\alpha(\mathbf{k}) = \sum_{\ell} \Omega_{\alpha,\ell 1}(E) \cos(\ell\phi) + \Omega_{\alpha,\ell 2}(E) \sin(\ell\phi), \qquad (2D) \tag{4.21}$$

where $E = E(\mathbf{k})$ and $\phi$ is the angle of $\mathbf{k}$ from a fixed axis. Time reversal invariance requires $\mathbf{\Omega}(\mathbf{k}) = -\mathbf{\Omega}(-\mathbf{k})$, so $\Omega_{\alpha,\ell m} = 0$ for even $\ell$. Each of these $\Omega_{\alpha,\ell m}$ components can be considered as an independent fluctuating field.

The timescale for field reversal will depend on $\ell$. In order to change the sign of $\cos\theta$, $\theta$ needs to be changed by $180°$, however $\cos 3\theta$ only requires a $60°$ change in $\theta$ in order to change sign. Thus a scattering event which changes the angle of the momentum by $60°$ will change the sign of the fluctuating field corresponding to $\Omega_3$, but not that of $\Omega_1$. Clearly the effective time for field reversal $\tau_\ell$ depends on $\ell$.

Quasi-elastic scattering simplifies the problem considerably, for it allows division of the carrier population into subpopulations characterized by the same energy $E$. The decoherence is then evaluated for each of these subpopulations, and the final stage is a properly weighted average of the decoherence rate over the carrier energy.

$$\frac{1}{T_1} = \frac{1}{n} \int D(E) f(E) [1 - f(E)] \sum_{\ell} \tau_\ell(E) \Omega_\ell^2(E) dE, \tag{4.22}$$

where $f(E)$ is the Fermi occupation function, $D(E)$ is the density of states and $n$ is the electron density. The $f(E)[1-f(E)]$ is familiar from quasi-elastic scattering theory, for an initial state is occupied with probability $f(E)$, and a final state is unoccupied with probability $1 - f(E)$. Several of the fluctuating fields have been grouped together, as

$$\Omega_\ell^2(E) = \sum_{\alpha,m} \Omega_{\alpha,\ell m}^2(E). \tag{4.23}$$

We now proceed to describe a successful method of calculating $\Omega_\ell(E)$ and $\tau_\ell(E)$.

### 4.3.1  Magnitude of the Fluctuating Field

$\Omega_\ell(E)$ can be obtained from any suitable electronic structure model of a semiconductor. Recently a flexible and accurate model was applied to calculate $\Omega_\ell(E)$ for bulk and quantum well systems [49], via a non-perturbative calculation in a fourteen-band basis [50]. This basis, shown schematically in Fig. 4.9, is the minimum required to generate spin splitting nonperturbatively. The constituents of the basis are two conduction antibonding $s$ states ($\Gamma_6$), six valence (bonding) $p$ states ($\Gamma_7 + \Gamma_8$), and six antibonding $p$ states ($\Gamma_7 + \Gamma_8$). Such a basis has, for example, been used to analyze spin-splitting in heterostructures [51]. The Hamiltonian is well-known, and can be found in Refs. [31,50]. The parameters that enter this Hamiltonian include the zone-center energies of the constituent bulk semiconductors and the momentum matrix elements between bands, which are obtained from the conduction band mass, the heavy-hole mass, and the $g$-factor.

For quantum wells the electronic structure is obtained by expressing the electronic states as spatially-dependent linear combinations of the fourteen states in the basis. The full Hamiltonian is projected onto this restricted basis set, which produces a set of fourteen coupled differential equations for the spatially-dependent coefficients of the basis states (generalized envelope functions). These equations are then solved in Fourier space in a similar method to that of Winkler and Rössler. [52] Further details are available elsewhere [50].

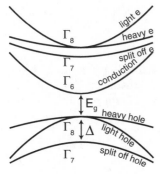

**Fig. 4.9.** Schematic of the fourteen-band basis used to calculate the crystal magnetic field in bulk and quantum well structures. The double group representations of the bands at zone center are shown, as well as the common names for bulk bands. The band gap between the conduction and valence bands $E_g$ and the energy splitting $\Delta$ between the valence band edge and the split-off band are also indicated. Spin splittings of each band are not shown, for they remain very small on this scale.

### 4.3.2    Calculation of the Effective Time for Field Reversal

As the scattering cross section $\sigma(\theta, E)$ is dependent on energy $E$ as well as angle $\theta$, the time for field reversal depends on energy, according to

$$\tau_\ell^{-1}(E) = \int_{-1}^{1} \sigma(\theta, E)(1 - P_\ell(\cos\theta))d\cos\theta \qquad (3D) \qquad (4.24)$$

$$\tau_\ell^{-1}(E) = \int_{0}^{2\pi} \sigma(\theta, E)(1 - \cos[\ell\theta])d\theta \qquad (2D). \qquad (4.25)$$

Unless the scattering process is completely angle-independent, typically the field reversal time decreases with increasing $\ell$.

The scattering cross section $\sigma(\theta, E)$ is taken from standard expressions for ionized impurity (II), neutral impurity (NI – such as arises from quantum well interface roughness), or optical phonon (OP) scattering. The $\tau_\ell$'s differ for different mechanisms – these are shown for 3D and 2D in Table 4.2 below.

**Table 4.2.** Relationship between $\tau_\ell(E)$ (for odd $\ell$) and $\tau_1(E)$ in two and three dimensions for ionized impurity, neutral impurity, and optical phonon scattering

|     | 2D | 3D |
| --- | --- | --- |
| II | $\tau_\ell = \tau_1/\ell^2$ | $\tau_\ell = \left[\sum_{k=0}^{(\ell-1)/2} \frac{(-1)^k(2\ell-2k)!}{2^\ell k!(\ell-k)!(\ell-2k-1)!}\right]^{-1} \tau_1$ |
| NI | $\tau_\ell = \tau_1$ | $\tau_\ell = \tau_1$ |
| OP | $\tau_\ell = \tau_1/\ell$ | $\tau_\ell = \left[\sum_{k=0}^{(\ell-1)/2} \frac{(-1)^k(2\ell-2k)!}{2^\ell k!(\ell-k)!(\ell-2k)!} \sum_{m\,\text{odd}}^{(\ell-2k)} \frac{1}{m}\right]^{-1} \tau_1$ |

The scattering cross section $\sigma(\theta, E)$ could be calculated using Monte Carlo techniques for particular structures (as in [53]). Instead, for simplicity, we use the experimentally-observed scattering mechanism and mobility for each structure. From that we can determine the transport scattering time $\tau_1(E)$ from

$$\mu = (e/mn) \int D(E)f(E)[1 - f(E)]\tau_1(E)EdE. \qquad (4.26)$$

### 4.3.3    Spin Decoherence Times in Bulk III-V Semiconductors

For bulk semiconductors the relevant electronic states for spin decoherence are near the bulk band edge, and thus perturbative expressions for $\Omega_\ell^2(E)$ for these bulk semiconductors ($\Omega_1^2(E) = 0$, $\Omega_3^2(E) \propto E^3$) [29] are identical to those obtained from a full fourteen-band calculation within numerical accuracy. Shown in Fig. 4.10 are calculated $T_2$'s for GaAs, InAs, and GaSb assuming II scattering. Solid with squares and solid lines respectively represent the results of experiments [7] and the non-perturbative theory for bulk

GaAs at the electron density $n = 1.0 \times 10^{16}$ cm$^{-3}$. The agreement with experimental measurements for GaAs at the higher temperatures is quite good, whereas for low temperatures other spin relaxation mechanisms are expected to dominate.

Also shown are results for bulk InAs at $n = 1.7 \times 10^{16}$ cm$^{-3}$ and bulk GaSb at $n = 1.49 \times 10^{18}$ cm$^{-3}$, which are indicated with dashed and dot-dashed lines respectively. The difference in slope between GaSb and GaAs occurs because GaSb is degenerate for this density. The tabulated mobilities [54] for InAs and GaSb extend only to 77 K, so at lower temperatures $\tau_3(E)$ was assumed to have the same value as at 77 K. The smaller $T_2$'s in InAs and GaSb are due partly to the larger conduction spin splitting, which originates from a larger ratio of the spin-orbit coupling $\Delta$ to the band gap $E_g$ (see [55] on perturbative expansions of spin splittings). The agreement between calculated and measured $T_2$'s in Fig. 4.10 indicates that the spin splitting of bulk GaAs is well described by our model.

For a cubic system $T_1 = T_2$, so these theoretical calculations are appropriate for $T_1$ as well. A recent experimental measurement of $T_1$ for InAs at room temperature [56] is shown as the solid diamond, indicating good agreement with calculations.

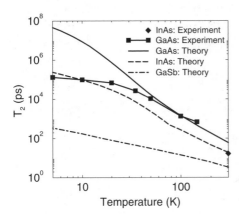

**Fig. 4.10.** Spin coherence times in GaAs, InAs, and GaSb as a function of temperature.

### 4.3.4 Spin Decoherence in III-V (001) Quantum Wells

**D'yakonov-Kachorovskii Theory.** Many comparisons are made of experiments on quantum wells with a simple theory due to D'yakonov and Kachorovskii (DK). [57] The DK theory for (001) quantum wells is derived as follows. First, negligible penetration of the electronic states into the barriers is assumed, so $E_1 \ll \Delta E_c$, where $E_1$ is the confinement energy of the first quantum well state and $\Delta E_c$ is the conduction band offset. Then the perturbative expressions [58] $\Omega_1^2(E) \propto E(4E_1 - E)^2$ and $\Omega_3^2(E) \propto E^3$ are used. Further-more the energies of relevant states are assumed to be $\ll E_1$, and thus (i) the contribution from $\Omega_3(E)$ is ignored, and (ii) it is assumed that $\Omega_1^2(E) \propto E$. It is not generally recognized that the conditions $kT \ll E_1 \ll \Delta E_c$ are quite restrictive and are difficult to satisfy at room temperature.

An elegant result under these approximations is that

$$\frac{1}{T_1} = \frac{1}{n} \int D(E)f(E)[1 - f(E)]\tau_1(E)\Omega_1^2(E)dE$$

$$\propto \int D(E)f(E)[1 - f(E)]\tau_1(E)EdE = \mu. \tag{4.27}$$

Thus the spin decoherence rate under the DK assumptions is proportional to the mobility *independent of the dominant scattering mechanism* [see (4.26)]. In addition, $T_1^{-1}$ is proportional to $E_1^2$. These trends are not supported by recent experimental measurements [30,59] on 75Å $n$-doped GaAs/Al$_{0.4}$Ga$_{0.6}$As MQWs at 300 K [shown in Fig. 4.11a,b (filled circles)]. In both cases the experimental trends are weaker than the predicted theoretical ones.

**Accurate Calculations of Spin Coherence Times in Heterostructures.** Calculations are shown in Fig. 4.11a,b using the fourteen-band heterostructure model of [49] assuming OP (solid line) and NI (dashed line) as the dominant processes determining the mobility. The results agree with

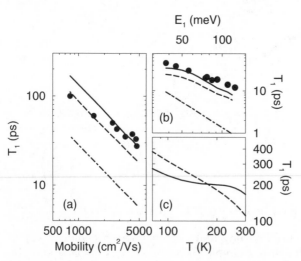

**Fig. 4.11.** $T_1$ as a function of (**a**) mobility, (**b**) confinement energy, and (**c**) temperature, for 75Å GaAs/Al$_{0.4}$Ga$_{0.6}$As MQWs at room temperature. Closed circles represent the results of experiments [30]. The non-perturbative theory results with OP scattering (*solid lines*) and NI scattering (*dashed lines*) are shown, as well as the DK theory results (*dot-dashed lines*).

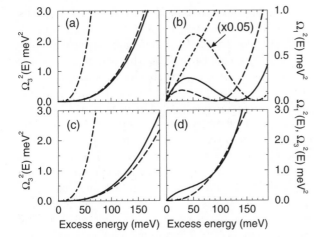

**Fig. 4.12.** $\Omega_1^2(E)$ and $\Omega_3^2(E)$ for several structures. (a) $\Omega_3^2(E)$ for bulk GaAs (*solid*), InAs (*dashed*), and GaSb (*dot-dashed*). (b) $\Omega_1^2(E)$ for GaAs (*solid*), In-GaAs (*long dashed*), and GaSb (*dot-dashed*) quantum wells described in Table 4.3. The *short-dashed line* is the DK approximation for the GaAs quantum well. (c) $\Omega_3^2(E)$ for the same three structures as (b). (d) $\Omega_1^2(E)$ (*solid*) and $\Omega_3^2(E)$ (*dashed*) for a thin-layer InAs/GaSb superlattice.

experiment if one assumes a shift from OP to NI scattering as the mobility drops – this is the origin of the unusual experimental dependence of $T_1$ on the mobility. The weaker dependence of $T_1$ on $E_1$ in the general theory versus DK theory in Fig. 4.11b is due to wavefunction penetration into the barriers and non-perturbative effects. We emphasize as well the key role of temperature studies of the mobility in analyzing the temperature dependence of spin coherence. In Fig. 4.11c the calculated $T_1$ for one sample with a given room-temperature mobility is presented as a function of temperature for NI and OP scattering. In particular the OP results appear relatively insensitive to temperature from 100–250 K – this is due to the rapid decrease in the mobility from OP scattering with increasing temperature. This may play a role in the weak temperature dependence seen in [6].

The origin of the improved agreement with experiment can be explored by comparing the more accurate results for the fluctuating magnetic fields with those of DK theory. Figure 4.12 compares the energy dependence of $\Omega_1^2(E)$ and $\Omega_3^2(E)$ for several additional material systems. The cubic dependence of $\Omega_3^2(E)$ for the three bulk semiconductors is confirmed in Fig. 4.12a. Fig. 4.12b, however, shows that for quantum wells $\Omega_1^2(E)$ is only linear (short dashed line for the GaAs MQW) for a small energy range ($\sim 20$ meV) above the band edge before it begins to deviate. More energetic states than this certainly contribute to the spin coherence times at room temperature. The wider the well the lower the energy where $\Omega_1^2(E)$ deviates from linear behavior, as it approaches a bulk-like $E^3$ behavior. $\Omega_3^2(E)$ for these structures is shown in Fig. 4.12c, and is comparable in magnitude to $\Omega_1^2(E)$. As the wells become narrower, even the perturbative expressions for $\Omega_3$ and $\Omega_1$ break down. Figure 4.12d shows $\Omega_1^2(E)$ and $\Omega_3^2(E)$ for a thin-layer InAs/GaSb superlattice, indicating very different behavior from the other structures, poorly reproduced by even the general forms of the perturbative expression.

The broad applicability of these accurate calculations is shown in Table 4.3, which presents calculations and experimental measurements of $T_1$ for several material systems. The order of magnitude discrepancy between DK calculations and measurements occurs here as well. Given the uncertainty in experimental mobilities and densities, the agreement of our calculations with experiment for both NI and OP scattering is good for all systems, and is much better than DK theory. Note that OP and NI scattering calculations in the full theory differ from each other by factors of of up to 2 (due to differences in $\tau_\ell(E)$), whereas all scattering mechanisms produce the same result in DK theory. As expected, for several systems the $T_1$'s are much shorter at higher electron densities, for as the carrier distribution is spread further from zone center the effective crystal magnetic fields increase. The DK approximation (i) can be evaluated by comparing $OP_1$ to OP and $NI_1$ to NI, where calculations using all terms up to $\ell$ are designated $OP_\ell$ and $NI_\ell$. The difference is up to 40%. Approximation (ii), however, produces a discrepancy between the DK result and both $NI_1$ and $OP_1$ which can greatly exceed an order of magnitude.

Interface bonding asymmetry (native interface asymmetry, or NIA), which arises in non-common-atom structures, could play a role in systems II, IV, or V. The NIA spin splitting for perfect interfaces (imperfect interfaces reduce

**Table 4.3.** Spin decoherence times $T_1$ (ps) for several structures, I: a 75Å GaAs/Al$_{0.4}$Ga$_{0.6}$As MQW [30], II: a 70Å In$_{0.53}$Ga$_{0.47}$As/97Å InP MQW [60], III: an 80Å GaSb/80Å AlSb MQW [61], IV: a 51Å GaAs$_{0.19}$Sb$_{0.81}$/80Å AlSb MQW [61], and V: a 21.2Å InAs/36.6Å GaSb superlattice. Calculated times are shown for a given total electron density (n.d. indicates nondegenerate) using DK theory (DK), and the nonperturbative theory with optical phonon (OP) and neutral impurity (NI) scattering. The subscript $\ell$ indicates that only terms up to $\Omega_\ell$ were used in the calculation.

| | System | Density (cm$^{-3}$) | $\mu$ (cm$^2$/Vs) | Exp | DK | $OP_1$ | OP | $NI_1$ | NI |
|---|---|---|---|---|---|---|---|---|---|
| I | GaAs/AlGaAs | $2.7 \times 10^{17}$ | 800 | 100 | 27 | 151 | 120 | 162 | 111 |
| II | InGaAs/InP | n.d. | 6700 | – | 1.45 | 53 | 37 | 52 | 32 |
| | | $3.0 \times 10^{18}$ | 6700 | 2.6 | 0.21 | 6.0 | 4.9 | 6.9 | 4.0 |
| III | GaSb/AlSb | n.d. | 2000 | – | 0.59 | 1.9 | 1.8 | 1.5 | 1.4 |
| | | $2.8 \times 10^{18}$ | 2000 | 0.52 | 0.09 | 0.64 | 0.55 | 0.88 | 0.53 |
| IV | GaAsSb/AlSb | n.d. | 2000 | – | 0.09 | 0.53 | 0.52 | 0.44 | 0.43 |
| | | $3.4 \times 10^{18}$ | 2000 | 0.42 | 0.01 | 1.9 | 1.4 | 1.7 | 0.87 |
| V | InAs/GaSb | n.d. | 3000 | – | 0.38 | 0.77 | 0.77 | 1.7 | 1.6 |

the NIA contribution) has been calculated for System II in [43]. By comparing with [43] we find the spin splitting of this quantum well is dominated by BIA. For system V, however, further experiments and calculations of the NIA contribution to the spin decoherence time clearly indicate that it dominates over the BIA contribution [44].

The above is a *quantitatively accurate* non-perturbative nanostructure theory for electron spin relaxation in bulk and quantum well zincblende semiconductors based on a fourteen band model. The calculated electron spin lifetimes in III-V semiconductor bulk and quantum well materials are in agreement with experimental measurements, indicating the importance of accurate band structure calculations for zincblende type nanostructures.

## 4.4 Spin Transport

Spin device physics requires not just the theory of spin coherence times described above, but also a theory of spin transport. Such a theory should properly address regimes of tunneling, ballistic transport, and diffusive transport. This chapter, however, will focus on the mobility and diffusion properties in the diffusive transport regime. Considerable progress has been made in understanding spin transport in metals [62–65], however semiconductors behave very differently. The origin of the differences in spin diffusion between semiconductors and metals are

(1) the much greater spin coherence times in semiconductors,
(2) the ineffectiveness of screening in semiconductors relative to metals, and
(3) the possibility of controlling whether carriers in a band are degenerate or not by small perturbations (e.g. electric fields or doping).

The first of these was explored in detail in Sect. 4.3 and will be described further in Chap. 5. The focus here will be on elements (2) and (3).

Generation of the characterized inhomogeneous spin polarized perturbations required to measure spin transport can be performed optically with circularly polarized light, either with a spin grating [66] or direct lateral drag [8]. Observation of the decay of a spin grating, generated from two beams of linearly polarized light with crossed polarizations, provides information about the electron spin diffusion constant, and in the structure of [66] incoherent spin diffusion was observed in the presence of a background of unpolarized holes. In the direct lateral drag measurement of [8] a coherent spin packet was generated by a localized pump pulse, and was later detected with a probe pulse spatially localized elsewhere. The geometry of [8] is an optical analogue to the classic Haynes-Shockley experiment, in which holes were injected electrically from an emitter, and detected with a collector [67]. From the time-of-flight measurements of Haynes and Shockley the mobility and diffusion constant of minority holes in $n$-doped germanium were extracted. Through measurements of lateral drag in a transverse magnetic field in [8], the motion of local packets of *coherent* electron spin was determined in the low-field limit and in the absence of any background hole population. For this reason, we focus on the quantitative results of [8] in our theoretical analysis.

We note [68] reports an attempt to measure vertical spin transport, however the experimental geometry was constructed to probe incoherent spin transport in an intermediate-field regime over a fixed distance in the presence of both photoexcited electrons and holes. Proper theoretical analysis of this experimental situation is therefore more involved, requiring simultaneous solution of drift-diffusion equations for both electrons and holes in an intermediate-strength field.

In this section we focus on the low-field transport properties of doped and undoped semiconductors which have a localized perturbation of spin-polarized carriers. First, nonmagnetic semiconductors are considered. In undoped semiconductors the diffusion and mobility of spin packets are very similar to those of charge polarization packets. There are, however, qualitatively different diffusion and mobility properties for spin packets in *doped* systems compared to the traditional charge polarization packets [70]. Specifically, consideration of the consequences of (2) and (3) above explains the anomalously high diffusion rates of spin packets observed in [8].

We also describe spin diffusion in spin-polarized semiconductors. The diffusion and mobility of spin packets are found to differ by orders of magnitude depending on whether they are polarized parallel or antiparallel to the spin polarization of the equlibrium carriers. This work may assist in understanding spin transport within metallic ferromagnetic semiconductors, such as GaMnAs [17], and semimagnetic semiconductors, such as BeMnZnSe [71]. Both the $p$-doped GaMnAs and the $n$-doped BeMnZnSe have been used in spin-dependent devices [16,14,13].

### 4.4.1   Drift-Diffusion Equations

The drift-diffusion equations for carrier motion in a semiconductor describe the combined motion of electrons and holes in the presence of electric and quasi-electric fields, including the effects of space charge fields on carrier motion. In order to include the possibility of spin imbalance in both the conduction and valence bands, four currents are required:

$$\mathbf{j}_{e\uparrow} = en_\uparrow \mu_{e\uparrow} \mathbf{E} + eD_{e\uparrow} \nabla n_\uparrow, \tag{4.28}$$

$$\mathbf{j}_{e\downarrow} = en_\downarrow \mu_{e\downarrow} \mathbf{E} + eD_{e\downarrow} \nabla n_\downarrow, \tag{4.29}$$

$$\mathbf{j}_{h\uparrow} = ep_\uparrow \mu_{h\uparrow} \mathbf{E} - eD_{h\uparrow} \nabla p_\uparrow, \tag{4.30}$$

$$\mathbf{j}_{h\uparrow} = ep_\downarrow \mu_{h\downarrow} \mathbf{E} - eD_{h\downarrow} \nabla p_\downarrow. \tag{4.31}$$

The evolution in time and space of these four currents and the electric field comes from the four continuity equations,

$$\frac{\partial n_\uparrow}{\partial t} = -\nabla \cdot \mathbf{j}_{e\uparrow} + \Gamma_{e\uparrow\downarrow} n_\downarrow - \Gamma_{e\downarrow\uparrow} n_\uparrow + G_\uparrow - R_\uparrow n_\uparrow p_\downarrow \tag{4.32}$$

$$\frac{\partial n_\downarrow}{\partial t} = -\nabla \cdot \mathbf{j}_{e\downarrow} - \Gamma_{e\uparrow\downarrow} n_\downarrow + \Gamma_{e\downarrow\uparrow} n_\uparrow + G_\downarrow - R_\downarrow n_\downarrow p_\uparrow \tag{4.33}$$

$$\frac{\partial p_\uparrow}{\partial t} = -\nabla \cdot \mathbf{j}_{h\uparrow} + \Gamma_{h\uparrow\downarrow} p_\downarrow - \Gamma_{p\downarrow\uparrow} p_\uparrow + G_\uparrow - R_\downarrow n_\downarrow p_\uparrow \tag{4.34}$$

$$\frac{\partial p_\downarrow}{\partial t} = -\nabla \cdot \mathbf{j}_{h\downarrow} - \Gamma_{h\uparrow\downarrow} p_\downarrow + \Gamma_{h\downarrow\uparrow} p_\uparrow + G_\downarrow - R_\uparrow n_\uparrow p_\downarrow \tag{4.35}$$

and the Poisson equation relating the local deviation from equilibrium of the charge densities to the electric field,

$$\nabla \cdot \mathbf{E} = -\frac{e}{\epsilon}(\Delta n_\uparrow + \Delta n_\downarrow - \Delta p_\uparrow - \Delta p_\downarrow). \tag{4.36}$$

Here $\Gamma$ are rates for spin relaxation of electrons and holes, $G$ are generation rates for electron-hole pairs, $R$ are their recombination rates and $\epsilon$ is the dielectric constant.

### 4.4.2  Low-Field Motion of Spin Packets in Nonmagnetic Semiconductors

It can be seen from Sect. 4.4.1 that local charge imbalance plays a key role through the Poisson equation (4.36) on the motion of packets of spin-polarized carriers. For example, the presence of inhomogeneous space-charge fields produces dramatic differences in the diffusion of spin-polarized carrier packets depending on whether they move through doped or undoped (intrinsic) semiconductors. Whereas a spin packet in an intrinsic semiconductor must be a multiple-band disturbance, involving inhomogeneous distributions of both electrons and holes, in a doped semiconductor a single-band disturbance is possible.

In the low-field regime local variations in the conduction electron density

$$\Delta n(\mathbf{x}) = \Delta n_\uparrow(\mathbf{x}) + \Delta n_\downarrow(\mathbf{x}) \tag{4.37}$$

must be approximately balanced by a local change in the valence hole density

$$\Delta p(\mathbf{x}) = \Delta p_\uparrow(\mathbf{x}) + \Delta p_\downarrow(\mathbf{x}). \tag{4.38}$$

Exceptions require large space-charge fields, such as occur when donor or acceptor concentrations vary substantially over short distances. In metals, by contrast, these density variations are screened out on length scales of Angstroms. The approximate local charge neutrality constraint in semiconductors,

$$\Delta n(\mathbf{x}) \sim \Delta p(\mathbf{x}), \tag{4.39}$$

has key implications for the motion of packets of increased carrier density [72,67,73]. If such a packet moves, both the conduction electrons and valence holes which comprise it must move together. The motion of holes in semiconductors tends to be much slower (due to their lower mobility) than that of electrons, so hole mobility and diffusion tends to dominate the properties of a packet consisting of both electron and hole density variations. In the absence of spin polarization this disturbance is referred to as a charge packet (shown in Fig. 4.13a for an $n$-doped system).

Spin packets in semiconductors are also subject to these constraints. Consider a spin packet which involves an increase in the density of spin-up electrons, or $\Delta n_\uparrow(\mathbf{x}) > 0$. In undoped semiconductors it is not possible for the population of the other spin species to be substantially decreased, for the thermally generated background of conduction electrons is quite small. Hence an increase in the population of one spin species of carrier implies an increase in the total population of that carrier, so $\Delta n_\uparrow(\mathbf{x}) > 0$ implies $\Delta n(\mathbf{x}) > 0$. The increase in the total electron density then implies a local increase in the hole density to maintain $\Delta n(\mathbf{x}) \sim \Delta p(\mathbf{x})$. This multiple-band disturbance is shown in Fig. 4.13b. Even if the holes in the packet are not spin polarized themselves, their presence affects the motion of the spin-polarized electrons just as occurred in the charge packet.

In a doped semiconductor, however, there is a substantial background of conduction electrons, so $\Delta n_\downarrow(\mathbf{x})$ can be significantly less than zero. Thus one can create a spin packet through a spin imbalance in the conduction band $(\Delta n_\downarrow(\mathbf{x}) = -\Delta n_\uparrow(\mathbf{x}))$, without excess electrons or holes $(\Delta n(\mathbf{x}) = 0 = \Delta p(\mathbf{x}))$ [7]. This single-band spin packet (Fig. 4.13c) *does not drag a local inhomogeneous hole density with it,* and thus its mobility and diffusion properties are very different from those of a spin packet in the undoped semiconductor.

We now describe the implications for mobility and diffusion of these two types of packets. Ignoring generation, recombination, and spin relaxation in (4.32–4.35), we find the motion of electrons is only influenced by the holes through the space-charge field coming from the Poisson Equation. The importance of this field can be seen in Fig. 4.14. A charge packet (Fig. 4.14a) in the presence of an electric field will begin to separate (Fig. 4.14b) – elec-

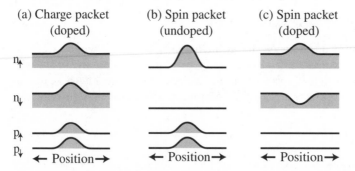

**Fig. 4.13.** Spin subband density profile of (**a**) charge polarization packets in an $n$-doped semiconductor, (**b**) spin polarization packets in a undoped semiconductor, and (**c**) spin polarization packets in an $n$-doped semiconductor.

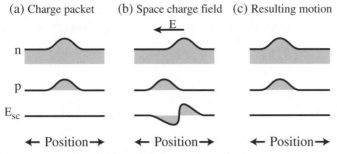

(a) Charge packet    (b) Space charge field   (c) Resulting motion

**Fig. 4.14.** (a) charge polarization packet, (b) initial response of that packet to an electric field. (c) motion of carriers in response to space-charge field in (b) results in the packet following the low-conductivity minority carriers (holes).

trons move one way and holes the other. The space-charge field pulls them together and the high-conductivity carriers (here the electrons) move towards the low-conductivity carriers (here the holes) to reduce the space-charge field.

For a small applied field the motion of the charge packet can be calculated by setting $\Delta n_\uparrow(\mathbf{x}) = \Delta n_\downarrow(\mathbf{x})$ and $\Delta p_\uparrow(\mathbf{x}) = \Delta p_\downarrow(\mathbf{x})$ and then eliminating the space-charge field from (4.28)–(4.31). Finally, setting $\Delta n(\mathbf{x}) = \Delta p(\mathbf{x})$ one finds that the motion of a charge packet (Fig. 4.13a) involves dragging both a conduction and valence disturbance, and is described by the ambipolar mobility and diffusion constant, [73]

$$\mu_a = \frac{(n-p)\mu_e\mu_h}{n\mu_e + p\mu_h}, \quad D_a = \frac{n\mu_e D_h + p\mu_h D_e}{n\mu_e + p\mu_h}, \tag{4.40}$$

where $D_e$, $\mu_e$ and $D_h$, $\mu_h$ are the diffusion constants and mobilities for electrons and holes respectively. For $n$-doping ($n \gg p$), $D_a \sim D_h$ and $\mu_a \sim \mu_h$, so the diffusion and mobility of the charge packet is dominated by the *holes*.

The mobility and diffusion constants of the spin packet of Fig. 4.13c, however, involve dragging only a spin-up and a spin-down conduction disturbance, so $\Delta p_\uparrow(\mathbf{x}) = \Delta p_\downarrow(\mathbf{x}) = 0$. Again, eliminating the space-charge field from (4.28), (4.29) and then setting $\Delta n_\uparrow(\mathbf{x}) = -\Delta n_\downarrow(\mathbf{x})$ Ref. [70] found

$$\mu_s = \frac{(n_\downarrow + n_\uparrow)\mu_{e\downarrow}\mu_{e\uparrow}}{n_\downarrow\mu_{e\downarrow} + n_\uparrow\mu_{e\uparrow}}, \quad D_s = \frac{n_\downarrow\mu_{e\downarrow}D_{e\uparrow} + n_\uparrow\mu_{e\uparrow}D_{e\downarrow}}{n_\downarrow\mu_{e\downarrow} + n_\uparrow\mu_{e\uparrow}}. \tag{4.41}$$

For the nonmagnetic semiconductor of [8], with $n_\uparrow = n_\downarrow$, $\mu_{e\uparrow} = \mu_{e\downarrow}$, and $D_{e\uparrow} = D_{e\downarrow}$, (4.41) predict $\mu_s = \mu_e$ and $D_s = D_e$. Thus the mobility of the spin packet is predicted to be the same as that of the electron sea background.

### 4.4.3  Diffusion and Mobility of Packets in GaAs

Because the diffusion and mobility of spin and charge packets in doped semiconductors are determined by the properties of a single carrier species, we can relate the mobility $\mu$ of a packet to the diffusion constant $D$ describing the spread of the packet with an expression [73] derived for a single species,

$$eD = -\mu \frac{\int_0^\infty N(E)f_0(E)dE}{\int_0^\infty N(E)\left(\partial f_0(E)/\partial E\right)dE}, \tag{4.42}$$

$$= 2\mu kT \frac{F_{1/2}[E_F/kT]}{F_{-1/2}[E_F/kT]}. \tag{4.43}$$

Here $N(E)$ is the density of states of the band with the zero of energy chosen so that the band edge is $E = 0$ and $f_0(E)$ is the Fermi function. Equation (4.42) comes from considering diffusion to be driven by a gradient in the chemical potential caused by the increase in density. Equation (4.43) holds for a bulk parabolic band, where

$$N(E) = (2E)^{1/2}m^{3/2}/(\pi^2\hbar^3) \tag{4.44}$$

$$F_n(\xi) = \int_0^\infty x^n[\exp(x - \xi) + 1]^{-1}. \tag{4.45}$$

In the low density limit

$$(\partial f_0(E)/\partial E) = -f_0(E)/kT, \tag{4.46}$$

where $T$ is the temperature and $k$ is Boltzmann's constant, and so $eD = \mu kT$, which is Einstein's relation. For degenerate systems, however, $eD/kT\mu > 1$ because of the faster increase of the chemical potential with density (Fermi pressure). Whereas $eD/kT\mu$ is adjustable by light doping of a semiconductor, in a metal this value is more difficult to change (point (3) above).

Figure 4.15 shows $eD/kT\mu$ obtained from (4.43) for a spin packet (solid line) and a charge packet (dashed line) in $n$-doped GaAs at $T = 1.6$ K. The quantitative difference in $eD/kT\mu$ for spin packets, which are dominated by conduction electron properties, and for charge packets, which are dominated by valence hole properties, occurs because the conduction band electrons have a very different mass ($m_e = 0.067m_0$, where $m_0$ is the free electron mass) from the valence band holes ($m_h = 0.53m_0$), and $N(E) \sim m^{3/2}$. This mass dependence of $N(E)$ implies $eD/kT\mu \gg 1$ for spin packets (due to conduction band degeneracy) at much lower densities ($(m_e/m_h)^{3/2} \sim 0.045$) than have been predicted for charge packets [74]. At $n = 10^{16}$ cm$^{-3}$, $eD/kT\mu \sim 12$, which is in good agreement with the "more than one order of magnitude" enhancement seen in [8].

For the $p$-doped semiconductor the charge packet is dominated by the diffusion and mobility properties of the conduction electrons, whereas the spin packet is dominated by the properties of the valence holes. Thus for typical band-edge masses, the charge packet is over an order of magnitude more mobile than the spin packet, precisely the opposite case as for an $n$-doped semiconductor.

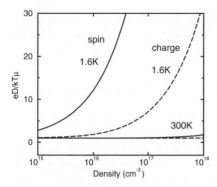

**Fig. 4.15.** (*Solid*) Ratio of diffusion to mobility for spin packets in GaAs at 1.6 K and 300 K as a function of background conduction electron ($n$) density. (*Dashed*) Same for charge packets as a function of packet density. The diffusion and mobility of the charge packet is dominated by the valence holes, whereas that of the spin packet is dominated by the conduction electrons.

### 4.4.4   Influence of Many-Body Effects on Low-Field Spin Diffusion

The relationships above between the diffusion constant and mobility are quantitative for noninteracting electrons. Many-body effects, however, reduce the diffusion constant for a given electron density [75], due to reductions in the spin stiffness (primarily from exchange) and also due to spin Coulomb drag. The larger reduction is due to spin Coulomb drag, which is a frictional effect driven purely by the correlations between electrons with opposite spins. The effect of these interactions on the mobility, however, is negligible.

For interacting electrons [75] the generalization of the Einstein relation is

$$e^2 D_{\alpha\beta} = \sum_\gamma \sigma_{\alpha\gamma} S_{\gamma\beta}, \tag{4.47}$$

where $\alpha$, $\beta$, and $\gamma$ label spin directions, $D$ and $\sigma$ are now tensors, and

$$S_{\gamma\beta} = \frac{\partial^2 f(n_\uparrow, n_\downarrow, T)}{\partial n_\gamma \partial n_\beta} \tag{4.48}$$

is the spin stiffness tensor [minus the inverse spin susceptibility matrix]. Many-body effects introduce off-diagonal components into $S_{\alpha\beta}$ and $\sigma_{\alpha\beta}$ as well as change the overall value of the spin stiffness. Here an off-diagonal conductivity ($\sigma_{\uparrow\downarrow}$) indicates that an external force applied only to the up electrons produces as well a current of down electrons. Similarly, an off-diagonal spin stiffness ($S_{\uparrow\downarrow}$) indicates that the chemical potential of up electrons depends on the density of down electrons (not the case for noninteracting electrons).

Whereas the reduction in the ratio of diffusion constant to mobility due to electron interactions is only about 20% at 1.6 K for the $n = 10^{16}$ cm$^{-3}$ GaAs material considered in Sect. 4.4.3, the importance of these corrections can be considerably greater, particularly as the temperature and carrier density are decreased.

### 4.4.5  Motion of Spin Packets in Spin-Polarized Semiconductors

We now turn to the behavior of spin and charge packets in a spin-polarized semiconductor, where equilibrium densities, mobilities and diffusion constants can differ for the two spin densities. In $n$-doped ferromagnetic and semimagnetic semiconductors the motion of spin packets polarized antiparallel to the equilibrium carrier spin polarization is predicted to be an order of magnitude faster than for parallel polarized spin packets. These results are reversed for $p$-doped semiconductors [70].

For example, consider a 100% spin-polarized $n$-doped semiconductor, such as BeMnZnSe, which has spin subbands split by $sp - d$ exchange and is fully spin polarizable in $\sim 1$ T [71]. For this semiconductor in equilibrium $n_\uparrow > 0$, but $n_\downarrow$, $p_\uparrow$, and $p_\downarrow$ are all approximately zero. Here $\uparrow$ is defined as the low-energy spin direction in the magnetic field. As shown in Fig. 4.16 a single-band spin polarization packet is only possible for a spin packet polarized antiparallel to the equilibrium carrier spin polarization. This restriction occurs because $\Delta n_\uparrow(\mathbf{x}) < 0$ is possible, but not $\Delta n_\downarrow(\mathbf{x}) < 0$. Thus a packet with spin polarized parallel to the equilibrium spin (Fig. 4.16a) must consist of both electron and hole perturbations ($\Delta n_\uparrow(\mathbf{x}) > 0$ and $\Delta p(\mathbf{x}) > 0$) and would have diffusion and mobility properties dominated by the minority holes. The antiparallel spin packet (Fig. 4.16b), however, can be a single band disturbance with $\Delta n_\uparrow(\mathbf{x}) < 0$ and $\Delta n_\downarrow(\mathbf{x}) > 0$. Such a spin packet would have diffusion and mobility properties entirely determined by those of the majority electrons, and thus over an order of magnitude faster. We show in Fig. 4.17 the different ratios of diffusion constant to mobility for spin packets polarized parallel and antiparallel to the equilibrium carrier spin polarization. Here $N(E)$ is half that in the spin unpolarized state, $m_e = 0.16m_0$, and $m_h = 0.74m_0$ [71].

The behavior of spin packets in a spin-polarized $p$-doped semiconductor, such as ferromagnetic GaMnAs, is completely the opposite. Here a spin packet polarized parallel to the equilibrium carrier spin polarization would require a conduction electron component. The minority carriers (the electrons) would

(a) Parallel          (b) Antiparallel

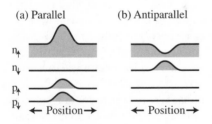

$n_\uparrow$
$n_\downarrow$
$p_\uparrow$
$p_\downarrow$
← Position →          ← Position →

**Fig. 4.16.** Spin subband density profile of spin polarization packets polarized **(a)** parallel and **(b)** antiparallel to the equilibrium carrier spin polarization of an $n$-doped spin-polarized semiconductor.

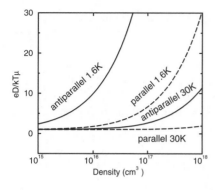

**Fig. 4.17.** (*Solid*) Ratio of diffusion to mobility for spin packets polarized antiparallel to the equilibrium carrier spin polarization of the semiconductor $Be_xMn_yZn_{1-x-y}Se$ at 1.6 K and 30 K as a function of conduction electron density (Be doping). (*Dashed*) Same, but for parallel spin packets as a function of packet density.

determine the mobility and diffusion constant of such a packet. A spin packet polarized antiparallel to the equilibrium carrier spin polarization could consist entirely of holes, however, and would have a much smaller mobility and diffusion constant.

### 4.4.6  High-Field Spin Transport in the Diffusive Regime

Recently it has been pointed out [76] that there is a second relevant diffusive regime in semiconductors which has no analogue in metals. If the density dependence of the conductivity is treated appropriately in (4.28)–(4.36) then an equation for diffusion is obtained that includes a term depending on the electric field:

$$\nabla^2(n_\uparrow - n_\downarrow) + \frac{e\mathbf{E}}{k_\mathrm{B}T} \cdot \nabla(n_\uparrow - n_\downarrow) - \frac{(n_\uparrow - n_\downarrow)}{(L^{(s)})^2} = 0, \tag{4.49}$$

where $L^{(s)}$ is the intrinsic spin diffusion length. Here the term depending on electric field will dominate over the more common diffusive term for sufficient fields $E > E_c$. For a typical spin diffusion length, $L^{(s)} = 10^4$ nm, $E_c = 25$ V/cm at $T = 300$ K and $E_c = 0.25$ V/cm at $T = 3$ K in non-degenerate $n$-doped GaAs. Hence the fields required to enter the high-field regime are quite small. In this high-field regime there are two spin diffusion lengths (upstream and downstream relative to the electric field). Spin polarization can travel downstream over distances much greater than the intrinsic spin diffusion length.

## 4.5  Spin Transport in Inhomogeneous Structures

In previous sections we have considered the spin coherence times of homogeneous spin perturbations in homogeneous materials and the spin transport properties of inhomogeneous spin perturbations in homogeneous materials. We now turn to inhomogeneous materials. In Chaps. 1–3 some recent advances in electrical spin injection from magnetic materials to nonmagnetic materials have been considered. The range of devices include the proposal of spin filters [77,78] and the unipolar spin diodes and transistors of [79] as well as initial demonstrations in resonant tunneling diode geometries [16], and spin-LED's [13–15]. The drift-diffusion equations described before have also been used for calculations of transport in nonmagnetic [80] p-n junctions and magnetic p-n junctions [81]. Here we analyze an alternative situation, the effect of the ferromagnet/semiconductor boundary on spin injection into a nonmagnetic semiconductor.

### 4.5.1    Transport Across the Ferromagnet/ Semiconductor Boundary

We will treat this subject only briefly to outline general considerations. Two special cases of the ferromagnet/semiconductor interface have been considered theoretically in the literature, depending on how the interface is treated. In [82] (also see Chaps. 2–3) the interface is considered to be perfectly transparent, and the injected spin density is found to be small. In [83] and later [84], by contrast, a tunnel barrier is introduced at the interface, and a large injected spin density can be achieved. Both cases are appropriate for particular situations, but the general case is considerably richer. For a true semiconductor-metal interface the carrier density in the semiconductor responds strongly to the presence of the interface, requiring a self-consistent determination of the conductivity in the interface region. A typical self-consistent conduction-band profile, as shown below in Fig. 4.18a, incorporates a high-resistance region at the interface. This is the Schottky barrier. The voltage is drawn for the case of electrons flowing from the semiconductor to the ferromagnet, however the barrier is present for flow either direction. In principle if there is a spin-selective transport across or over the barrier in either direction (e.g. due to the difference in the spin-polarized density of states in the ferromagnet), a spin polarization of the carriers in the semiconductor will ensue.

Such a high-resistance region should, according to the arguments of [83], yield a large injected spin density, however our findings differ. Shown in Fig. 4.18b are spin coherence times for carriers excited above the conduction band minimum in GaAs. As the carriers involved in carrying current across the Schottky barrier are typically very energetic carriers, it is quite possible that they will lose their spin coherence before crossing the interface, thus leading to a weak injected spin density.

Shown in Fig. 4.18c are band alignments for a highly-doped semiconductor contiguous to a ferromagnet. The presence of the Schottky barrier and the highly-doped region leads to the development of a tunnel barrier. Again, whereas this satisfies the conditions of [83], the spin coherence time for carriers in highly-doped GaAs is very short. Thus only a weak injected spin signal might be expected in this case as well.

To see how the Schottky barrier arises in the typical electrochemical formulation for transport, consider the following. For a steady-state current

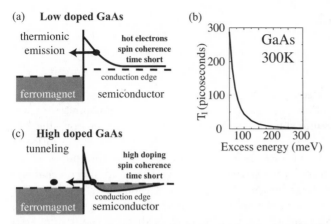

**Fig. 4.18.** (**a**) band alignment for Schottky barrier between ferromagnet and lightly $n$-doped semiconductor. (**b**) coherence times for carriers in GaAs at 300 K as a function of energy above the conduction band minimum. (**c**) band alignment for Schottky barrier between ferromagnet and heavily $n$-doped semiconductor.

distribution, with $\nabla \cdot \mathbf{j} = 0$, it is possible to write the current as the gradient of a hydrodynamic potential: $\mathbf{j} = -\nabla \xi$. A local version of Ohm's law relates this hydrodyamic potential to the electrochemical potential according to

$$\mathbf{j}(\mathbf{r}) = -\sigma(\mathbf{r})\nabla\mu(\mathbf{r}). \tag{4.50}$$

It is important to recognize that in situations where the carrier density can change from point to point and when the conductivity depends strongly on the density (such as in a lightly-doped semiconductor), the conductivity must be determined self-consistently in (4.50). It is the neglect of this effect which renders the results of [82–84] only applicable to special cases where the Schottky barrier vanishes.

Hence we find that for most semiconductor-ferromagnetic interfaces, neither the analysis of [82] or of [83,84] properly applies. The clean interface naturally develops a high-resistance region, however spin coherence times near the interface are quite short for III-V semiconductors.

One possible route around this is to attempt spin injection in the high-field regime (described in Sect. 4.4.6). In this regime the amplitude of the spin injection can be enhanced by several orders of magnitude, yielding moderate injection amplitudes for ferromagnet-semiconductor spin injection [76]. Also predicted from the high-field regime is a pronounced increase in the spin injection amplitude with increasing current. This may help explain some of the puzzling characteristics of experimental reports of spin injection from a ferromagnetic metal into a semiconductor [85–87].

## 4.6   Conclusion

In the course of this chapter we have outlined several themes in the dynamics of mobile spin in semiconductors. These include:

(1) the interplay between carrier spin and motion, as incorporated in the spin-orbit interaction and the presence of inversion asymmetry in the semiconductor. Spin-orbit interaction itself gives rise to the optical selection rules (described in Chap. 2) that can be used to generate optically spin-polarized distributions. As we have described here, the spin-orbit interaction with inversion asymmetry generates the dominant mechanism for spin decoherence in both bulk and confined geometries.
(2) the interplay between carrier spin and charge. The sensitivity of carriers to the electric field can be used to manipulate directly the flow of spin through a material. It also influences the mobility and diffusion of spin within a material as well as the transport of spin from one material to another.

These themes will continue to be developed in future work, and will form two constituents of any fundamental theory of spin coherence in semiconductors.

A further theme, not explored here, is that of the interplay between non-magnetic semiconducting host materials and dilute magnetic atoms embedded within them. The presence of magnetic atoms, and their influence on the local electronic structure of the semiconductor, will incorporate and modify both of the above themes. Mn, for example, is a dopant in GaAs, so a complete theory of the alloy GaMnAs requires proper consideration of the doping contributions of Mn to the material.

### Acknowledgments

We would like to acknowledge discussions with D. D. Awschalom, C. Ullrich, and G. Vignale. This work was supported in part by DARPA/ARO through grant DAAD19-01-0490 and the National Science Foundation through grant ECS-0000556.

# References

1. S. Datta and B. Das, Appl. Phys. Lett. **56**, 665 (1990).
2. D. D. Awschalom and J. M. Kikkawa, Phys. Today **52**, 33 (June 1999).
3. L. Sham, Science, **277**, 1258 (1997).
4. J. Wagner, H. Schneider, D. Richards, A. Fischer, and K. Ploog, Phys. Rev. B **47**, 4786 (1993).
5. S. A. Crooker, D. D. Awschalom, J. J. Baumberg, F. Flack, and N. Samarth, Phys. Rev. B **56**, 7574 (1997).
6. J. M. Kikkawa, I. P. Smorchkova, N. Samarth, and D. D. Awschalom, Science **277**, 1284 (1997).
7. J. M. Kikkawa and D. D. Awschalom, Phys. Rev. Lett. **80**, 4313 (1998).
8. J. M. Kikkawa and D. D. Awschalom, Nature (London) **397**, 139 (1999).
9. J. A. Gupta, R. Knobel, N. Samarth, and D. D. Awschalom, Science **292**, 2458 (2001).
10. J. M. Kikkawa and D. D. Awschalom, Science **287**, 473 (2000).
11. A. Malinowski and R. T. Harley, Solid State Commun. **114**, 419 (2000).
12. G. Salis, D. T. Fuchs, J. M. Kikkawa, D. D. Awschalom, Y. Ohno, and H. Ohno, Phys. Rev. Lett. **86**, 2677 (2001).
13. R. Fiederling, M. Keim, G. Reuscher, W. Ossau, G. Schmidt, A. Waag, and L. W. Molenkamp, Nature **402**, 787 (1999).
14. Y. Ohno, D. K. Young, B. Beschoten, F. Matsukura, H. Ohno, and D. D. Awschalom, Nature **402**, 790 (1999).
15. B. T. Jonker, Y. D. Park, B. R. Bennett, H. D. Cheong, G. Kioseoglou, and A. Petrou, Phys. Rev. B **62**, 8180 (2000).
16. H. Ohno, N. Akiba, F. Matsukura, A. Shen, K. Ohtani, and Y. Ohno, Appl. Phys. Lett. **73**, 363 (1998); H. Ohno, Science **281**, 951 (1998).
17. H. Ohno, A. Shen, F. Matsukura, A. Oiwa, A. Endo, S. Katsumoto, and Y. Iye, Appl. Phys. Lett. **69**, 363 (1996); T. Hayashi, M. Tanaka, T. Nishinaga, H. Shimada, H. Tsuchiya, and Y. Otuka, J. Cryst. Growth **175**, 1063 (1997); A. Van Esch, L. Van Bockstal, J. De Boeck, G. Verbanck, A. S. van Steenbergen, P. J. Wellmann, B. Grietens, R. Bogaerts, F. Herlach, and G. Borghs, Phys. Rev. B **56**, 13103 (1997).
18. D. DiVincenzo, Science **270**, 255 (1995).
19. D. Loss and D. P. DiVincenzo, Phys. Rev. A **57**, 120 (1998).
20. B. E. Kane, Nature **393**, 133 (1998).
21. A. Imamoğlu, D. D. Awschalom, G. Burkard, D. P. DiVincenzo, D. Loss, M. Sherwin, and A. Small, Phys. Rev. Lett. **83**, 4204 (1999).
22. G. A. Prinz, Science **282**, 1660 (1998); *ibid* **283**, 330 (1999).
23. *Principles of Magnetic Resonance*, C. P. Slichter, Springer, 3rd edition, New York, 1992.
24. G. L. Bir, A. G. Aronov, and G. E. Pikus, Zh. Eksp. Teor. Fiz. **69**, 1382 (1975)[Sov. Phys. JETP **42**, 705 (1976)].
25. G. Fishman and G. Lampel, Phys. Rev. B **16**, 820 (1977).
26. *Optical Orientation*, edited by F. Meier and B. P. Zakharchenya (North-Holland, Amsterdam, 1984), Vol. 8.
27. Y. Yafet in *Solid State Physics Vol. 14*, (Academic Press, New York, 1963), p. 1.
28. G. Dresselhaus, Phys. Rev. **100**, 580 (1955).

29. M. I. D'yakonov and V. I. Perel', Sov. Phys. JETP **33**, 1053 (1971); Sov. Phys. Solid State **13**, 3023 (1972).
30. R. Terauchi, Y. Ohno, T. Adachi, A. Sato, F. Matsukura, A. Tackeuchi, and H. Ohno, Jpn. J. Appl. Phys. **38**, Pt. 1, No. 4B., 2549 (1999).
31. P. Y. Yu and M. Cardona, *Fundamentals of semiconductors, Physics and materials properties*, 2nd ed. (Springer, New York 1999).
32. see, *e. g. Group Theory and Quantum Mechanics*, M. Tinkham, McGraw-Hill, New York, 1964.
33. E. I. Rashba, Sov. Phys. Solid State **2**, 1109 (1960); Yu. A. Bychov and E. I. Rashba, J. Phys. C **17**, 6093 (1984).
34. P. Pfeffer and W. Zawadzki, Phys. Rev. B **52**, R14332 (1995); P. Pfeffer, *ibid* **55**, R7359 (1997); P. Pfeffer and W. Zawadzki, *ibid* **59**, R5312 (1999).
35. E. A. de Andrada e Silva, Phys. Rev. B **46**, 1921 (1992).
36. E. A. de Andrada e Silva, G. C. La Rocca, and F. Bassani, Phys. Rev. B **50**, 8523 (1994).
37. N. S. Averkiev and L. E. Golub, Phys. Rev. B **60**, 15582 (1999).
38. W. H. Lau and M. E. Flatté, J. Appl. Phys. in press.
39. O. Krebs and P. Voisin, Phys. Rev. Lett. **77**, 1829 (1996).
40. P. V. Santos, P. Etchegoin, M. Cardona, B. Brar, and H. Kroemer, Phys. Rev. B **50**, 8746 (1994).
41. L. Vervoort, R. Ferreira, and P. Voisin, Phys. Rev. B **56**, R12744 (1997).
42. E. L. Ivchenko, A. Yu. Kaminski, and U. Rössler, Phys. Rev. B **54**, 5852 (1996).
43. L. Vervoort, R. Ferreira, and P. Voisin, Semicond. Sci. Technol. **14**, 227 (1999).
44. J. T. Olesberg, W. H. Lau, M. E. Flatté, C. Yu, E. Altunkaya, E. M. Shaw, T. C. Hasenberg, and T. F. Boggess, Phys. Rev. B **64**, 201301(R) (2001).
45. A. Mysyrowicz, D. Hulin, A. Antonetti, A. Migus, W. T. Masselink, and H. Morkoç, Phys. Rev. Lett. **56**, 2748 (1986).
46. S. Schmitt-Rink and D. S. Chemla, Phys. Rev. Lett **57**, 2752 (1986).
47. A. V. Khaetskii and Y. V. Nazarov, Phys. Rev. B **61**, 12639 (2000).
48. J. S. Sandhu, A. P. Heberle, J. J. Baumberg, and J. R. A. Cleaver, Phys. Rev. Lett. **86**, 2150 (2001).
49. W. H. Lau, J. T. Olesberg, and M. E. Flatté, Phys. Rev. B **64**, 161301(R) (2001).
50. J. T. Olesberg, Ph. D. thesis, University of Iowa, 1999.
51. L. Wissinger, U. Rössler, R. Winkler, B. Jusserand, D. Richards, Phys. Rev. B **58**, 15375 (1998).
52. R. Winkler and U. Rössler, Phys. Rev. B **48**, 8918 (1993).
53. A. Bournel, P. Dollfus, E. Cassan, and P. Hesto, Appl. Phys. Lett. **77**, 2346 (2000).
54. O. Madelung, *Semiconductors-Basic Data*, 2nd ed. (Springer, New York, 1996).
55. M. Cardona, N. E. Christensen, and G. Fasol, Phys. Rev B **38**, 1806 (1988).
56. T. F. Boggess, J. T. Olesberg, C. Yu, M. E. Flatté, and W. H. Lau, Appl. Phys. Lett. **77**, 1333 (2000).
57. M. I. D'yakonov and V. Yu. Kachorovskii, Sov. Phys. Semicond **20**, 110 (1986).
58. E. L. Ivchenko and G. E. Pikus, *Superlattices and Other Heterostructures*, 2nd ed. (Springer, New York, 1997).
59. The experimental results have been adjusted from [30], for the authors defined an effective spin flip time for a single spin, $\tau_s = 2T_1$, and plotted their results for $\tau_s$. The DK calculation (dot-dashed line) is of $T_1$, so in addition to the errors in trends, the discrepancy in the magnitude of $T_1$ (using our values of the confinement energy) is about a factor of 4 in Fig. 4.11a,b.

60. A. Tackeuchi, O. Wada, and Y. Nishikawa, Appl. Phys. Lett. **70**, 1131 (1997); R. Tackeuchi, T. Kuroda, S. Muto, Y. Nishikawa, and O. Wada, Jpn. J. Appl. Phys. **38**, Pt. 1, No. 8, 4680 (1999).

61. K. C. Hall, S. W. Leonard, H. M. van Driel, A. R. Kost, E. Selvig, and D. H. Chow, Appl. Phys. Lett. **75**, 3665 (1999); **75**, 4156 (1999).

62. M. Johnson and R. H. Silsbee, Phys. Rev. Lett. **55**, 1790 (1985); M. Johnson and R. H. Silsbee, Phys. Rev. B **35**, 4959 (1987); M. Johnson and R. H. Silsbee, Phys. Rev. B **37**, 5312 (1988).

63. P. C. van Son, H. van Kempen, and P. Wyder, Phys. Rev. Lett. **58**, 2271 (1987).

64. T. Valet and A. Fert, Phys. Rev. B **48**, 7099 (1993).

65. P. M. Levy and S. Zhang, Phys. Rev. Lett. **79** 5110 (1997).

66. A. R. Cameron, P. Riblet, and A. Miller, Phys. Rev. Lett. **76**, 4793 (1996).

67. J. R. Haynes and W. Shockley, Phys. Rev. **81**, 835 (1951).

68. D. Hägele, M. Oestreich, W. W. Rühle, N. Nestle, and K. Eberl, Appl. Phys. Lett. **73**, 1580 (1998).

69. S. S. P. Parkin, R. Bhadra, and K. P. Roche, Phys. Rev. Lett. **66**, 2152 (1991).

70. M. E. Flatté and J. M. Byers, Phys. Rev. Lett. **84**, 4220 (2000).

71. B. König, U. Zehnder, D. R. Yakovlev, W. Ossau, T. Gerhard, M. Keim, A. Waag, and G. Landwehr, Phys. Rev. B **60**, 2653 (1999).

72. W. van Roosbroeck, Bell. Syst. Tech. J. **29**, 560 (1950); Phys. Rev. **91**, 282 (1953).

73. See, e.g., *Semiconductors,* R. A. Smith, Cambridge University Press, New York, 1978.

74. J. F. Young and H. M. van Driel, Phys. Rev. B **26**, 2147 (1982).

75. I. D'Amico and G. Vignale, Europhys. Lett. **55**, 566 (2001).

76. Z. G. Yu and M. E. Flatté, cond-mat/0201425.

77. J. C. Egues, Phys. Rev. Lett. **80**, 4578 (1998).

78. Y. Guo, B.-L. Gu, Z.-Q. Li, J.-Z. Yu, and Y. Kawazoe, J. Appl. Phys. **83**, 4545 (1998).

79. M. E. Flatté and G. Vignale, Appl. Phys. Lett. **78**, 1273 (2001).

80. I. Žutic, J. Fabian, and S. Das Sarma, Phys. Rev. B **64**, 121201 (2001).

81. I. Žutic, J. Fabian, and S. Das Sarma, Phys. Rev. Lett. **88**, 066603 (2002).

82. G. Schmidt, D. Ferrand, L. W. Molenkamp, A. T. Filip, and B. J. van Wees, Phys. Rev. B **62**, R4790 (2000).

83. E. I. Rashba, Phys. Rev. B **62**, R16 267 (2000).

84. D. L. Smith and R. N. Silver, Phys. Rev. B **64**, 045323 (2001).

85. P. R. Hammar, B. R. Bennett, M. J. Yang, and M. Johnson, Phys. Rev. Lett. **83**, 203 (1999)

86. H. J. Zhu, M. Ramsteiner, H. Kostial, M. Wassermeier, H.-P. Schönherr, and K. H. Ploog, Phys. Rev. Lett. **87**, 016601 (2001).

87. A. T. Hanbicki, B. T. Jonker, G. Itskos, G. Kioseoglou, and A. Petrou, Appl. Phys. Lett. **80**, 1240 (2002).

# 5 Optical Manipulation, Transport and Storage of Spin Coherence in Semiconductors

David D. Awschalom and Nitin Samarth

## 5.1 Introduction

The drive to build a framework for coherent semiconductor spintronic devices provides a strong motivation for understanding the coherent evolution of spin states in semiconductors [1,2]. The fundamental aim in this context is to discover regimes in which carefully prepared quantum states based upon spin can evolve coherently long enough to allow the storage, manipulation and transport of quantum information in devices. Such devices might exploit, for instance, the interference between two coherently-occupied spin states whose time variation occurs at a frequency $\Delta E/h$, where $\Delta E$ is their energy separation. Since typical spin splittings in semiconductors are in the range of meV, the rapidly varying oscillations of a classical observable such as the spin orientation (magnetization) can occur at GHz-THz frequencies, providing the basis for ultrafast devices. Another possibility is that this quantum interference may actually be used as part of a calculation within the context of quantum computing algorithms [3]. It is hence crucial to develop experimental tools that probe spin coherence in semiconductors and that allow one to map out schemes for its manipulation, storage and transport. The previous chapter formulated the theoretical foundations underlying coherent spin dynamical phenomena in semiconductors and introduced specific mechanisms that may be responsible for spin relaxation and spin decoherence, pointing out the important physical distinctions between longitudinal and transverse spin relaxation times ($T_1$ and $T_2$, respectively) [4]. We note that it is the latter timescale that is of direct relevance to coherent spin devices and hence we focus on experimental techniques that probe the transverse spin relaxation time in semiconductors.

These experiments are based upon measurements of the time-resolved Faraday effect, a technique that has its genesis in early experiments that studied magnetization dynamics in bulk magnetic semiconductors [5]. Later variations on this experimental theme have been applied to magnetic semiconductor quantum wells, providing insights into diverse phenomena such as exciton "spin beats" [6,7], conduction electron spin resonance (at THz frequencies) and electron paramagnetic resonance of $Mn^{2+}$ ions (at GHz frequencies) [8]. Since these initial experiments focused on systems containing magnetic ions, the observed electron spin decoherence times were typically quite short ($\sim 5$ ps), owing to spin-flip processes resulting from the carrier-ion exchange interaction. The more recent application of the time-resolved Faraday effect to conventional (non-magnetic) semiconductors, however, reveals a series of surprising discoveries. First, the conduction electron spin coherence times in n-doped semiconductors are found to be quite long, ranging from several nanoseconds at room temperature in n-ZnSe [9] to hundreds of nanoseconds in n-GaAs at low temperatures [10]. Further, the long spin coherence times can be exploited for driving a coherent spin polarization across macroscopic distances in homogeneous semiconductor crystals [11] and even across heterointerfaces [12,13]. Strikingly, the long spin coherence times appear to be insensitive to defect densities in samples [14]. The optical techniques developed for measuring electron spin coherence have also given rise to other variations that enable the coupling of electron spin coherence to nuclear spins in an all-optical form of nuclear magnetic resonance [15] and also permit the all-optical coherent manipulation of electron spins in a manner akin to nuclear spin echoes [16]. In this chapter, we provide an overview of all these recent developments in coherent electronic spin dynamics in semiconductors, beginning with an introduction to the basic measurement techniques. This is followed by a discussion of the measurement of spin coherence in bulk semiconductor crystals, focusing principally on extensive studies of GaAs. The next section looks at spin coherence in nanostructures, specifically in semiconductor quantum dots. (We note that spin coherence in quantum wells has already been addressed earlier in Chap. 4.) Since the transport of coherent spin information is of great interest to eventual applications, we then discuss experiments that probe the transfer of spin coherence in both homogeneous and inhomogeneous systems. The next section examines how the excitation of coherent electron spin populations can couple with local moments (paramagnetic ions) and with nuclear spins, resulting in time-resolved spin resonance measurements. The final sections of the chapter discuss different approaches for manipulating spin coherence, first by using the optical Stark effect for coherently tipping coherent spin states and then by exploiting a coupling between ferromagnetic layers and semiconductors in hybrid heterostructures.

## 5.2 Experimental Techniques for Measuring Spin Coherence in Semiconductors

Ideally, an experiment aimed at measuring electron spin coherence in a semiconductor would proceed by preparing a coherent superposition of spin states of a single electron (e.g. the spin "up" and spin "down" states in the presence of an external magnetic field). Some appropriate measurement would then interrogate the coherent time evolution of this superposition state and examine the influences of decohering factors produced by the coupling of this state with the environment. Such a *gedanken* experiment is easily imagined in semiconductors because of the electric dipole selection rules underlying optical transitions across the energy gap (see previous chapter). As a result of these selection rules, a circular polarized photon can excite a spin polarized electron from the lowest energy heavy hole state to the lowest energy conduction band state in a direct gap semiconductor such as GaAs [9,17]. If an external magnetic field is applied perpendicular to the direction of the propagation of the circularly polarized light pulse, then the promoted electron is in a coherent superposition of the basis states defined by the magnetic field. The coherent evolution of this superposition results in a phase difference between the two energy eigenstates that varies linearly in time as $\Delta Et/\hbar$. One may view this semi-classically as the Larmor precession of the classical spin vector in a plane normal to the applied field. A projection of the classical spin vector along its initial direction oscillates as a cosine; incoherent evolution of the spins wavefunction results in a decay of the amplitude of the oscillation. The latter "decoherence" can arise from a number of possible events such as spin-spin scattering due to interactions with holes, local magnetic impurities, or other electrons, spin-orbit scattering due to phonons or impurities, or finite lifetime effects caused by optical recombination with a hole. Another way in which one can prepare a coherent superposition of spin states is by using a linearly polarized optical pulse that propagates parallel to the external magnetic field. Again, this creates a coherent superposition of the basis spin states defined by the magnetic field and this superposition evolves at a frequency that corresponds to the spin splitting between the basis states [6,18].

The single electron experiment described above is of course much harder to carry out in practice; instead, a more tractable measurement deals with an ensemble of electron spins ($\sim 10^{15}$ electrons per $cm^{-3}$) excited by an optical pulse from an ultrafast laser. In this initial state, all the optically-excited spins add constructively to yield a net magnetization that precesses at the Larmor frequency. If all conduction electrons are identical and non-interacting, then the decay of the net magnetization reflects the intrinsic spin decoherence of individual electrons and – in the case of an exponential decay – this may be characterized by the transverse spin relaxation time $T_2$. However, in addition to such homogeneous decoherence, there may be inhomogeneous effects such as variations in $g$-factors or local magnetic fields that lead to an extra decay (or "dephasing") of the spin polarization. These result in a spreading in the relative orientation of spins within the electron distribution, even when all spins are evolving coherently. Hence, we refer to the experimentally measured spin decay as $T_2^*$, indicating that the measured decay time only establishes a lower bound on $T_2$ unless inhomogeneous effects can be eliminated. We note that an additional complication arises in the presence of electron-electron spin interactions: since intra-electronic interactions of the form $\mathbf{s}_i \cdot \mathbf{s}_j$ commute with both the total electron spin and its equations of motion, they can introduce a hidden decoherence of individual spins that leaves the net electron magnetization undisturbed and is not detectable in the measurement of the Larmor precession unless electron spins are isolated. Thus, measurements of electron Larmor precession in semiconductors can only reveal the rate at which spin coherence is lost to outside the electronic system, a process we will refer to as environmental decoherence.

We now address the problem of detecting spin motion once an initial spin polarized state has been prepared. In general, spin dynamical processes in semiconductors can be interrogated using a variety of time-resolved optical techniques in which an optical pulse prepares a spin-oriented state whose time evolution is then followed by an optical measurement at some later time. For instance, the spin scattering and decoherence of excitons can be studied by measuring the time-resolved circular polarization of photoluminescence (PL) emitted during electron–hole recombination; the electronic magnetization may then be inferred through the photon spin conservation. This particular method was first applied to the measurement of electron spin precession in undoped GaAs quantum wells and bulk crystals [17]. These experiments detected the time-resolved intensity of the co- and counter-circularly polarized components of the PL along the axis of optical excitation from a GaAs/AlGaAs quantum well after optical excitation with circularly polarized light. The experiment – carried out in the Voigt geometry (light propagation normal to the external magnetic field) – monitors the projection of the injected spin parallel or anti-parallel to their initial orientation, respectively. Periodic oscillations are seen in the luminescence intensity as a function of time, corresponding to the Larmor precession of excited spins and starting in phase for parallel detection (co-circularly polarized) detection and out of phase for antiparallel detection (counter-circularly polarized detection). The oscillation periodicity determines the $g$-factor; in this case, the measured value of the $g$-factor identifies the spin precession with an electron rather than a bound electron–hole pair (an exciton). We note however that such a technique does not convey any information about processes that may be evolving after electron–hole recombination has taken place. Instead, we resort to other techniques based upon dynamic measurements of the Faraday and Kerr effects to reveal information about coherent spin evolution that occurs on time scales much longer than the radiative recombination times.

Time resolved Faraday/Kerr spectroscopy uses pump-probe techniques to prepare/interrogate a spin population using a stream of short optical pulses at a wavelength tuned just above the absorption edge of a semiconductor. Typically, these optical pulses are provided by a pulsed laser source such as a Ti-sapphire laser that generates a 76 MHz train of $\sim$ 120 fs pulses. A beam splitter divides the pulse train into pump and probe pulses whose polarization can be appropriately set or modulated. As described below, the pump pulse can be either circularly or linearly polarized, depending on the specific experiment. The probe pulse is always linearly polarized and a mechanical delay line produces pump-probe time delays in the range $t = 0 \to 3$ ns. For longer time delays ($\Delta t > 3$ns), a dual laser system is used with an electronic control system which allows the repetition rate of the lasers to be locked with a fixed time interval between pulses. An extension of this dual laser technique allows the energies of the pump and probe pulses to be tuned independently. The optical pulses are normally incident on the sample and focused to a $\sim 50\ \mu\mathrm{m}$ diameter spot; the angle of rotation of the linearly polarized probe pulse is then measured using an optical bridge as a function of the pump-probe time delay. In a transmission geometry, we refer to this measurement as the time resolved Faraday rotation (TRFR), while the equivalent measurement in a reflection geometry is called the time-resolved Kerr rotation (TRKR).

TRFR measurements of coherent electron spin dynamics in semiconductors were first applied to probe exciton "beats" in magnetic semiconductor quantum wells [6]. The measurements are performed in the Faraday geometry wherein the pump and probe pulses propagate along the direction of an external magnetic field. In this case, a coherent superposition of spin states is excited using a linearly polarized optical pulse; the Faraday rotation of a time-delayed linearly polarized probe pulse then interrogates the net magnetization along the field direction. In this geometry, coherent exciton beats are observed at a frequency given by the exciton spin splitting (i.e. the $g$-factors of both the electrons and holes are involved) and the decoherence is very rapid (a few picoseconds) due to the rapid spin scattering of holes. Although these Faraday geometry measurements yield important insights into many body effects such as exciton-exciton correlations [7], the rapid decoherence of the hole states obscures valuable information about the electron spin coherence. This can be unmasked by carrying out the TRFR measurement in the Voigt geometry (see Fig. 5.1).

In this scheme, the pump-probe Faraday rotation measurements employ a normally-incident, *circularly-polarized* pump pulse, where the magnetic field, $B$, is applied in-plane. The circularly-polarized pump pulse initiates an electric-dipole allowed transition across the semiconductor band gap, hence exciting electron–hole pairs that are spin polarized along the optical path. Whereas holes spin-relax rapidly due to valence band mixing or are pinned in certain quantum geometries, the electron spins continue to precess around

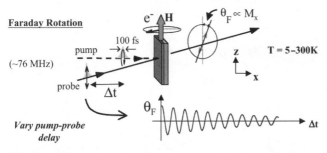

**Fig. 5.1.** Schematic view of pump-probe Faraday rotation measurement carried out in the Voigt geometry.

the magnetic field at a characteristic Larmor frequency. This spin motion is interrogated by the Faraday/Kerr rotation of the time-delayed linearly polarized probe pulse through an angle proportional to the net electron spin along the probe path, which is normal to both the sample and the field. The result is a Faraday rotation of the form:

$$\theta_F = A \exp\left(-\frac{\Delta t}{T_2^*}\right) \cos\left(\frac{g\mu_B B \Delta t}{\hbar}\right), \tag{5.1}$$

whose frequency in the pump-probe delay, $\Delta t$, directly measures the electronic $g$-factor and whose decay yields the transverse spin lifetime, $T_2^*$. An example of such a measurement is shown in Fig. 5.2 wherein the TRFR is measured in a modulation-doped $ZnSe/Zn_{1-x}Cd_xSe$ single quantum well that contains a two dimensional electron gas (2DEG). The frequency of the measured oscillations corresponds to an electronic $g$-factor for $g = 1.1$; the decay of these oscillations occurs over a time scale of several nanoseconds at low temperatures and provides a measure of the spin dephasing time $T_2^*$.

When the spin dephasing time becomes longer than several ns, the measurement of TRFR in the time domain requires long pump-probe mechanical delays that are cumbersome and prone to misalignment. Hence, a complementary technique called resonant spin amplification (RSA) has been developed to extract spin lifetimes that exceed the pulse repetition interval ($t_{rep} \approx 13$ ns). Under such conditions, the spin signals from successive excitations can constructively or destructively interfere. The total spin polarization is then resonantly enhanced whenever $g\mu_B B t_{rep}/\hbar$ is a multiple of $2\pi$, a condition that is met periodically in the applied field. Hence, by measuring the Faraday rotation at a fixed pump-probe delay and sweeping the magnetic field, one obtains oscillations that are periodic in B at a frequency proportional to both $\Delta t$ and $g$. An example of such data is shown in Fig. 5.3.

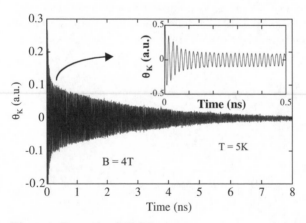

**Fig. 5.2.** Example of TRFR measurement in a modulation-doped ZnSe/(Zn,Cd)Se quantum well.

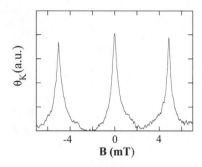

**Fig. 5.3.** Example of resonant spin amplification measured in an n-doped ZnSe epilayer ($n = 5 \times 10^{16}$ cm$^{-3}$). The data shows $\theta_K$ at $\Delta t = -10$ ps taken at small magnetic field variations around $B = 0$, showing the central resonance and the $g\mu_B B t_{\rm rep}/\hbar = \pm 2\pi$ peaks. The spacing between the resonant peaks measures the electronic $g$-factor, while the width of the peaks is related to $T_2^*$.

Synchronous injection of electron spins in phase with their precession yields a tenfold increase in spin polarization and a series of sharp field-dependent resonances. Since $t_{\rm rep}$ can be actively stabilized to $\pm 1.5$ ps and the magnetic field is measured to $10^{-5}T$, this technique is capable of resolving spin lifetimes as long as 5 µs and $\Delta g/g \approx 0.1$. Furthermore, these parameters are easily measured for low magnetic fields, which is impractical in conventional spin beat and Hanle effect measurements. A more quantitative treatment of the RSA data is given in the next section when we discuss spin coherence measurements on bulk GaAs.

## 5.3 Electron Spin Coherence in Bulk Semiconductors

Measurements of $T_2^*$ in bulk semiconductors have been carried out on single crystals of GaAs [10], ZnSe [9] and GaN [14] using both the TRFR/TRKR and RSA techniques. Although there are variations in the details, the general behavior of the spin coherence in all these direct band gap semiconductors is quite similar. This is particularly interesting given that the materials systems studied encompass defect densities that span many orders of magnitude ($\sim 10^4$ cm$^{-2}$ in GaAs to $\sim 10^{10}$ cm$^{-2}$ in GaN). Here, we focus our discussion on measurements of n-GaAs single crystals that have served as templates for the most extensive studies of spin coherence thus far. The measurements employ Si-doped GaAs single crystals with doping densities ranging from insulating to metallic behavior, with room temperature carrier densities $n = 1 \times 10^{16}$ cm$^{-3}$, $1 \times 10^{18}$ cm$^{-3}$, and $5 \times 10^{18}$ cm$^{-3}$ and mobilities $\mu = 5400$, 2300, 1340 cm$^2$/Vs, respectively. Control measurements are carried out on a semi-insulating GaAs sample, hereafter referred to as $n = 0$. We note that since increased excitation densities $N_{\rm ex}$ degrade $T_2^*$ in nondegenerate samples, $N_{\rm ex}$ is kept several orders of magnitude below n for the doped samples. For $n = 0$, the laser energy (1.540 eV) includes an additional 20 meV which does not significantly affect the free electron spin lifetime and suppresses the excitation of excitons whose precession complicates data analysis. All wafers are mechanically thinned down to $\sim 50$ µm for transmission measurements and mounted strain-free in a magneto-optical cryostat.

The oscillatory temporal evolution of the TRFR is shown in Fig. 5.4 for the different doping concentrations at $T \sim 5$ K and $B = 4$ T. The TRFR measurements show a striking feature: the free carrier spin lifetime is a non-monotonic function of carrier density, abruptly increasing between the $n = 0$ and $n = 1 \times 10^{16}$ cm$^{-3}$ samples and steadily decreasing thereafter. The coincidence of the maximum spin lifetime with the metal-insulator transition is not yet understood. However, such behavior appears to be generic since it is also observed in GaN and ZnSe. The inset shows $T_2^*$ vs field, spanning over 2 orders of magnitude and ranging from nearly 10 ns for $n = 1 \times 10^{16}$ to less than 40 ps for $n = 5 \times 10^{18}$. For $n = 1 \times 10^{16}$, the zero-field polarization exhibits virtually no decay in the 1 ns measurement interval, making its lifetime difficult to quantify using the time domain measurements; as described below, RSA measurements can be used to accurately measure these long spin lifetimes and show that the range in lifetimes is over 3 orders of magnitude. In addition to the variation of $T_2^*$ with doping, the data also show a change in the spin precession frequency. This effect is related to an energy dispersion in the GaAs conduction band g factor that manifests itself because of shifts in the absorption edge with doping.

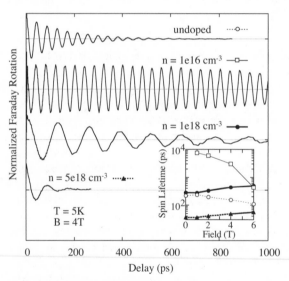

**Fig. 5.4.** TRFR for undoped and n-type GaAs at $B = 4$ T. Data are normalized just after zero pump-probe delay. Plots are offset for clarity, with zeros marked by *dotted lines*. The inset shows $T_2^*$ vs field. Data are taken at $T \sim 5$ K with $N_{ex} \sim 2 \times 10^{14}$ cm$^{-3}$, $2 \times 10^{14}$, $1.4 \times 10^{15}$, $3 \times 10^{15}$ cm$^2$ for $n = 0$, $10^{16}$, $10^{18}$, $5 \times 10^{18}$ cm$^{-3}$, respectively.

The TRFR measurements on these GaAs crystals also show that $T_2^*$ increases with magnetic field for the two highest carrier densities and decreases with field for the two lowest ones. This behavior may be understood by considering processes that introduce a broadening, $\langle \Delta \Phi^2 \rangle / 2$ to the spin direction $\Phi$ [8]. If the resulting distribution is Gaussian, the spin polarization decays according to $\exp - \langle \Delta \Phi^2 \rangle / 2$, and $T_2^{*-1}$ equals the broadening rate $\Gamma$. As discussed in the previous chapter, contributions to $\Gamma$ may arise from spin-orbit scattering during collisions with phonons or impurities (the Elliot-Yafet (EY) mechanism [22]) or from precession about anisotropic internal magnetic fields (the Dyakanov-Perel (DP) mechanism [23]). In the former case, $\Gamma \propto \Gamma_P$, where $\Gamma_P$ is the momentum scattering rate. In the latter case, the broadening accumulates between collisions, so $\Gamma \propto (\Gamma_P)^{-1}$. The application of a field adiabatically scatters electron momentum at the cyclotron frequency $\omega_c$ and suppresses DP relaxation by randomizing the internal field axis between collisions when $\omega_c \approx \Gamma_P$ [24]. The data suggest a DP mechanism for the two highest doping levels. Electron–hole spin scattering, important in p-type GaAs, is not a favored mechanism in these n-type samples because the number of holes injected yields a spin relaxation that is too slow by several orders of magnitude and is independent of $n$ [25]. An additional contribution to the spin relaxation rate may arise from a spread in electronic g factors, $\Delta g$, which results in an inhomogeneous dephasing of $\phi$ given by $\Delta \phi \approx \Delta g \mu_B B t / \hbar$ [26]. However, such a process would lead to an inverse relationship between $T_2^*$ and $B$ which is not observed. Moreover, for $n = 10^{16}$, the measured spin lifetimes at 6 T imply that $\Delta g < 0.005$. Using the dispersion $g = -0.44 + 6.3E$ eV [27], this implies that spin precession occurs within a carrier energy width that is far less than its initial distribution, as determined by the excitation spectrum.

As evidenced in Fig. 5.4, the spin coherence times in n-GaAs can be long enough that time domain measurements of TRFR cannot be used to extract reliable data. This regime is instead best probed using the RSA technique that allows precise measurements of spin lifetimes well in excess of $t_{\mathrm{rep}}$. Figures 5.5a,b show low-field studies of TRFR at fixed $\Delta t$. Dotted lines indicate zero Faraday rotation (theory fits are offset for clarity). While one ordinarily sees almost no field dependence at $\Delta t = 10$ ps, the data comprise a sequence of resonances, individually resolved in Fig. 5.5c, which increase in amplitude and decrease in width as the field approaches zero. This behavior may be understood qualitatively as arising from increased spin lifetimes near $B = 0$, which require an increasing number of successive pump pulses to be in phase. Resonances are consequently narrower in field and concurrently amplified further because more spin polarization survives from each preceding spin excitation.

The RSA measurements can be made quantitative by fitting the observed spin polarization $M_s$ to the sum of exponentially decaying oscillations,

$$M_s(\Delta t, B) = \sum_n \Theta(\Delta t + n t_{\mathrm{rep}}) A e^{-(\Delta t + n t_{\mathrm{rep}})/T_2^*}$$

$$\times \cos[g \mu_B B (\Delta t + n t_{\mathrm{rep}})/\hbar], \tag{5.2}$$

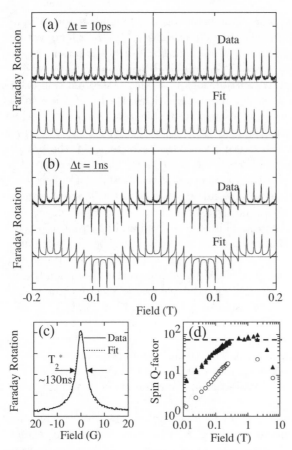

**Fig. 5.5.** TRFR for $n = 10^{16}$ cm$^{-3}$ at small magnetic fields with $\Delta t = 10$ ps (**a**), and 1 ns (**b**). Fits are derived from (5.1) and are offset for clarity. Zeros are indicated by *dotted lines*. (**c**) An expanded view of the central resonance in (**a**). The fit given by the *dotted line* corresponds to a spin lifetime of 130 ns. (**d**) The spin Q factor vs field on a log-log plot. $N_{ex} = 2 \times 10^{14}$ cm$^{-3}$ (*solid triangles*) and $3 \times 10^{15}$ cm$^{-3}$ (*open circles*), respectively. The *dotted line* indicates a thermal ceiling (described in the text). Data are taken at $T = 5$ K.

where $n$ represents a sum over all laser pulses and the step function $\Theta$ ensures that only preceding pump pulses contribute to $M_s$ at any particular delay $\Delta t$. Data are fit by adjusting $T_2^*$, $A$, and $g$ separately for each resonance (except at zero field, where $g$ is interpolated). The results are shown in Figs. 5.5a,b, where excellent agreement for both delays (and over the entire interval from 210 ps to 1 ns) is obtained with identical fitting parameters. Hence, $T_2^*$ may be extracted for each resonance, and is seen to increase by tenfold from 1 to 0 T, reaching 130 ns at zero field [Fig. 5.5c].

Figure 5.5d shows the quality factor ($Q$) of the electron spins in the same sample, computed from the fits shown in Figs. 5.5a,b according to the relation $Q = g\mu_B B T_2^*/\hbar$. Also shown are data for a higher excitation density, which appears to follow the same trend but with a uniformly lower $Q$. Three regimes are found: (a) a low-field region obeying a power law $Q \propto B^{0.8}$, (b) an intermediate regime where $Q$ is independent of $B$, and (c) a high-field region in which $Q$ decreases. Since dephasing of spin direction produces an effective lifetime which is inversely proportional to B, the resulting $Q \propto g/\Delta g$ is field independent. The plateau at $Q \approx 80$ seen in Fig. 5.5d suggests the possibility that spin dephasing limits the value of $Q$ in this region, with a variation of $\Delta g/g = 0.28\%$. Assuming that the electrons occupy an energy bandwidth $kT$, the $g$-factor dispersion used above implies a maximum value of $Q = 88$, shown as a dotted line in Fig. 5.5d.

The temperature dependence of $T_2^*$ helps distinguish between spin scattering processes [26,28]. For $n = 0, 1 \times 10^{18}, 5 \times 10^{18}$, measurements of $T_2^*$ at 4 T reveal only a weak temperature dependence from which the operative relaxation mechanisms are difficult to identify. However, a much more striking temperature dependence is observed for $n = 1 \times 10^{16}$, (Fig. 5.6) for $B = 0$ and $B = 4$ T. For T $\leq 50$ K, the spin lifetime becomes strongly field dependent, splitting into high- and low-field regimes. Dashed lines indicate EY and DP predictions for isotropic charged impurity scattering [29], where $\Gamma_P$ is estimated from the measured mobility as $e/m^*\mu$, and contributions to the electron kinetic energy from doping have been included. The low-field behavior shows good agreement with DP for $T \geq 30K$, below which a weaker temperature dependence $T_2^* \approx (kT)^{-1/2}$ is suggestive of electron-electron scattering. Estimates taking $N = n + N_{ex}$, however, show that this mechanism is actually too strong and may require a more explicit consideration of doping effects [30]. These data support a transition to the EY mechanism below 30 K, accompanied by a strong field dependence which suppresses the high-field spin lifetimes.

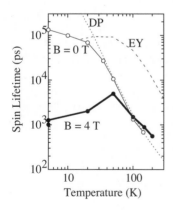

**Fig. 5.6.** Temperature dependence of $T_2^*$ in n-GaAs for $n = 10^{16}$ at $B = 0$ and $B = 4$ T. The excitation density is $N_{ex} = 2 \times 10^{14}$ cm$^{-3}$. *Dashed lines* indicate DP and EY predictions.

The TRKR and RSA techniques have also been employed to initiate and monitor electronic spin coherence in ZnSe [9] and GaN [14] epilayers. Our studies in II-VI materials reveal that these nanosecond spin lifetimes persist to room-temperature. As alluded to above, these effects can be seen in a variety of semiconductors with widely ranging material properties. Specifically, GaN has a large band gap ($\sim 3.54$ eV), high defect density ($\sim 10^8$ cm$^{-3}$), and Wurtzite crystal symmetry. Additionally, and somewhat counter-intuitively given its defect density, GaN has been shown to have both a sharp absorption edge and bright luminescence [31]. This combination of characteristics makes GaN particularly well suited to studies of the importance of defects in determining the spin lifetime, a question of both scientific and technological relevance. The dominant defect in epitaxial GaN is a charged threading dislocation, nucleated at the interface with the lattice-mismatched substrate (typically sapphire or SiC) and extending throughout the thickness of the film [Fig.5.7a] [32]. Transport measurements tracking the free-carrier mobility as a function of the density of these defects has revealed that they act as strong momentum scattering centers [33,34], and as such should serve as a probe of the relevance of spin-orbit scattering in limiting the spin lifetime.

In fact, TRFR data reveal that the spin lifetime in GaN is quite long [Fig. 5.7c] and persists to room temperature [Fig. 5.7b] [14]. While the decay envelope of TRFR in GaAs can be well fit by a double exponential (with the fast component correlating with the carrier lifetime and the slow component with $T_2^*$ [10]), the decay of the TRFR in Fig. 5.7c is comprised of at least three distinct lifetime components. The fastest component can be correlated with the carrier lifetime, but in addition to an intermediate lifetime (ranging from $\sim 100$ ps to 10 ns) there is a long-lived component ($\sim 10$ ns) whose origin is unclear and which depends relatively weakly on temperature and magnetic field. However, if one concentrates on the intermediate lifetime component, which we label $\tau$, striking similarities with GaAs emerge. As can be seen in Fig. 5.8a,b both the field and temperature dependence show not only the onset of field dependence at low temperatures, but also the non-monotonic dependence on carrier density corresponding to the metal-insulator transition found in GaAs. The presence of these long-lifetime effects, and the fact that they bear a striking qualitative similarity to what has been observed in GaAs, argue that the spin scattering mechanism responsible for this behavior in both these systems is the same and further that it is insensitive to momentum scattering in the GaN system. A possible explanation for this insensitivity can be found in the small spin-orbit coupling [35] and short charge diffusion length [33] found in GaN.

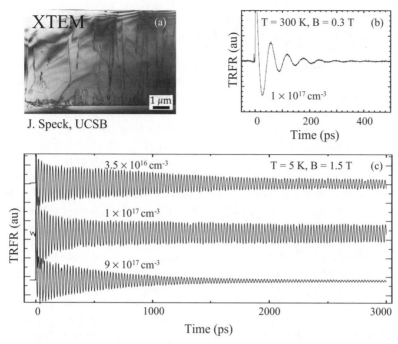

**Fig. 5.7.** (a) Cross-sectional transmission electron microscopy (XTEM) image of an MOCVD GaN epilayer. *Dark vertical lines* are charged threading dislocations. (b) TRFR taken at a magnetic field of 0.3 T and a temperature of 300 K in a sample with $n = 1 \times 10^{17}$ cm$^{-3}$. (c) TRFR measurements in n-doped GaN epilayers at $T = 5$ K and $B = 1.5$ T. Carrier densities, $n$, are room temperature values.

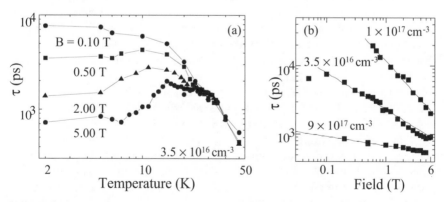

**Fig. 5.8.** (a) Spin scattering time $\tau$ vs. temperature at various fields for a sample with $n = 3.5 \times 10^{16}$ cm$^{-3}$. (b) $\tau$ vs. magnetic field on a log-log plot at $T = 5$ K for all samples. Solid lines are power-law fits to $\tau_2 \propto B^{-\alpha}$, yielding $\alpha = 0.52$, 0.93, and 0.13, as shown.

## 5.4   Electron Spin Coherence
## in Semiconductor Quantum Dots

Research into semiconductor quantum dots (QDs) is driven by their promise in exploring carrier behavior in the mesoscopic regime between bulk and molecular systems, and a variety of technological applications which exploit their size-tunable optical properties [36]. Recent interest in manipulating semiconductor spins for applications ranging from spin-polarized magneto-electronics to quantum computation is based on the ability to control and maintain spin coherence over practical length and time scales. To this end, QDs have been suggested as potential elements for such devices due to control over the structural and electronic environment of localized carriers [37]. In particular, it is expected that the confinement of carriers to nanometer-sized clusters and the attendant quantization of the energy level spectrum will reduce the number and efficiency of operative spin scattering mechanisms, resulting in longer spin lifetimes than comparable bulk semiconductors.

Chemically synthesized nanocrystals are one class of semiconductor QD that can be controllably fabricated with nearly spherical shapes in sizes ranging from 1.5–10 nm in diameter, with relatively narrow size distributions of $\sim 5-10\%$ [38]. For CdSe QDs, this range of size spans the strong-confinement regime where the size of the QD is smaller than the bulk exciton Bohr diameter ($\sim 12$nm). This figure of merit determines the relative importance of quantum confinement and electrostatic effects in calculating electron–hole energy levels. While the Coulomb attraction between electrons and holes is the glue that keeps ordinary excitons together, carriers in strongly-confining QDs are kept together simply by the physical size of the QD itself. As the simplest approximation, an energy level spectrum for the electrons and holes can be computed from the particle in a sphere quantum mechanics problem. While this treatment is a crude approximation, it does qualitatively reproduce some features of the optical absorption in these QDs, like the blueshift in optical absorption energy with decreasing size. Electron–hole pair states can be simply visualized as products of envelope wavefunctions analogous to atomic hydrogen. Fig.5.9 shows a typical linear absorption spectrum from 5.7 nm QDs, with some of the inhomogeneously-broadened energy levels labeled by the small-but-significant size distribution.

Time-resolved Faraday rotation measurements of the spin dynamics in these quantum dots revealed relatively long spin lifetimes that persist up to room temperature [39]. Figure 5.10 shows a comparison of spin precession data taken in 4 nm QDs at $T = 6$ and $T = 280$ K. The two main features of these data are the surprising persistence of nanosecond-scale spin lifetimes up to room temperature, and the multiple frequency precession, yielding two distinct peaks in the Fourier transform spectrum (inset). Such extended spin precession lifetimes at room temperature would be unusual for the corresponding nominally undoped bulk semiconductors, and might be attributable to the increased stability of quantum confined energy levels.

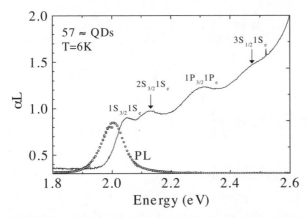

**Fig. 5.9.** Absorption and PL spectra of chemically synthesized CdSe QDs. Inhomogeneously broadened peaks in the absorption spectrum can be assigned to exciton states via calculations of energy position and oscillator strengths.

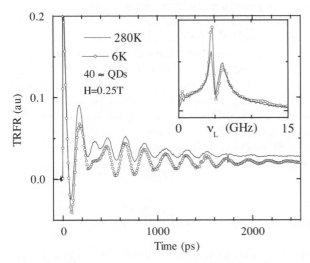

**Fig. 5.10.** Spin precession in 4 nm QDs at $T = 6$ K and $T = 280$ K, showing only a modest decrease in spin lifetime with temperature.

A series of measurements have been made to assign the distinct spin precession signatures to electron (the lower frequency component in Fig. 5.10) and exciton spins (the upper frequency) [40]. The main possibilities to choose from included electron, hole, and exciton spins, each of which can respond to magnetic fields with a characteristic Lande $g$-factor. While all of these species have been seen in higher dimensional semiconductor structures, observation of the latter two was only possible in specially designed samples at low temperatures and excitation powers [41]. In contrast, the spin dynam-

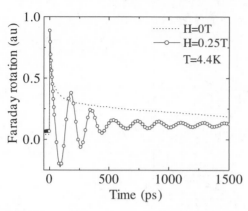

**Fig. 5.11.** Electron spin precession in a semiconductor-doped glass, showing nanosecond spin lifetimes at low temperature.

ics in these quantum dots are remarkably robust against energy relaxation, temperature, and excitation intensity. The observation of coherent spin dynamics in QDs is not limited to only the highest quality samples, but can be generally observed in samples with even the largest distributions in size and composition [50]. This was recently demonstrated in a class of $CdS_xSe_{1-x}$ QDs grown in glass matrices, where quantum size effects and composition have long been exploited to produce commercial optical filters. The sample presented in Fig. 5.11 consisted of $\sim$ 6-10 nm diameter QDs, with a size distribution of $\sim$ 25%, and a mean composition of $x \sim 0.5$. Despite the larger size distribution and possible composition fluctuations, comparable nanosecond scale spin lifetimes (here attributed solely to electrons) were also measured in these QDs.

## 5.5   Coherent Spin Transport in Semiconductors

Thus far in this chapter, we have shown that time-resolved optical techniques permit the measurement of spin coherence in semiconductors and their hetero- and nanostructures. The observation of relatively long spin coherence times in n-doped semiconductors opens up new possibilities in transporting coherent spin information across semiconductor devices. In this section, we describe the first steps towards this goal, demonstrating how the TRKR technique can be modified to measure macroscopic displacement of coherent spin precession in both homogeneous and inhomogeneous systems.

### 5.5.1   Lateral Drag in GaAs

We now discuss how time-resolved optical studies of non-local Faraday rotation in n-type bulk GaAs reveal macroscopic lateral transport of coherently precessing electronic spins over distances exceeding 100 µm. The ability to drag these spin packets by their negative charge, without a substantial increase in spin decoherence, is a consequence of the rather weak coupling between spin coherence and orbital motion in this system. The measurements focus on Si-doped GaAs wafers ($n = 1 \times 10^{16}$ cm$^{-3}$) that are mechanically polished to a thickness of 30 µm and then uniformly contacted on one face with standard AuGeNi except in a $\sim$ 560 µm linear gap where optical studies are performed. The sample is contacted with voltage leads and mounted strain-free in a magneto-optical cryostat for TRFR measurements in the Voigt configuration. The gap orientation induces electric fields along the direction of the magnetic field. The spatio-temporal profile of optically-excited spin polarization is measured using a variation on the pump-probe TRFR technique described earlier. Here, the pump and probe beams can be focused onto spatially distinct regions of the sample; a stepper motor is used to adjust the lateral pump-probe separation, $\Delta x$, along the direction of the in-plane electric field. Hence, measurements of the Faraday rotation of the probe beam map both the time- and spatial-variation of the spin coherence.

Figure 5.12a shows RSA data near zero field in the n-doped sample, obtained by scanning the magnetic field at $\Delta t = 50$ ps. The inset shows the associated temporal behaviour of the spin polarization taken slightly off the B = 0 T resonance, with the offset at zero time delay arising from the injection of new spins on the arrival of the pump pulse. We note that pump-probe pairs reappear at 76 MHz, and past spin injections leave an imprint of negative polarization at $t < 0$. Spatial scans of the pump-probe overlap are taken at a fixed delay with the pump maximally focused, as shown in Fig. 5.12b for a field of B = 0 T. Since there is no precession at this field, spin accumulates from consecutive pump pulses, and the spatial profile at any given delay is broadened because of spin diffusion. The spin injection profile is extracted by taking the difference signal between scans obtained immediately before and after spin injection ($\Delta t = \pm 10$ ps). The upper part of Fig. 5.12b compares the spatial profile obtained by this method (solid line) to the wider profile taken at $\Delta t = -10$ ps (dotted line). The spatial resolution of these measurements is limited to $\sim$ 18 µm by the full-width at half-maximum (FWHM) of the former.

A macroscopic displacement of the electron spin polarization is obtained by the application of an in-plane electric field. The lower portion of Fig. 5.12b shows that an electric field of 16 V cm$^{-1}$ produces a lateral displacement and an asymmetry in the spin distribution. This asymmetry stems from a separation of the zero-field spin resonance into constituent "spin packets" created by distinct pump events. Under the influence of an electric field, packets created at different times drift variable distances that are proportional to their ages. Hence, these spins no longer constructively reinforce each other at the resonance magnetic fields. The data in Fig. 5.12b clearly show that the measured spin polarization is that of free electrons with a drift distance that is

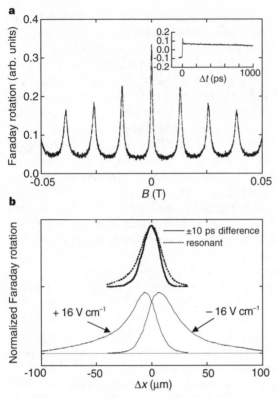

**Fig. 5.12.** (a) RSA measurements for a 30-μm-thick, $1 \times 10^{16}$ cm$^{-3}$ Si-doped GaAs sample. Data are taken with a 100-μm pump diameter and a tightly focused probe. The *inset* shows TRFR versus pump probe delay taken at $B = 0.0030$ T. (b) Spatial scans of the pump-probe interaction with 12-μm FWHM beams, normalized and offset in amplitude for clarity. The *dotted line* is the resonant response at $\Delta t = 0$ ps and $B = 0$. The *thick solid line* shows the unbroadened spin injection profile, obtained by subtracting the corresponding data at $\Delta t = -10$ ps. The lower data are taken at $B = 0$ with an applied electric field $E = 16$ V cm$^{-1}$.

linear in electric field, and that the spins are carried by negative charges. Since there are no indications of any spin polarization travelling opposite to the electron spins, the assumption that hole spins scatter rapidly in these systems appears to be valid. Finally, we note that control measurements on insulating (undoped) samples show no changes in spin profile at fields up to 400 V cm$^{-1}$.

The spin drift can be measured more precisely by varying B at each displacement. We first assume that the total spin response $M$ is the sum of non-interacting spin packets and is given by:

$$M(\Delta x, \Delta t) = \Sigma_n m_n(\Delta x, \Delta t) = \Sigma_n m(\Delta x, \Delta t + n t_{\text{rep}}).  \tag{5.3}$$

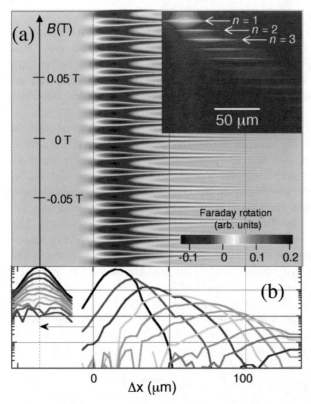

**Fig. 5.13.** (a) Faraday rotation versus magnetic field and displacement. The field range is $-0.1 \rightarrow +0.1$ T and $\Delta x$ values are taken from the axis of (b). The inset shows the spectral power of the $m_n$ pulses over the range $-11$ μm $< x < 137$ μm. The first three pulses are labelled by their index, $n$. (b) Amplitudes of the first ten $m_n$ constituents versus displacement, obtained by explicit fits to the data in (a) and shown on a logarithmic scale. The $m_n$ profiles at zero bias are also shown, displaced $-36$ μm for clarity. Sample and excitation remain unchanged from Fig. 5.12a.

Each constituent of the spin resonance has a different periodicity in the applied magnetic field, given by $[g\mu_B(\Delta t + nt_{\mathrm{rep}})]/\hbar$. Hence, the spatial extent of the various $m_n$ components can be extracted by Fourier decomposition. Figure 5.13 shows the Faraday rotation versus magnetic field profile taken at a displacement of $x = 54$ μm from the injection point and a pump-probe delay of $\Delta t = -10$ ps. An applied electric field of $-37$ V cm$^{-1}$ creates a measurable spin polarization at this lateral position, and harmonic analysis (Fig. 5.13a inset) reveals that the oscillatory behaviour arises from Larmor precession of the third, fourth and fifth most recent pump pulses. Field scans over a range in $\Delta x$ allow the tracking of the spatial position of successive spin packets, as indexed by their injection time. Figure 5.13b shows a two-dimensional assembly of field scans similar to Fig. 5.13a, taken from $\Delta x = -54$ μm to

+137 μm. A narrowing of the spin resonances and an increasing periodicity in field accompanies lateral displacement, reflecting increased pulse ages with increased $x$. Figure 5.13b inset shows the corresponding harmonic power versus position, indicating that pulse ages increase in steps of $t_{rep}$. The data can thus be used to mark the positions of individual spin packets, and indicate that spin drift is linear in time and corresponds to a drift mobility $\mu_d = 3 \times 10^3$ cm$^2$ V$^{-1}$ s$^{-1}$.

A quantitative measure of spin drift and diffusion over a 130-ns interval following spin injection is obtained by fitting the oscillatory data to obtain the amplitude of each spin packet at every position. The profiles of the ten most recent spin injections at zero and $-37$ V cm$^{-1}$ are compared in Fig. 5.13c, where the former are laterally displaced for clarity. To remove artifacts from a strongly field-dependent transverse spin lifetime in the zero-bias sample, these data are obtained by fitting only the $B = 0$ T spin resonance. The logarithmic amplitude scale indicates that the zero-bias spins decay exponentially with a characteristic time $T_2^* = 29$ ns. Fitting the broadening of the spin packet to $\sqrt{D_s(t + t_0)}$, where $t = \Delta t + n t_{rep}$ is the age of the nth pulse and $t_0$ adjusts for its initial width, a spin diffusion constant $D_s$ is obtained that exceeds the electron diffusion constant, $D_e = \mu_d kT/e$, by more than one order of magnitude. As explained in Chap. 4, this surprising discrepancy arises because the transport of spin packets is dominated by conduction electron properties, while the transport of charge packets is dominated by the valence band properties [42]. The significant difference in the effective mass of electrons and holes hence effectively results in a large effective spin diffusion constant.

## 5.5.2   Transport Across Heterointerfaces in ZnSe/GaAs

As we have just seen, time-resolved optical techniques show that coherent spin transport can occur across macroscopic length scales approaching $\sim 100$ μm in a homogeneous semiconductor crystal. Since designs for coherent semiconductor devices are likely to involve more complex semiconductor systems (such as p-n junctions and heterojunction bipolar transistors), it is also important to examine whether the coherent transmission of spin information is possible across heterogeneous systems. We now discuss experiments that provide insights into this problem [12,13]. The system chosen for these measurements (GaAs/ZnSe) is a closely lattice-matched heterostructure ($\frac{\delta a}{a} \approx 0.25\%$) in which the two semiconductors have very different band gaps ($E_g$) and electronic $g$-factors: at 4.2 K, $E_g = 1.5$ eV, $g = -0.44$ for GaAs and $E_g = 2.82$ eV and $g = 1.1$ for ZnSe. As a result, optical pulses resonantly tuned to the very different energy gaps of the two semiconductors may be used to selectively excite spin coherence in one semiconductor and unambiguously measure its time evolution in either of the two different heterojunction constituents. For instance, circularly polarized optical excitation at the GaAs absorption threshold creates spin polarized carriers only in that layer since

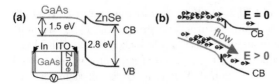

**Fig. 5.14.** (a) Schematic conduction band (CB) and valence band (VB) diagram of the n-GaAs/n-ZnSe heterostructure. The actual offset between the different layers is not known for the samples measured here since the band bending and interface potential depends sensitively on growth details. (b) Schematic representation of the spin transfer in the conduction band with (*bottom*) and without (*top*) bias.

the optical pulses propagate freely through the larger bandgap ZnSe layer. Subsequently, a time-delayed linearly polarized probe resonant with either the GaAs or the ZnSe bandgap interrogates the spin dynamics in that specific layer. An additional feature that aids this experiment is the band alignment between ZnSe and GaAs. Although the conduction band offset at ZnSe/GaAs heterojunctions ($\sim 250 - 750$ meV) is highly sensitive to interface conditions and hence not well known for our structures, it is clear that the band offset is largely taken up by the valence band; hence, electrons created in GaAs easily migrate into ZnSe, while holes created in GaAs are blocked by the sizeable band offset (Fig. 5.14a). Finally, the use of doped GaAs substrates and transparent indium tin oxide contacts on the upper ZnSe surface allows the application of an electric field so that we can study the bias driven spin transport across the heterojunction (Fig. 5.14b).

The ZnSe/GaAs heterostructures used in these studies are grown by molecular beam epitaxy on a variety of GaAs substrates: semi-insulating, n-doped ($n \sim 3 \times 10^{16}$ cm$^{-3}$) and also p-doped ($p \sim$ cm$^{-3}$). ZnSe epilayers are n-doped with Cl and transparently contacted by 40nm of indium-tin-oxide. The epilayers are 100, 150, 200 and 300 nm thick. We note that the critical thickness for misfit dislocation formation in ZnSe on GaAs is $\sim 150$ nm; hence, the structures include heterostructures that have a wide range of defect density. The n-doping in the ZnSe epilayers is in the range $5 - 15 \times 10^{17}$ cm$^{-3}$ and is provided by Cl donors.

The two-color pump-probe optical experiments use pump and probe pulses tuned to different energies. Electron spins oriented along the sample normal are created using a circularly-polarized pump pulse tuned to either the ZnSe or GaAs absorption threshold (1.52 and 2.80 eV, respectively, at 5 K); we will denote these pump pulses as $P_{\mathrm{ZnSe}}$ or $P_{\mathrm{GaAs}}$, accordingly. Whereas primary absorption of $P_{\mathrm{ZnSe}}$ occurs in ZnSe, $P_{\mathrm{GaAs}}$ passes through the ZnSe epilayer unimpeded since it is far below the band gap, exciting spins in the GaAs substrate. The Kerr rotation of a linearly-polarized probe pulse at either the ZnSe absorption threshold then records the dynamics of the normal component of the total ZnSe spin, $S_x(\Delta t)$, where $\Delta t$ is the pump-probe temporal interval. As mentioned earlier, the probed layer selectivity is obtained by tuning its energy within the range 2.70-2.80 eV to obtain maximum Kerr effect from the ZnSe epilayer at each temperature. The probe and $P_{ZnSe}$ are split from the 100 fs output of a frequency-doubled, mode-locked Ti:sapphire laser and focused to a $\sim$ 80μm diameter spot. $P_{GaAs}$ emerges from an additional Ti:sapphire cavity and focuses to $\sim$ 20μm. Whereas $\Delta t$ is set using a conventional delay stage for $P_{ZnSe}$, for $P_{GaAs}$, $\Delta t$ is set with 3 ps resolution by actively synchronizing pulse trains from both Ti:sapphire lasers. Excitation power densities are 750 and 5 W cm$^{-2}$ for $P_{GaAs}$ and $P_{ZnSe}$, respectively.

We first discuss the measurements of spin transfer in the absence of external electric and magnetic fields. Figure 5.15a shows a low-temperature TRKR measurement of spin accumulation in ZnSe originating from $P_{GaAs}$. Data are shown at $B = 0$ T as the pump energy $E_P$ is tuned through the GaAs absorption threshold. The data obtained using $P_{ZnSe}$ are identical to the measurements discussed in the section on measurements of bulk crystals: a signal appears suddenly at $\Delta t = 0$ and then decays exponentially at a rate characteristic of $T^*_{2ZnSe}$. In contrast, pumping in the GaAs layer (with $P_{GaAs}$) and probing in the ZnSe layer produces a ZnSe magnetization $S_x$ that grows over the first few hundred picoseconds and then decays with a spin lifetime appropriate to the ZnSe host. This suggests spin migration from GaAs to ZnSe, a suspicion confirmed by the sudden growth of $S_x$ as $E_P$ is tuned from below to above the GaAs absorption threshold (Fig. 5.15b). The absence of signal below this threshold precludes a contribution from two-photon absorption in ZnSe. Conversely, the probes selectivity in measuring only ZnSe spins is seen in Fig. 5.15c, where $\theta_K$ (and hence $S_x$) is maximized at the ZnSe absorption threshold. An estimate of the fraction of GaAs spins entering ZnSe is 2.5–10% at $B = 0$ T, based on the Kerr response obtained by introducing a known number of spins directly into the ZnSe. This means that most of the optically excited electron spins remain trapped in the GaAs. As we will see later, these spins can be "liberated" using an external electric field that significantly enhances the spin transfer efficiency.

The spin arrival distribution can be estimated by extrapolating the exponential decay of the transferred spins back into the time of their arrival. This is shown in Fig. 5.16a, where the dashed line fits the exponential decay from 2000 ps to 9000 ps and the shaded region marks the discrepancy between this fit and the measured spin profile. Figure 5.16b shows this difference on a log scale for both substrates, where the data are normalized by the extrapolated spin amplitude and equal the fraction of spins yet to cross the interface. The data are well described by $e^{-\Delta t/\tau}$, where $\tau$ is the accumulation time of spin transfer and equals 210 and 440 ps for SI and n-GaAs substrates, respectively. Generally, $\tau^{-1} = \tau_0^{-1} + T^*_{2GaAs}{}^{-1}$, where $\tau_0$ is the spin accumulation time for non-decaying GaAs spins and $T^*_{2GaAs}$ is the substrate transverse spin lifetime. The observed decrease in $\tau$ for insulating substrates may then arise from a drop in $T^*_{2GaAs}$ relative to the doped substrate, or from a doping dependence of band alignment, band bending, and spin diffusion. Presently, the relative

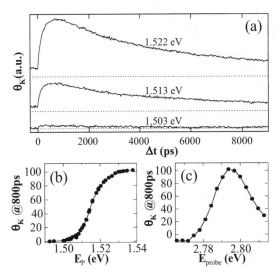

**Fig. 5.15.** (a) TRKR data for spin excitation in the SI GaAs substrate at $T = 5$ K and $B = 0$ T. The *dashed lines* mark $\theta_K = 0$. (b) Pump energy dependence of $\theta_K$ at a fixed delay $\Delta t = 800$ ps using a probe energy of 2.80 eV. (c) Probe energy dependence of $\theta_K$ at $\Delta t = 800$ ps and $E_P = 1.52$ eV.

contributions of electron transport and pure spin diffusion are unknown. In any event, since the spin lifetime in the n-GaAs substrate is nearly three orders of magnitude longer than $\tau$, it appears that thermal relaxation may act to extinguish either process after a few hundred picoseconds. This might occur by reducing the carrier energy in GaAs below an interface potential or by moving spins into low energy states with reduced diffusion coefficients. While the laser bandwidth $\sim 13$ meV provides some initial kinetic energy, one might expect additional contributions from band bending at the interface. Interestingly, no significant changes in $\tau$ are observed as $E_P$ is increased to as much as 40 meV above the GaAs band edge.

We now examine the changes that occur upon the application of a transverse magnetic field B (Fig. 5.16c) that induces coherent spin precession at the Larmor frequency, Fig. 5.15c shows the evolution of $S_x$ at 50 mT, where spin precession results in an oscillatory profile, $S_x(\Delta t) = Ae^{-\Delta t/T_2^*} \cos(\omega\Delta t + \phi)$. For $\Delta t \gg \tau$, the data due to $P_{\text{GaAs}}$ resembles that obtained from $P_{\text{ZnSe}}$. As with the data at $B = 0$ T, we extrapolate this post-transfer behavior to earlier $\Delta t$ (dashed line in Fig. 5.15), and the difference between this extrapolation and the measured data shows a non-zero accumulation time. Significantly, values of $\omega$ and $T_2^*$ determined from the post-transfer fit reflect the $g$-factor and spin lifetime of the ZnSe layer, providing a clear indication of spin transfer. In addition to the coherent spin precession, the TRKR measurements at $B > 0$ also provide insights into the dephasing of the spin system during interlayer spin transport. This dephasing arises from conventional inhomo-

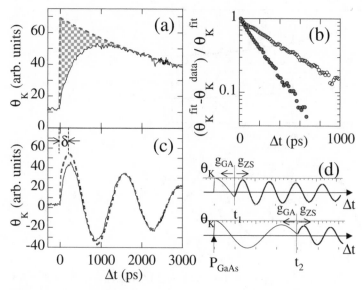

**Fig. 5.16.** (a) TRKR at $B = 0$ T (*solid line*) for spins excited in an n-GaAs substrate. The dashed line shows the fit described within the text. The decay of the shaded amplitude (fit data) gives a measure of the spin accumulation time in ZnSe. (b) A plot of the fraction of spins yet to cross into ZnSe. Both SI (*solid*) and n-doped (*open*) substrates are shown, with $\tau = 210$ ps and 440 ps, respectively. (c) Data is similar to (a) but with $B = 50$ mT. The temporal phase shift, $\delta$, is indicated.

geneous effects as well as from a field-dependent suppression of $T_2^*$. Spin evolution is particularly inhomogeneous during spin transfer because GaAs and ZnSe have different $g$-factors ($g_{GaAs}$ and $g_{ZnSe}$, respectively). Hence, we expect significant spin dephasing phenomena to occur only during spin accumulation. As we will show later in this section, a successful phenomenological model of the coherent spin transfer can be developed to understand the data, by considering inhomogeneous effects arising from differences between $g_{GaAs}$ and $g_{ZnSe}$. In Fig. 5.16c, $\phi < 0$, indicating that the angle of transferred spins is retarded compared to that of spins directly excited within ZnSe ($\phi = 0$). This phase shift shows that the $g$-factor prior to spin transfer is lower than afterwards.

The temperature dependence of spin accumulation provides further insight into the incident carrier kinetics. Figure 5.17a shows that at $B = 0$ T, $\tau$ is roughly constant up to 100 K and decreases sharply thereafter. These changes may reflect a temperature dependence of the interface potential, the substrate spin lifetime, or perhaps an increase in the mean carrier velocity, but more extensive study is necessary to confirm these trends as $\tau$ appears to vary with probe energy. Of significant importance to technology, Fig. 5.17b

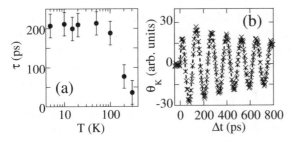

**Fig. 5.17.** (a) Temperature dependence of $\tau$ at $B = 0$ T. (b) TRKR measured (*crosses*) and fit (*dashed line*) at $T = 300$ K and $B = 0.5$ T. All data taken on a SI substrate.

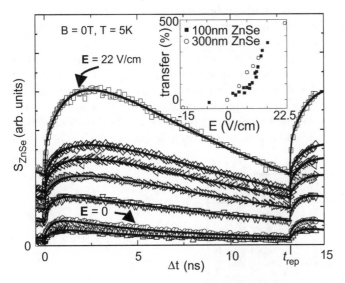

**Fig. 5.18.** Changes in the efficiency of the spin transfer with bias. Time evolution of the spin transfer at different biases with $B = 0$ T. The time interval is long enough to include two successive pump pulses. The 300 nm-thick $n = 1.5 \times 10^{18}$ cm$^{-3}$ ZnSe epilayer was used with $E = -16.5, 0, 4.5, 7.5, 9.8, 12, 22$ V/cm (at higher voltages heating occurred). No offset added. (*Inset*) Percentile change in the spin transfer from the reservoir to the epilayer, calculated using the sum of the three mechanisms amplitudes $(A + B + C)$ as a measure of the spin transfer at different biases, and comparing them to the unbiased case.

shows that the coherent transfer of spins from GaAs to ZnSe occurs even at 300 K, where we observe no significant decrease in the measured signal.

We now examine the effect of an external electric field on the coherent spin transfer, beginning with measurements at $B = 0$ T where bias-dependent changes in net spin transfer are more evident in the absence of spin precession. In Fig. 5.18, time scans show an increase in the amplitude of the spin signal in

ZnSe as the applied bias is increased, while a reverse bias reduces the number of spins that cross the interface. The data also show a striking offset to the spin polarization. This is due to the long spin lifetimes in GaAs that acts as a "reservoir" whose spin population decays only slightly before receiving a boost from the next pump pulse. Hence, the GaAs layer sources a spin current that never reaches zero for $E > 0$. The inset in Fig. 5.18 shows the change in the total spin polarization transferred from the reservoir to the epilayer, where modest electric fields increase spin transfer nearly 500%.

An alternative way in which to witness the remarkable changes brought about by an external electric field is through the RSA spectra measured under different bias conditions. Figure 5.19 shows the RSA spectra for GaAs (top red) and ZnSe (bottom blue) layers, obtained by exciting and detecting spins in the same layer (degenerate pump-probe). Also shown are resonance spectra obtained by exciting in the GaAs layer and detecting spins that have crossed into the ZnSe epilayer (two-color pump-probe) under different bias conditions. Note that as the bias increases, the resonance spectrum transforms from that characteristic of ZnSe to one whose spacing more closely resembles that of GaAs. The data show that an electric field can be used to control the magnetic response of the ZnSe layer during spin injection, and vice-versa. As we will show below, this unusual multi-functional behavior results from concurrent changes in the amplitude and orientation of the spin current itself. These data

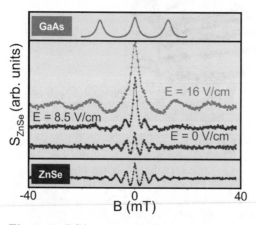

**Fig. 5.19.** RSA scans taken by sweeping the applied magnetic field at a fixed delay of 10 ps at $T = 5$ K in a 100nm-thick $n = 5 \times 10^{17}$ cm$^{-3}$ ZnSe epilayer sample. The *top red* (*bottom blue*) spectrum corresponds to the pump and probe pulses tuned to the GaAs (ZnSe) band gap. For the GaAs spectrum a 400 Wcm$^{-2}$ pump and a 40 Wcm$^{-2}$ probe are used, while the ZnSe spectrum is taken with 80 Wcm$^{-2}$ and 30 Wcm$^{-2}$ for the pump and probe respectively. The measurement of spins transferred from the GaAs substrate to the ZnSe epilayer at different biases is shown in the middle. All the interface transfer data is taken using a 200 Wcm$^{-2}$ 1.52 eV pump and a 30 Wcm$^{-2}$ 2.79 eV probe.

suggest that when $E > 0$, spins trapped in the GaAs spin reservoir are pulled into the ZnSe epilayer, continuously sourcing a spin current whose amplitude is determined by the spin of the GaAs resorvoir. These concepts are made more quantitative by developing a phenomenological model for the measured TRKR in the presence of both external magnetic and electric fields.

Our model for coherent spin transfer assumes that the Larmor angle in each material grows as $\omega \Delta t$, where $\omega$ is now material dependent. It can be shown that the total spin contribution in the ZnSe layer is given by:

$$S_{\text{ZnSe}}(\Delta t) = A \cdot \sum_{n=0}^{\infty} e^{-\frac{\Delta t + n \cdot t_{\text{rep}}}{T^*_{\text{2ZnSe}}}} \cos(\omega_{\text{ZnSe}} \Delta t + \phi)$$

$$-e^{-\frac{\Delta t + n \cdot t_{\text{rep}}}{\tau_{\text{eff}}}} \cos(\omega_{\text{GaAs}} \Delta t + \phi) \tag{5.4}$$

where $\phi = \tan^{-1}(\tau_{\text{eff}}(\omega_{\text{GaAs}} - \omega_{\text{ZnSe}}))$, $T^*_{\text{GaAs}}$ and $T^*_{\text{ZnSe}}$, are the GaAs and ZnSe spin lifetimes, $\omega_{\text{GaAs}}$ and $\omega_{\text{ZnSe}}$ their Larmor frequencies (which contain the $g$-factor explicitly), $t_{\text{rep}}$ is the laser repetition time, $\tau_{\text{eff}}^{-1} = \tau^{-1} + T^*_{\text{GaAs}}^{-1}$ is the effective spin accumulation time, and $\tau$ is the spin accumulation time. This equation is obtained by summing over spins crossing the interface at different times $t_i$ with an exponentially decreasing probability $\propto e^{-t_i/\tau}$. During spin accumulation, the dephasing process phase shifts the spin precession and suppresses the spin amplitude, $A \equiv S \cdot \left(1 + (\tan \phi)^2\right)^{-1/2}$, as the applied magnetic field increases. In contrast to spin dephasing due to local field fluctuations or $g$-factor dispersion, this inhomogeneous process ceases once spin accumulation in the ZnSe is complete ($\Delta t \gg \tau$). We can model the experimental spin precession data by using the functional form for $S_x(\Delta t)$ to extrapolate spin behavior from $B = 0$ T to non-zero fields. A fit to the $B = 0$ T data determines the zero-field values of $S$, $\tau$ and $T^*_2$. Using nominal values of $g_{\text{GaAs}} = -.44$ and $g_{\text{ZeSe}} = 1.1$, determined from spin precession measurements on these systems, (5.4) then predicts how $S_x(\Delta t)$ might evolve as the magnetic field increases (assuming that A, $\tau$ and $T^*_2$ are field-independent). Theoretical plots based upon this model exhibit many qualitative similarities with the actual data, including a decrease in the spin amplitude and a negative phase shift of spin precession. More detailed fits to such data actually reveal the presence of parallel fast and slow transfer mechanisms, A and B, each described by functions $S_A$ and $S_B$ of the form of (5.4) but with amplitudes $A$ and $B$, respectively, and corresponding accumulation times $\tau_{effA} \approx \tau_A \approx 20$ ps and $\tau_{effB} \approx \tau_B \approx 500$ ps (where we used $T^*_{\text{2GaAs}} \gg \tau_{A,B}$). Long spin lifetimes require summation over prior pump pulses as well.

In the presence of an electric field, modeling of the TRKR measurements reveals the onset of a fundamentally new mode of spin transfer. Without an external bias (Figs. 5.20a,b for $E = 0$), spin accumulation in the epilayer lasts only a few hundred ps and can leave a large spin polarization trapped in the reservoir. Once transferred, spins precess about $B$ at a rate determined by the ZnSe $g$-factor, and decay with a time reflecting the ZnSe spin lifetime. In

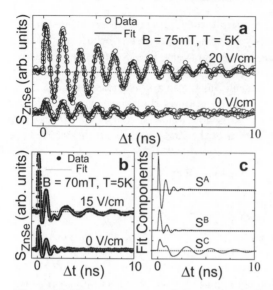

**Fig. 5.20.** Time progression of the spin transfer from the reservoir to the epilayer with and without a bias. An offset is added for clarity; the *dotted lines* mark the zeros. (**a**) 300 nm-thick $n = 1.5 \times 10^{18}$ cm$^{-3}$ ZnSe epilayer. (**b**) 100 nm-thick $n = 5 \times 10^{17}$ cm$^{-3}$ ZnSe epilayer. (**c**) Decomposition of the fit to the biased scan in (**b**) into its three different conduction channels.

epilayers with lifetimes longer than 4 ns, the appearance of a second $g$-factor is revealed by the presence of beats (Fig. 5.20a); in contrast, in epilayers with shorter spin lifetimes (Fig. 5.20b), after a few ns, the spin signal is present only under a biased condition and exhibits a precession rate different from that of ZnSe. A Fourier transform of the biased time scans shows that the second $g$-factor is within 2% of the nominal $g_{GaAs} \approx 0.44$ value for GaAs.

The biased spin transfer data may be understood by adding a persistent spin flow (mechanism C), described by (5.4) with an amplitude $C$ and $\tau_C = \infty$. In this case, spin transfer never turns off, and $\tau_{Ceff} = \tau_C \approx T_{2GaAs}^*$ represents not the spin accumulation time but a decay in the persistent spin current that mirrors spin polarization decay in the reservoir. The total spin signal in the epilayer is therefore described by $S_{ZnSe}(\Delta t) = S^A + S^B + S^C$, shown in Fig. 5.20. Fig. 5.20c shows the dynamics of the three parallel channels contributing to spin injection: the spins that cross over tens of ps (A), over a few hundred ps (B), and the persistent spin current that appears with a bias (C). Fitting variables are the amplitudes $A$, $B$ and $C$ (the latter is taken as non-zero only for positive biases), and the risetimes $\tau_A$ and $\tau_B$. The other values are fixed and obtained from complementary measurements, such as resonance spectra or time scans from degenerate pump-probe arrangements. Note that the persistent spin current contribution is only sensitive to spin that has arrived within a time $T_{2ZnSe}^*$. Hence, for $t > T_2^*$, the epilayer polarization reflects the precession and the lifetime of the spin current itself, which follows the reservoir spin dynamics. This explains why the average spin polarization in the epilayer is characterized by $\omega_{GaAs}$ and $T^*_{2GaAs}$ although spins reside in ZnSe (and each spin has $\omega_{ZnSe}$ and $T^*_{2ZnSe}$).

Similar physics also arises from the built-in interfacial electric fields in a p-n junction, and is demonstrated by coherent spin transfer across a p-GaAs/n-ZnSe heterojunction. As a control, an undoped GaAs substrate is placed together with the p-type substrate in the growth chamber, and 300 nm of n-ZnSe ($n = 1.5 \times 10^{18}$ cm$^{-3}$) is deposited simultaneously on both. A $\sim 4000\%$ increase in spin transfer due to the heterojunction voltage is observed, compared to the control (degenerate pump-probe measurements of the ZnSe epilayers are nearly identical, with a slight change in spin lifetime). We note that unlike the n-doped GaAs spin reservoirs, the spin lifetimes in the p-doped GaAs substrate are extremely short (a few ps) so that persistent spin currents do not explain the observed increase. Instead, the data suggest an enhancement of spontaneous transfer mechanisms similar to A and B, but with a non-exponential spin accumulation profile.

## 5.6   Spin Coherence and Magnetic Resonance

### 5.6.1   Electron Paramagnetic Resonance in II-VI Magnetic Semiconductor Quantum Structures

As we have seen in earlier sections of this chapter, the utility of the TRFR technique in the Voigt geometry lies in its ability to measure induced sample magnetizations that persist long after the injected carriers have recombined. This feature can be exploited for revealing the coherent dynamics of magnetic ions embedded in semiconductor quantum structures [8]. These studies are carried out on quantum wells derived from the II-VI magnetic semiconductors, wherein an sp-d exchange interaction exists between the extended band states of the semiconductor and the localized d-states of the magnetic ions ($Mn^{2+}$) [43]. The principal effect of this exchange interaction is an enhancement of the spin splitting of the conduction and valence band states that may be viewed as an effective enhancement of the electron and hole $g$-factors. As a result, TRFR measurements reveal electron spin precession at THz frequencies rather than at the slower GHz frequency observed in the conventional semiconductor systems described elsewhere in this chapter. The presence of the magnetic ions however leads to rapid spin-flip scattering and hence produces spin decoherence times of only $\sim 10$–$20$ ps. This is shown, for instance, in Fig. 5.21a where we witness the final decay over a time scale of $\sim 20$ ps of the THz electron spin precession in a modulation doped ZnSe/(Zn,Cd,Mn)Se 2DEG sample [44]. We note parenthetically that similar data is observed in insulating magnetic semiconductor quantum well samples.

Suprisingly, TRFR measurements at time scales longer than the electron spin decoherence time reveal a remarkable result: even after the signal from

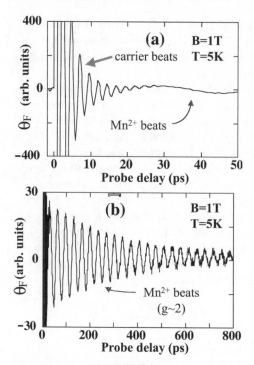

**Fig. 5.21a,b.** TRKR measurements in a "magnetic" 2DEG fabricated by modulation doping of a single ZnSe/(Zn,Cd,Mn)Se quantum well.

electron spin precession disappears, an additional oscillation is observed, persisting for hundreds of picoseconds (Fig. 5.21b), with a period and decay time implying the free induction decay of coherent $Mn^{2+}$ spins. The photoexcited carriers have imparted a net transverse magnetization to the ensemble of local moments, which subsequently and collectively precess at microwave frequencies about the applied field. The signal reverses sign with opposite circular pump, while the precession frequency scales linearly with applied field, is sample and temperature independent (over the range 2–100 K), and corresponds to that expected for $g_{Mn} = 2.0$. The observed decay time of these magnetic oscillations agrees well ($\pm 10\%$) with ESR measurements of the dephasing time $(T_2)_{Mn}$ of $Mn^{2+}$ spins in related bulk magnetic semiconductors [45], confirming the realization of an all-optical time-domain spin-resonance experiment that is easily capable of probing small numbers of spins.

A possible mechanism for initiating the observed spin resonance signal lies in the impulsive coherent rotation of $Mn^{2+}$ moments about the transient exchange field $H_{exch} \parallel \hat{z}$ generated by the hole spins. To first order, we exclude the electrons from consideration since any contribution averages away due to their much faster precession. The sample magnetization $M_{Mn}$, oriented initially along the applied field $H_x$, is rotated away from the x axis upon application of a torque $\parallel \hat{y}$ by the strong exchange field of the holes, which persists for the hole spin relaxation time, acting on each ion. A similar process has been invoked to explain Raman data in magnetic semiconductor quantum wells, where up to 15 Mn spin-flip Stokes lines are observed [46]. After the holes equilibrate, the perturbed Mn moment, which has been rotated by up to a half degree from the $x$-axis in these experiments, then precesses freely. Strong evidence for this mechanism is seen from extrapolating the Mn beats back to zero delay, which shows the oscillations build up sinusoidally as predicted in this model. The effect of the hole exchange field is analogous to the radio-frequency tipping pulses used in NMR studies to initiate free-induction decays in nuclear moments. The variation of $(T_2)_{\mathrm{Mn}}$ with temperature and spin distribution is in qualitative agreement with exchange narrowing models [45], which relate the transverse relaxation time to anisotropic and isotropic spin-spin interactions, as well as to static and dynamic spin-spin correlations.

### 5.6.2   All-Optical Nuclear Magnetic Resonance in Semiconductors

Electron and nuclear spins in a semiconductor are coupled through the hyperfine interaction. Electrons in the conduction band have s-type wave functions, and the average hyperfine energy can be written as $A\langle \boldsymbol{I} \rangle \cdot \boldsymbol{S}$, where $\langle I \rangle$ is the average nuclear spin [24]. Comparing this expression with the electron spin Hamiltonian $H = g\mu_{\mathrm{B}} \boldsymbol{B} \cdot \boldsymbol{S}$, we see that the nuclear polarization acts as an additional magnetic field for the electron spins with nuclear field $B_N = A\langle I \rangle/(g\mu_{\mathrm{B}})$. In particular, for small values of g, the nuclear field can dominate the electron spin dynamics.

Since TRFR directly measures the electron Larmor frequency, it can be employed as a sensitive in situ magnetometer for nuclear spin polarization. In these experiments, dynamic nuclear polarization (DNP) [47] is used to achieve nuclear spin polarization that exceeds the thermal equilibrium values by orders of magnitude. DNP occurs when the electron spin distribution is brought out of equilibrium, such as when excited with circularly polarized light. These non-equilibrium electron spins attempt to thermally equilibrate, in part by flipping nuclear spins, thereby transferring their angular momentum to the nuclear system. Combining DNP and Larmor magnetometry enables optical detection of nuclear magnetic resonance with direct and accurate measurement of nuclear spin polarization at arbitrary magnetic fields. This is in marked contrast to the conventional approach of optically detected NMR, that relies on dc measurements of the photoluminescence polarization reduced by the Hanle effect.

Our all optical NMR technique is also quite different from traditional NMR experiments wherein the nuclear resonance is induced by externally applied radio-frequency magnetic fields. Instead, the all optical technique induces the resonance by using modulated light pulses interacting indirectly with the nuclei through the electron system. Polarization modulation as well

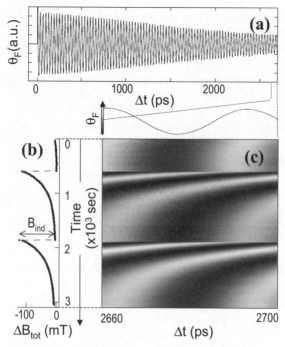

**Fig. 5.22.** (a) TRFR $\theta_F$ in $n \sim 3 \times 10^{16}$ cm$^{-3}$ GaAs at $B = 5.19$ T. (b)Relative magnetic field changes $\Delta B_{\text{tot}}$ versus laboratory time. The sample was translated at 600 and 1860 s. Faraday rotation data used to extract $\Delta B_{\text{tot}}$. (c) A horizontal cut through the grayscale image represents an individual scan of $\theta_F$ versus $\Delta t$, as shown at top (a). (a) through (c) were taken at $T = 5$ K, a photon energy $E_{\text{g}} = 1.50$ eV, and a total pump pulse energy $E_{\text{P}} = 13$ mJ cm$^{-2}$ [15].

as intensity modulation can be used, each of which is responsible for different nuclear spin transitions as discussed below. Resonant depolarization of the nuclear spin is observed through electron Larmor magnetometry. This all-optical NMR technique can be performed in either bulk semiconductors or in nanostructures. Moreover, applying the all-optical scheme in a quantum-confined electronic structure leads to spatial confinement of the nuclear excitation and therefore to localization of all the NMR mechanisms of initial polarization, excitation, and detection.

Figure 5.22 shows how DNP is observed in a bulk GaAs sample through Larmor magnetometry [15]. The sample is an n-doped bulk GaAs, and TRFR is measured repeatedly as a function of laboratory time at $T = 5$ K and $B = 5.19$ T. When the sample is physically translated, TRFR measures a previously unilluminated part of the material, and a reduction in Larmor precession frequency is observed. The nuclear field subsequently builds up within a time frame of $\sim 30$ minutes due to the relatively long nuclear spin-lattice relaxation time.

In this case, electron spins are injected perpendicular to the magnetic field (along the optical beam), resulting in an equal number of up and down spins. At a field of 5.19 T, this distribution differs from thermal equilibrium due to the Zeeman splitting, causing the electron spins to relax, and as a result, a nuclear polarization develops. If electron spins are excited parallel to the magnetic field (Faraday geometry), the system is further away from equilibrium compared to the Voigt geometry. Nuclear polarization can be achieved far more efficiently in the former geometry, but in order to resolve electron spin precession, the latter must be employed. An efficient polarization of nuclear spins in the Voigt geometry is achieved by exploiting an anisotropic $g$-factor in a quantum well (QW).

In a QW, nuclear polarization can be enhanced considerably by tilting the sample with respect to the magnetic field [48,49]. Quantum-confinement in such a structure gives rise to a spin splitting that depends on the direction of applied field. In these systems, the $g$-factor is described by a tensor, rather than a scalar. The anisotropy results in a precession axis tilted away from the applied field, determined by a product $gB$ (Fig. 5.23). The component of the electron spin perpendicular to the precession axis has negligible contribution to nuclear polarization, since it averages out as it precesses. On the other hand, if the precession axis is tilted away from $B$, a considerable component of the electron spins is not precessing even in the Voigt geometry. If the non-precessing part has a component along $B$, this component can induce nuclear polarization.

It is possible to determine all three components of the g-tensor using TRFR. In a [110] oriented QW, we expect the principal axes to be the growth axis [110](z), the [001] direction (y) and the [1$\bar{1}$0] direction (x). If such a sample is mounted so that $B$ is in the $x - z$ plane, the effective $g$-factor is given by $g = \sqrt{g_x^2\cos^2\alpha + g_z^2\sin^2\alpha}$, where $\alpha$ is the angle between $B$ and $\hat{x}$. At temperatures where nuclear polarization is negligible, this model describes the observed TRFR data very well. In addition, analysis of the amplitudes of precessing and non-precessing components reveal that the initial orientation

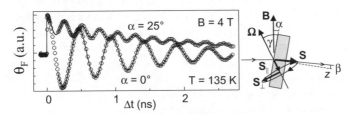

**Fig. 5.23.** (*Left*) TRFR data showing the effect of anisotropic $g$-factor. At large angles $\alpha$, there is a considerable component that is not precessing. (*Right*) Illustration showing the orientation of the precession axis. The injected electron spins precess about an axis whose direction depends on $\alpha$ [49].

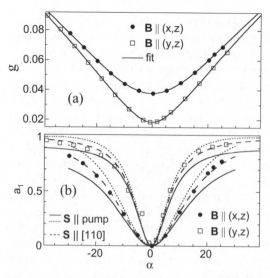

**Fig. 5.24.** (a) The electron g factor vs. $\alpha$ is obtained from the oscillation frequency of fits to TRFR data from a [110] GaAs/AlGaAs QW at 135K. *Filled circles* and *squares* are for the two orientations of the in-plane sample directions. The *solid line* fits the angle dependence of $g$, giving the two components $g_x$ and $g_z$ (*circles*) or $g_y$ and $g_z$ (*squares*). (b) The amplitudes $a_1$ of the non-precessing component, normalized in units of total injected spin, are compared to a geometric model assuming spin injection along the refracted pump beam direction, where same (*solid line*) and opposite (*dotted line*) signs for $g_x$, $g_y$ and $g_z$ are chosen. The *dashed lines* represent data assuming spin orientation strictly along the QW confinement axis [49].

of spins is directed along the growth direction rather than the direction of refracted light (Fig. 5.24).

The enhanced nuclear polarization due to the anisotropic $g$-factor, combined with the small absolute value of $g$ in this sample, results in electron spin dynamics that are dominated by nuclear spins at low temperatures. NMR can easily be detected by TRFR, with a resonance provided by the repetition rate of the mode-locked laser. A resonance condition occurs whenever $\gamma B = (2\pi)/t_{\mathrm{rep}}$, where $\gamma$ is the gyromagnetic ratio of a particular isotope of Ga or As, and $t_{\mathrm{rep}}$ is the repetition rate of the laser. By fixing the time delay and sweeping the field as shown in Fig. 5.25, changes in the nuclear field at resonance result in sharp peaks.

In addition to the resonances at full-field, resonances at half-field are also observed. Such resonances correspond to $\Delta m = 2$ transitions, which cannot be induced by magnetic dipole interaction ($m$ is the angular quantum number of the nuclei). These peaks are due to the interaction of the electric quadrupole moment of nuclear spins with electric field gradients created by the optical excitation of electron–hole pairs. This is experimentally verified by comparing the effect of circularly and linearly polarized pump pulses on the depth of the

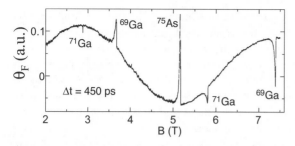

**Fig. 5.25.** All-optical NMR in a single GaAs/AlGaAs QW:. TRFR at $\Delta t = 450$ ps for $\alpha = 5^0$. B was swept from 7.5 to 2 T with 50 mT/min. Five distinct peaks are identified as full-field and forbidden half-field resonances of the different nuclei [48].

resonance. For a full-field resonance, changing the polarization from circular to linear decreases the depth of the resonance significantly, thus indicating that the injected electron spins are important, consistent with the picture of magnetic dipole interactions. On the other hand, for half-field resonances, polarization of the pump beam does not affect the resonance, but tuning the energy below band gap eliminates the peak. This suggests that the excited charge is responsible for the observed resonances. They cannot be explained by the interaction of created electric field gradients with nuclear quadrupole moments. This type of interaction is consistent with the occurrence of half-field resonances.

## 5.7   Coherent Manipulation of Spin in Semiconductors

Thus far in this chapter, we have seen how spin coherence in semiconductors may be prepared, probed and even driven across macroscopic distances using external fields. For true quantum coherent devices to be developed, we also need to construct schemes for the coherent manipulation of spins analogous to those employed in conventional spin resonance (both NMR and ESR) wherein pulsed techniques have evolved to an impressive degree of sophistication in part to isolate particular environmental contributions to nuclear and electron spin dynamics [26]. Such pulses are constructed by controlling the duration and intensity of an AC field $H_1$ applied perpendicular to the static field $H_0$. A canonical pulse sequence consists of a $\pi/2$ pulse to generate a non-equilibrium transverse spin polarization followed by a $\pi$-pulse that may enact a rephasing (spin echo) of transverse spin if inhomogeneous broadening dominates the ensemble spin dynamics. Current technology limits the number of systems to which pulsed-ESR experiments can be applied because the minimum achievable pulse length of $\sim$10ns should be much smaller than the spin coherence time. In order to apply pulse sequences to study conduction-band electron spin dynamics in a variety of semiconductors where spin lifetimes can vary from ∼3 ps [8] to ∼130 ns [10], a complementary ESR technique capable of much shorter pulse widths is desirable. Such an ability may also advance solid-state implementations of quantum computation [37]. In order for quantum algorithms to produce a meaningful improvement in computation speed, they must be rapidly enacted relative to the intrinsic decoherence time of the quantum bit (in this case an electron spin). The current figure-of-merit for the number of operations (on both single and multiple qubits) is ∼ $10^4$, again providing motivation for the development of rapid spin manipulation methods. Here we mention recent experiments that develop optical methods for producing coherent spin rotations on 100fs time scales [51]. The mechanism for this process relies on the generation of an effective magnetic field by a below-bandgap laser pulse through the optical Stark effect [52,53]. When the pulse is circularly polarized, initially degenerate states in the conduction band experience different Stark shifts due to optical selection rules, resulting in meV-scale spin splittings that correspond to effective field strengths of up to 20T. Any net torque between an existing electron spin population and the effective field then leads to an impulsive tip of the electron spin by angles up to ∼ $\pi/2$. Because the Stark shift only lasts for the duration of the laser pulse, the effective magnetic field is turned on for only ∼ 150fs, thus enabling ultrafast coherent control over electron spins in semiconductors. Figure 5.26 shows a schematic of the time-resolved Faraday rotation experiment extended to include a tipping pump (TP) which will act to rotate the electron spins. Because the TP must be tuned below-bandgap, the laser system in such experiments must be capable of producing at least two independently tunable beams, achievable in the visible spectrum with two synchronized optical parametric amplifiers.

In order for the magnetic field generated by the TP ($H_{Stark}$) to coherently rotate the electron spins as they precess about the static field, there must be a nonzero torque given by $\boldsymbol{\tau} \propto \boldsymbol{S} \times \boldsymbol{H}_{Stark}$. This condition is only met when the TP is incident on the sample at a time delay $\Delta_{TP}$ such that $\boldsymbol{S}(\Delta t_{TP}) \| \pm \hat{y}$, a situation that occurs at zero crossings in the detected Faraday rotation signal. Figure 5.27a shows a scan comparing the spin precession taken with and

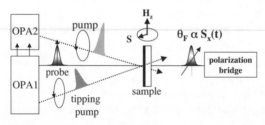

**Fig. 5.26.** Extension of the time-resolved Faraday rotation experiment for ultrafast manipulation of spin coherence. Pump and probe pulses are independently tunable across the visible spectrum, while the probe pulse consists of an ultrafast white light continuum.

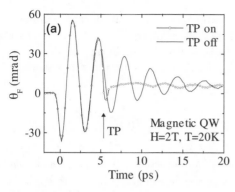
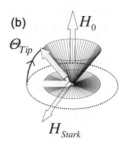

**Fig. 5.27.** (a) $\sim \pi/2$ spin rotations produced by the TP ($\Delta t_{TP} = 5.1$ ps) when positioned at a zero crossing in $\theta_F$. The pump initiates spin precession at $t = 0$ ps, and the probe time delay is scanned. (**b**) Schematic showing action of TP with nonzero torque.

without the TP in a magnetic QW sample [50]. The tipping pulse rotates the spin magnetization out of the $x - y$ plane; precession following the tipping event traces a cone about $H_0$ (Fig. 5.27b), thus giving a reduced value of the measured component $S_x$. A priori, the angle of rotation can be calculated from the relation: $\Theta_{tip} = \arccos(A/A_0)$, where $A$ and $A_0$ are the amplitudes of precession with and without the TP, respectively [50]. Such a calculation for the data in Fig. 5.27 yields a value for $\Theta_{tip}$ that exceeds $\pi/2$ radians, evidenced by the reversal in sign of the oscillations. The net rotation angle has been characterized as a function of tipping pump intensity and detuning, with dependencies qualitatively (but not quantitatively) consistent with expectations from measurements of optical Stark shifts. Measurements using two tipping pumps have established the coherent nature of the process by demonstrating reversible rotations [16]. Future experiments will seek to overcome limitations imposed by the semiconductor absorption linewidth. Stark shifts have been observed in bulk materials, quantum dots, and quantum wells, providing ample opportunity for the development of structures with tailored optical properties.

## 5.8   Spin Coherence in Hybrid Ferromagnet/Semiconductor Heterostructures

In the earlier section, we discussed one approach to the ultrafast coherent manipulation of electron spin coherence in semiconductors. The incorporation of ferromagnetic materials into semiconductor heterostructures creates yet more enticing possibilities for such control by coupling ultrafast techniques with local modifications in the spin environment. For instance, lithographically patterned ferromagnets should produce magnetic fields in nearby semiconductor regions to induce local modifications to the electron spin precession.

Such efforts may obviate the need for external magnetic fields and could lead to a new class of magnetoelectronic devices that rely on the coherent superposition of quantum spin states. In this section, we describe experiments that investigate the effect of ferromagnetic layers on electron spin coherence in adjacent semiconductors.

### 5.8.1  Ferromagnetic Imprinting of Nuclear Spins in Semiconductors

To examine how a ferromagnetic layer may be used to control coherent electron spins in a neighboring GaAs layer, we deposit epitaxial ferromagnetic films onto n-GaAs epilayers. Ultrafast optical pump-probe measurements on these samples reveal that the dynamics of coherent electron spins in the GaAs layer are strongly affected by the ferromagnet, but not through fringe fields or direct exchange interactions. Unexpectedly, the ferromagnet causes nuclear spins in the GaAs layer to become hyperpolarized and align with the magnetization [54]. These polarized nuclei, in turn, generate large effective magnetic fields on the coherent electron spins through the hyperfine interaction (see Sect. 5.6.2), leading to electron spin precession in the GaAs layer that is dominated by interactions with nuclei. Thus, ferromagnetic control of electron spin coherence is achieved by imprinting nuclear spins in the GaAs layer. Four ferromagnet/n-GaAs(100 nm) samples are prepared for this investigation, each with a different ferromagnetic layer: a single layer of Mn delta doping (henceforth termed single layer) [55], digital (Ga,Mn)As [55], (Ga,Mn)As [56], and MnAs [57]. Electron spin precession in the n-GaAs layer is measured at 5 K by TRFR in the Voigt geometry. To determine the relation between the ferromagnet's magnetization and the electron spin precession, TRFR time scans are measured for a series of applied magnetic fields ($B_{\mathrm{app}}$). As shown for the single layer sample, $B_{\mathrm{app}}$ is swept from 1000 G to 1000 G in steps of 10 G (Fig. 5.28a). The time interval between field steps is $\sim 40$ s, the time needed to obtain each TRFR scan. Near 250 G, we observe a sharp change in the Larmor frequency, corresponding to a magnetization reversal as revealed in the magnetic hysteresis loop obtained by magneto-optic Kerr effect (MOKE) measurements (Fig. 5.28c upper panel). Figure 5.28b shows the TRFR as the field sweep is reversed, and the sharp change appears near 250 G.

This measurement procedure is repeated for each of the samples, and the results are summarized in Figs. 5.28c–f. The Larmor frequency $\nu_{\mathrm{L}} = g\mu_{\mathrm{B}}B/h$ is determined by fitting the data at each applied field with (5.1). For all samples, the sharp change in $\nu_{\mathrm{L}}$ occurs when $B_{\mathrm{app}}$ matches the coercive field observed in the magnetic hysteresis loop. This is a clear indication that the ferromagnetic layer strongly influences the electron spin dynamics. The influence of the ferromagnet, however, is not through the expected mechanisms of fringe fields or direct exchange interactions with the magnetic moments.

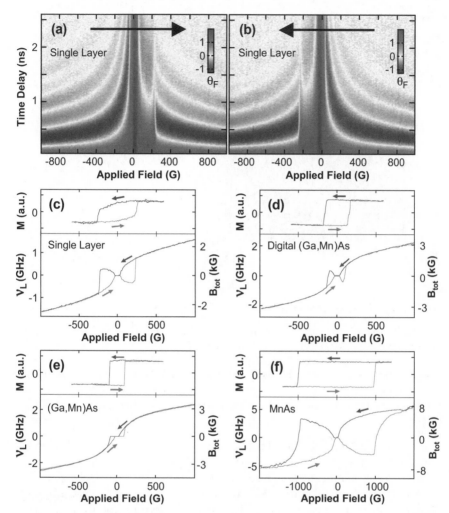

**Fig. 5.28.** (a) and (b) On the single layer sample, sequential TRFR time scans are taken at 5 K as the applied magnetic field is (a) ramped up from −1000 G to 1000 G in 10 G steps, and (b) ramped down from 1000 G to −1000 G. The amplitude of rotation $\theta_F$ is represented by the color. (c) *Top*: Magnetic hysteresis loop of the single layer sample at 5 K, measured by MOKE. *Bottom*: Larmor frequency as a function of field, obtained by fitting the TRFR time scans in (a) and (b) at each applied field. The *red curve* is from the up sweep (a), and the *blue curve* is from the down sweep (b). (d) through (f) MOKE hysteresis loop and Larmor frequency as a function of field at 5 K for the (d) digital (Ga,Mn)As sample, (e) (Ga,Mn)As sample, and (f) MnAs sample. The field step in (d–f) is 20 G. The total field $B_{\mathrm{tot}}$ is defined by the relation $\nu_{\mathrm{L}} = (g\mu_{\mathrm{B}}B_{\mathrm{tot}}/h)$.

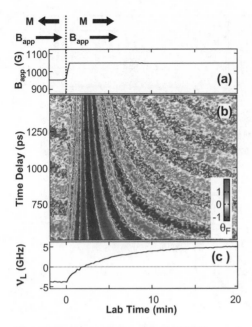

**Fig. 5.29.** Sequential scans of TRFR as a function of lab time after switching the magnetization at ∼1000 G. (**a**) The magnetic field $B_{\mathrm{app}}$ is ramped to induce a magnetization reversal. (**b**) The TRFR shows a continuous evolution for several minutes after magnetization reversal. (**c**) The Larmor frequency (obtained by fitting the TRFR data) shows a zero-crossing at $t_{\mathrm{L}} \sim 2$ minutes after switching, demonstrating that the nuclear polarization tracks the magnetization direction.

These mechanisms predict local fields that are proportional to the magnetization and therefore do not account for the disappearance of $\nu_{\mathrm{L}}$ at $B_{\mathrm{app}} = 0$ G, where there is significant remanent magnetization.

An alternative explanation is that the nuclear spins in the n-GaAs layer become polarized along the magnetization of the ferromagnetic layer. These polarized nuclei, in turn, generate effective magnetic fields $B_n \sim \langle I \rangle$ (via hyperfine interaction), where $\langle I \rangle$ is the average nuclear spin (see Sect. 5.6.2). Since the precession frequency is given by:

$$\nu_{\mathrm{L}} = (g\mu B/h)(B_{\mathrm{app}} + B_n), \tag{5.5}$$

the disappearance of $\nu_{\mathrm{L}}$ at $B_{\mathrm{app}} = 0$ G can be attributed to the depolarization of nuclear spins at zero field due to dipole-dipole interactions [58]. An intuitive understanding of the behavior of the nuclear spins is obtained by measuring their response to a magnetization reversal. The MnAs sample is prepared by saturating the magnetization at 2000 G and subsequently ramping the field to +950 G, which is just below the coercivity of ∼ 1000 G (Fig. 5.28f). At lab time $t_{\mathrm{L}} = 0$, the field is then quickly ramped to +1050 G and the magnetization quickly reverses (< 1s) to align with the applied field

(Fig. 5.29). Starting at $t_L = -2$ min, we take repeated TRFR time scans to obtain $\nu_L$ as a function of lab time (Fig. 5.29b,c). After the magnetization reversal, $\nu_L$ shows a continuous evolution over the next several minutes. This lab time dependence is characteristic of nuclear polarization (Sect. 5.6.2) and cannot be explained by fringe fields or direct exchange interactions. By interpreting $\nu_L$ as a magnetometer of the nuclear polarization, we find that $\langle I \rangle$ tracks the direction of the magnetization and requires $\sim 20$ min to reach steady-state (here, $\langle I \rangle / I \sim 13\%$) [58]). The fact that $\langle I \rangle$ tracks the magnetization explains why $\nu_L$ changes sign when $B_{app}$ crosses the coercive field of the ferromagnet (Fig. 5.28c–f).

Direct evidence that the GaAs nuclei are polarized is obtained through all-optical NMR, as discussed in Sect. 5.6.2. A photoelastic modulator (PEM) alternates the helicity of the pump beam at a given fundamental frequency (f = 40, 50, or 55 kHz), resulting in an effective ac-magnetic field through the hyperfine interaction. The nuclear spins depolarize when the resonance condition $\gamma B_{app} = nf$ is met, where $\gamma$ is the isotope-specific gyromagnetic ratio, and $n$ is an integer labeling the harmonics of the PEM modulation. On the single layer sample, we measure the Faraday rotation as a function of applied field for each PEM frequency at $\Delta t = 1500$ ps (Fig. 5.30). We observe peaks that shift with PEM frequency as expected from the resonance condition and can be associated with the isotopes $^{69}$Ga, $^{71}$Ga, and $^{75}$As, confirming that the nuclei are polarized.

In addition to the Mn-based samples discussed so far, investigations of ferromagnetic imprinting of nuclei have also been carried out with hybrid

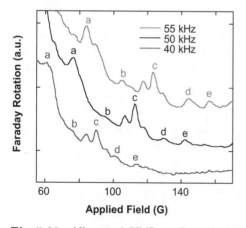

**Fig. 5.30.** All-optical NMR performed on the single layer sample at 5 K. The Faraday rotation is measured as a function of field at $\Delta t = 1500$ ps for three different fundamental frequencies of the PEM. The curves are vertically offset and scaled for clarity. Resonance peaks $b$ and $e$ are identified as the 2nd and 3rd harmonics of $^{69}$Ga, $a$ and $c$ are identified as the 2nd and 3rd harmonics of $^{71}$Ga, and $d$ is identified as the 2nd harmonic of $^{75}$As.

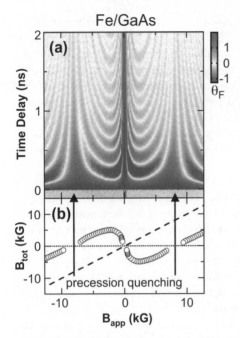

**Fig. 5.31.** (a) As in Fig. 5.28, TRFR time scans as a function of applied field for Fe/GaAs at $T = 5$ K. Spin precession quenching occurs at $\pm 8.5$ kG. (b) The total field $B_{tot}$ [obtained from data in (a)] vs. $B_{app}$ exhibits zero-crossings at $\pm 8.5$ kG due to cancellation of nuclear and applied fields.

Fe/GaAs structures. Remarkably, the nuclear spin polarization is of opposite sign in Fe/GaAs as compared to the Mn-based structures. That is, the nuclear field $B_n$ is generated along $B_{app}$ in the opposite direction to the magnetization. In fact, the nuclear field can be so strong as to cause the electron spin precession to reverse direction (i.e. $B_{tot}$ and $B_{app}$ have opposite signs). When the special condition is met where the nuclear and applied fields are equal in magnitude and opposite in sign ($B_{app} \sim 10$ kG), the spin precession is completely quenched (Fig. 5.31). Because the Fe and Mn-based ferromagnets generate nuclear fields of opposite sign, ferromagnetic imprinting cannot be explained by interactions of the nuclei with real magnetic fields generated by the ferromagnet. Instead, the following studies have revealed that the ferromagnet-nuclear coupling is mediated by spontaneous electron spin coherence.

### 5.8.2 Spontaneous Electron Spin Coherence in n-GaAs Produced by Ferromagnetic Proximity Polarization

Efforts to investigate the origin of ferromagnetic imprinting of nuclear spins have led to the discovery of spontaneous electron spin coherence via ferro-

magnetic proximity polarization. In this process, linearly-polarized optical
excitation in a semiconductor leads to electron spin coherence due to the
proximity of a ferromagnetic layer [59]. Since linearly-polarized light has zero
angular momentum, the ferromagnetic layer imparts angular momentum to
the electron spin system.

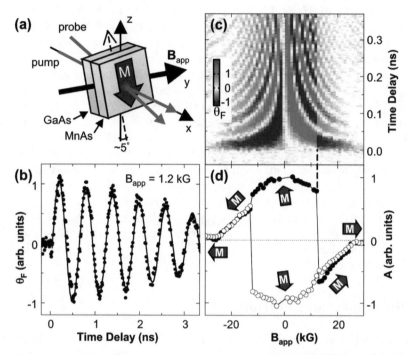

**Fig. 5.32.** Spontaneous spin coherence in the GaAs layer produced by the MnAs
layer (data taken with a linearly polarized pump). (**a**) Experimental geometry,
where M represents the magnetization direction, which aligns along the easy magnetization axis (*dashed arrow*) at low fields. (**b**) TRFR time scan for $B_{app} = 1.2$ kG
showing coherent spin precession. (**c**) TRFR time scans vs. $B_{app}$ featuring magnetization switching at $\sim 12$ kG. (**d**) TRFR amplitude A obtained from fits to data
in (**c**) demonstrating that the spontaneous polarization in the GaAs follows the
hysteresis of the ferromagnetic layer.

This effect is observed with TRFR using linearly-polarized pump and
probe beams in a Voigt geometry (Fig. 5.32a) with the sample magnetization
roughly perpendicular to the applied field. This unusual magnetization alignment is enabled by a strong in-plane uniaxial magnetic anisotropy for MnAs,
which pins the magnetization along a particular crystallographic axis at low
applied fields. The TRFR time scan at $B_{app} = 1.2$ kG exhibits spin precession
(Fig. 5.32b), indicating that electron spin coherence in GaAs is generated by
the linearly-polarized optical excitation near the ferromagnetic interface. The

field dependence of the spin precession (Fig. 5.32c,d) shows that the initial orientation of electron spin coherence tracks the ferromagnet's magnetization. Detailed measurements indicate that the spin coherence is established within 50 ps and possibly during the optical excitation. Spontaneous spin coherence is also observed for the Fe/GaAs samples, but the relative orientation of the electron spins and magnetization $M$ is opposite from the MnAs case: electron spins align parallel to $M$ for Fe, and antiparallel to $M$ for MnAs [59]. For both materials, the spin polarization reaches 1/3 of that due to circularly-polarized excitation, and the spin coherence lifetimes are several nanoseconds at low temperatures.

The connection between spontaneous spin coherence and ferromagnetic imprinting of nuclear spins is through the process of dynamic nuclear polarization (Sect. 5.6.2). In the traditional TRFR geometry, electron spins in GaAs become polarized by the ferromagnet (spontaneous spin coherence) and subsequently impart their spin angular momentum to the nuclear lattice during spin-relaxation. This model of electron spins mediating the ferromagnetic imprinting of nuclear spins is supported by the contrasting behaviors of Fe/GaAs and MnAs/GaAs: the opposite electron spin polarizations for these two structures explains the opposite nuclear polarizations observed.

While the physical origin of the spontaneous electron spin coherence is currently unclear, the opposite sign of polarization generated by these two materials provides flexibility in orienting regions of electron and nuclear spin in a semiconductor.

## 5.9   Summary and Outlook

In conclusion, we have shown how optical measurements prove to be a powerful tool for investigating electron and nuclear spin dynamics in semiconductors and nanostructures. Looking towards the future, another degree of freedom for optoelectronic devices is direct electrical control of spins in spin-engineered structures. A recent result demonstrates that electrical control of coherent electron spin precession is possible in a specially designed single $Al_x Ga_{1-x}As$ QW [60]. Exploiting the fact that the electron $g$-factor varies strongly with $x$ and changes sign (in bulk $Al_x Ga_{1-x}As$, $g = -0.44$ for $x = 0$ and $g = 0.40$ for $x = 0.3$) [61], the electron $g$-factor is continuously tuned by displacing an electron wave function. A parabolically-graded Al concentration is used for the structure design, since an applied bias does not change the shape of the potential, but is able to displace the minimum of the effective potential to a region with higher aluminum content. This allows for a continuous displacement of the complete electron wave function from one material to another without distortion, in contrast to a square quantum well where only the tail of the wave function is pushed into the barrier. As shown in Fig. 5.33, TRKR measurements in a series of spin-engineered structures show that gate-voltage mediated control of coherent spin precession is ob-

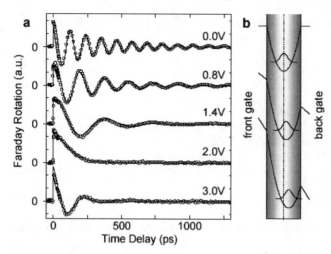

**Fig. 5.33a,b.** Voltage-controlled spin coherence. (**a**) TRKR measurements of the electron spin precession in the QW with minimum Al concentration of 7% at 5 K and $B = 6$ T. As a positive voltage $U_g$ is applied between back and front gate, the electron wave function is displaced towards the back gate into regions with more Al concentration [schematic in (**b**)], leading to an increase of $g$. At $U_g = 2$ V, no precession is observed, corresponding to $g = 0$. Circles are data points, and the solid lines are fits to the data [60].

tained over a 13 GHz frequency range in a fixed magnetic field of 6 T, including complete suppression of precession and a reversal of the sign of $g$. Such manipulations are possible up to room temperature.

We have discussed the detection of electron spin coherence in nanoscale structures, transport of spin-polarized coherent carrier populations across relevant length scales and heterointerfaces, and the manipulation of both electron and nuclear spins on unexpectedly fast time scales. Moreover, recent experiments suggest that the storage time of quantum information encoded in electron spins may be extended through their strong interplay with nuclear spins in the solid state. In the coming years, optical methods for spin injection, detection, and manipulation may also be developed that exploit the ability to precisely engineer the coupling between electron spin and optical photons. We envision that the merging of electronics, photonics, and magnetics will ultimately lead to new spin-based multifunctional devices. The success of these ventures depends on a deeper understanding of fundamental spin interactions in solid state materials as well as the roles of dimensionality, defects, and semiconductor band structure in modifying these dynamics.

## Acknowledgments

We thank R.J. Epstein, J.A. Gupta, E. Johnston-Halperin, Y. Kato, R.K. Kawakami, M. Poggio, G. Salis and G. Steeves for their help in the prepartion of this chapter, as well as the AFOSR, DARPA, NSF, ONR for their financial support.

# References

1. S. Wolf et al., Science **294**, 1488 (2001).
2. D. D. Awschalom and J. M. Kikkawa, Phys. Today **52**, 33 (1999).
3. D. Divincenzo, Science **270**, 255 (1995).
4. W. H. Lau, J. T. Olesberg, and M. E. Flatté, Phys. Rev. B (2001).
5. D. D. Awschalom, J. M. Halbout, S. von Molnar, T. Siegrist, F. Holtzberg, Phys. Rev. Lett. **55**, 1128 (1985).
6. J.J. Baumberg, D.D. Awschalom, N. Samarth, H. Luo, J.K. Furdyna, Phys. Rev.Lett. **72**, 717 (1994).
7. T. Ostreich, K. Schonhammer and L. J. Sham, Phys. Rev. Lett. **75**, 2554 (1995).
8. S. A. Crooker, D. D. Awschalom, J. J. Baumberg, F. Flack, and N. Samarth, Phys. Rev. B **56**, 7574 (1997); S. A. Crooker, J. J. Baumberg, F. Flack, N. Samarth, and D. D. Awschalom, Phys. Rev. Lett. (1996).
9. J. M. Kikkawa, I. P. Smorchkova, N. Samarth, and D. D. Awschalom, Science **277**, 1284 (1997).
10. J. M. Kikkawa and D. D. Awschalom, Phys. Rev. Lett. **80**, 4313 (1998).
11. J. M. Kikkawa and D. D. Awschalom, Nature (London) **397**, 139 (1999).
12. I. Malajovich, J. M. Kikkawa, D. D. Awschalom, J. J. Berry, and N. Samarth, Phys. Rev. Lett. **84**, 1015 (2000).
13. I. Malajovich, J. J. Berry, N. Samarth, and D. D. Awschalom, Nature (London) **411**, 770 (2001).
14. B. Beschoten, E. Johnston-Halperin, D.K. Young, M. Poggio, J.E. Grimaldi, S. Keller, S.P. DenBaars, U.K. Mishra, E.L. Hu, and D.D. Awschalom, Phys. Rev. B 63, R121202 (2001).
15. J. M. Kikkawa and D. D. Awschalom, Science **287**, 473 (2000).
16. J. A. Gupta, R. Knobel, N. Samarth, and D. D. Awschalom, Science **292**, 2458 (2001).
17. A.P. Heberle, J.J. Baumberg, K. Kohler., Phys. Rev. Lett. **75**, 2598 (1995).
18. S. Bar-Ad and I. Bar-Joseph, Phys. Rev. Lett. **68**, 349 (1992).
19. R.M. Hannak, M. Oestreich, A.P. Heberle, W.W. Ruhle, K. Kohler, Solid State Comm. **93**, 313 (1995).
20. M. Oestreich and W.W. Ruhle, Phys. Rev. Lett. **74**, 2315 (1995); M. Oestreich, et al., Phys. Rev. B **53**, 7911 (1996).
21. T. Amand, et al., Phys. Rev. Lett. **78**, 1355 (1997).
22. R. J. Elliot, Phys. Rev. **96**, 266 (1954).
23. M. I. D'yakonov and V. I. Perel', Sov. Phys. JETP **33**, 1053 (1971); Sov. Phys. Solid State **13**, 3023 (1972).
24. *Optical Orientation, Modern Problems in Condensed Matter Science*, edited by F. Meier and B. P. Zachachrenya (North-Holland, Amsterdam, 1984), Vol. 8.
25. G. Bir, A. Aronov, and G. Pikus, Zh. Eksp. Teor. Fiz. **69**, 1382 (1975) [Sov. Phys. JETP **42**, 705 (1976)].
26. A. Abragam, *The Principles of Nuclear Magnetism* (Clarendon, Oxford, 1961).
27. M. J. Yang et al., Phys. Rev. B **47**, 6807 (1993).

28. G. Fishman and G. Lampel, Phys. Rev. B **16**, 820 (1977); K. Zerrouati et al., Phys. Rev. B **37**, 1334 (1988).
29. A. G. Aronov, G. E. Pikus, and A. N. Titkov, Zh. Eksp. Teor. Fiz. **84**, 1170 (1983) [Sov. Phys. JETP **57**, 680 (1983)].
30. P. Boguslawski, Solid State Commun. **33**, 389 (1980).
31. S. Nakamura, Science 281, 956 (1998); S. F. Chichibu et al., Appl. Phys. Lett. **74**, 1460 (1999).
32. B. Heying, et al., Appl. Phys. Lett. **68**, 643 (1996); P.J. Hansen, et al., Appl. Phys. Lett. **72**, 2247 (1998).
33. D. C. Look and J. R. Sizelove, Phys. Rev. Lett. **82**, 1237 (1999).
34. H. M. Ng et al., Appl. Phys. Lett. 73, 821 (1998); N.G. Weinmann et al., J. Appl. Phys. 83, 3656 (1998).
35. W. E. Carlos, J. A. Freitas Jr., M. Asif Kahn, D. T. Olson, and J. N. Kuzina, Phys. Rev. B **48**, 17878 (1993).
36. A. P. Alivisatos, Science **271**, 933 (1996).
37. D. Loss and D. P. DiVincenzo, Phys. Rev. A **57**, 120 (1998).
38. C. B. Murray, D. J. Norris and M. G. Bawendi, J. Am. Chem. Soc. **115**, 8706 (1993).
39. J.A. Gupta, X.Peng, A.P. Alivisatos and D.D. Awschalom, Phys. Rev. B **59**, R10421 (1999).
40. X. Marie et al., Phys. Rev. B **60**, 5811 (1999).
41. J.A. Gupta, Al.L. Efros and D.D. Awschalom, in preparation.
42. M. E. Flatté and J. M. Byers, Phys. Rev. Lett. **84**, 4220 (2000).
43. D. D. Awschalom and N. Samarth, J. Mag. Magn. Mater. **200** (1999).
44. J. M. Kikkawa, I. P. Smorchkova, N, Samarth, and D. D. Awschalom, Physica E **2**, 394 (1998).
45. N. Samarth and J. K. Furdyna, Phys. Rev. B **37**, 9227 (1988); S. Rajagopalan, Ph. D. Thesis, Purdue University (1988).
46. J. Stühler et al., Phys. Rev. Lett. **74**, 2567 (1995).
47. G. Lampel, Phys. Rev. Lett. 20, 491 (1968).
48. G Salis et al., Phys. Rev. Lett. 86, 2677 (2001)
49. G. Salis et al., Phys. Rev. B 64, 195304 (2001).
50. J.A. Gupta and D.D. Awschalom, Phys. Rev. B **63**, 085303 (2001).
51. J. Preskill, quant-ph/9712048 at http://xxx.lanl.gov (1997).
52. C. Cohen-Tannoudji and J. Dupont-Roc, Phys. Rev. A **5**, 968 (1972).
53. M. Rosatzin, D. Suter, and J. Mlynek, Phys. Rev. A **42**, 1839 (1990).
54. R. K. Kawakami et al., Science **294**, 131 (2001).
55. R. K. Kawakami et al., Appl. Phys. Lett. **77**, 2379 (2000).
56. H. Ohno, Science **281**, 951 (1998).
57. M. Tanaka et al., Appl. Phys. Lett. **65**, 1964 (1994).
58. D. Paget, G. Lampel, B. Sapoval, and V. I. Safarov, Phys. Rev. B **15**, 5780 (1977).
59. R. J. Epstein et al., Phys. Rev. B **65**, 121202 (2002).
60. G. Salis et al., Nature (London) **414**, 619 (2001).
61. C. Weisbuch and C. Hermann, Phys. Rev. B **15**, 816 (1977).

# 6 Spin Condensates in Semiconductor Microcavities

Jeremy J. Baumberg

## 6.1 Introduction

Direct-gap semiconductors interact extremely strongly with light, absorbing energy in the promotion of electrons into the conduction band. The lifetime of the photoexcited electrons is several nanoseconds, set by competing processes of radiative and non-radiative recombination. Of increasing interest in the last decade, is the *phase* of the photoexcited electrons (or the *interband coherence*) induced by the oscillating optical field (Fig. 6.1). The time for this phase memory to be lost is much shorter than the carrier lifetime, typically less than 100 fs in bulk materials at room temperature, and is controlled by the range of possible phase scattering events accessible to the carriers. By freezing out the lattice vibrations at low temperatures, and quantum confining the carriers in volumes smaller than their de Broglie wavelengths, it is possible to reduce the phase scattering. Such confinement produces quasi-atomic energy levels whose separation restricts the events that can cause phase scattering. However even in fully-confining semiconductor quantum dots at liquid helium temperatures (see Chap. 9), the phase decay is only slowed by a factor of 200 [1]. Thus although such quantum dot systems have been suggested as all-solid-state elements for quantum computing applications, they are still prone to dephasing events which cause errors.

In this chapter we will explore the potential of an alternative semiconductor quantum system that displays features capable of enhancing electronic

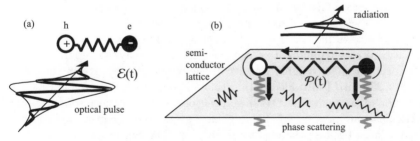

**Fig. 6.1.** (a) Optical pulse incident on a semiconductor creates an interband polarization $P(t)$ of Coloumb-bound electrons and holes which can (b) decay by reradiation, or phase scattering with a material excitation

phase memories. Although superficially similar to conventional semiconductor heterostructures, we will show how this system can produce radically different properties, one of which is a phase lifetime in excess of many nanoseconds. The extra ingredient employed is the *strong* coupling of light and matter, which can lead to peculiar quantum behaviours, as in quantum electrodynamics atom-cavity experiments such as micromasers [2,3]. In semiconductors, strong coupling mixes the photon and electronic excitations so completely that it becomes impossible to distinguish them. Instead quasiparticles built from their mixture become the new basis states, and their properties can differ considerably from those of their components. Thus it is possible to induce new properties for photoexcited electronic states in a semiconductor which open up alternative routes for coherent control and quantum coherent technologies. In Sect. 2 we will outline the origin of the strong coupling regime, and discuss the properties of these quasiparticles termed *polaritons*. An exhaustive review of polariton properties is beyond the scope of this chapter and the reader is referred to recent reviews [4,5]. Experimental evidence recently reported for the interactions between these quasiparticles is summarized in Sect. 3. This leads to a discussion in Sect. 4 of the properties of such quasiparticles at high densities, and predicts a new type of quasiparticle condensate analogous to Bose-Einstein atomic condensates. We then set out the evidence for this polariton condensate, and discuss how it might be usefully manipulated.

## 6.2    Polariton Properties

### 6.2.1    Strongly Coupled Microcavity Dispersion

When atoms emit light in a high-finesse cavity which is tuned to the same frequency as the atomic transition, the emission process changes its character. The enhanced electric field can stimulate emission even with only one photon in the cavity, however the photon has a strong probability of being reabsorbed. Energy can cycle repeatedly between atom and cavity field, at a rate known as the Rabi frequency which is determined by the atomic oscillator strength. With the advance of semiconductor epitaxy, it has become possible to devise monolithic solid samples which exhibit the same cyclic interplay between light and matter. Typical samples consist of two high reflectivity mirrors (composed of semiconductor multilayers) separated by a cavity of a thickness roughly the size of the optical wavelength in the material ($d \sim 250\,\text{nm}$). Embedded within this microcavity are a number of semiconductor quantum wells (QWs) at positions at which the resonant cavity electric fields are maximised. The oscillator strength of quantum wells is particularly large because the electron-hole pairs that can be photoexcited are bound together by the Coulomb interaction into excitons, which are further stabilised by the confined 2D planar geometry. In the samples we discuss, it is these excitons which dominate the optical response near the bandgap

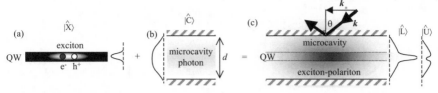

**Fig. 6.2.** Schematic wavefunctions of **(a)** exciton in a quantum well (QW), thickness $\simeq 10$ nm **(b)** cavity photon in a microcavity, thickness $\simeq 250$ nm **(c)** microcavity polariton. Light impinging at angle of incidence $\theta$ has in-plane momentum, $k_\parallel$

of the semiconductor. The most important condition engineered into these devices is that the emission wavelength of the excitons matches the resonant wavelength of the microcavity. At low temperatures, such a sample (shown schematically in Fig. 6.2) allows the cycling between light and exciton to become faster than both the phase scattering of the individual excitons and the photon decay from the cavity, and hence the strong coupling regime is attained.

The new quasiparticles, called 'exciton-polaritons' (henceforth abbreviated as polaritons), are mixtures of the cavity photon $|\hat{C}\rangle$, and exciton $|\hat{X}\rangle$ bases, given by

$$\begin{bmatrix} \hbar\omega_x & \hbar\Omega \\ \hbar\Omega & \hbar\omega_c \end{bmatrix} \begin{bmatrix} |\hat{X}\rangle \\ |\hat{C}\rangle \end{bmatrix} = E \begin{bmatrix} |\hat{X}\rangle \\ |\hat{C}\rangle \end{bmatrix} \tag{6.1}$$

The Rabi coupling strength, $\Omega$, produces an anticrossing between the energies of cavity photon $\omega_c$ and exciton $\omega_x = \omega_0 + \frac{\hbar k^2}{2m_x}$ (where the minimum exciton energy $\omega_0$ is several meV binding energy below the semiconductor band gap). At normal incidence, the detuning of the cavity from the exciton, $\Delta = \omega_c - \omega_x = 0$, and the two polariton branches (upper and lower) have equal admixtures of light and matter, $|\hat{L}\rangle = \frac{1}{\sqrt{2}}(|X\rangle + |C\rangle)$, $|\hat{U}\rangle = \frac{1}{\sqrt{2}}(|X\rangle - |C\rangle)$. It is important to note that this mixing changes as the angle of incidence onto the planar heterostructure changes, which corresponds to increasing the in-plane momentum, $k_\parallel = \omega \sin\theta/c$. The origin of this angular dispersion is the increase of cavity resonant frequency at increasing angles of incidence, $\omega_c = (\omega_0 + \Delta)\sqrt{1 + (k_\parallel/k_z)^2} \simeq (\omega_0 + \Delta)(1 + k_\parallel^2/2k_z^2)$ caused by the condition that the photon momentum perpendicular to the cavity must remain fixed, $k_z = 2\pi/d$. The exciton, being a much heavier particle, acquires little in-plane kinetic energy at these small momenta and thus the detuning between exciton and cavity modes increases as $k_\parallel$ increases, leading to their decoupling at angles greater than about 40° (Fig. 6.3). The two polaritons can be readily observed as strong dips in the near 100% reflectivity of the cavity mirrors, or as twin peaks in the sample transmission centered at $\omega_L = \omega_x - \Omega$, $\omega_U = \omega_x + \Omega$. The strength and position of each peak matches well with this model of coupled oscillators.

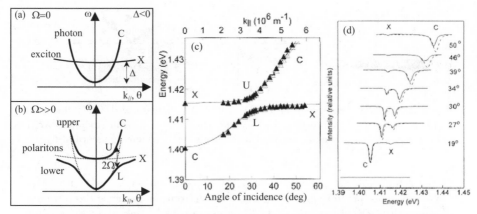

**Fig. 6.3.** In-plane dispersion relations at $\Delta \simeq -15$ meV for (**a**) cavity photons and excitons without interactions, and (**b**) coupled polaritons (**c**) Measured angular dispersion and transfer-matrix fit for TE/TM polarizations. (**d**) Reflectivity spectra for TE/TM at different angles of incidence (data courtesy of R.M. Stevenson)

**Polariton Traps.** The dispersion relation for the polaritons mapped in Figure 6.3b is critical for understanding the polariton interactions and merits further discussion. For $\Delta = 0$, the two polariton states are energetically split equally around the original energies by the Rabi frequency $\hbar\Omega = \sqrt{nf/2\varepsilon m_x d}$ where $n$ QWs of oscillator strength per unit area, $f \sim 8/\pi a_0^2 \sim 6.10^{12}$ cm$^{-2}$, are embedded in a microcavity of dielectric constant $\varepsilon$. As the cavity mode disperses strongly upwards in energy at greater angles, the exciton becomes less mixed in and tends to its unperturbed energy. This leads to the formation of a *polariton trap* in momentum space of depth $\Omega$ and momentum HWHM, $\Delta k \simeq \sqrt{3\varepsilon\omega_x\Omega}/c$ [6]. A near-stationary exciton can emit half a photon into the cavity producing an electric field which interferes with its own electron-hole polarization, and leads to a decrease in the total electromagnetic energy for the lower polariton branch $|\hat{L}\rangle$. Such a trap is rather different to magnetic or *rf* atomic traps which exist in real space. Instead it resembles a 'kinetic' trap for a macroscopic particle, because the lowest energy state has the particle at rest - this is where dissipation returns it. The polariton is very light in comparison to the exciton mass ($m_L = 4\pi^2\hbar/\omega_0 d^2 \sim 10^{-5}m_e$ [5]) so the trap is narrow and deep which encourages the polaritons to come to a *complete* standstill. The sequestration of polaritons in a trap near $k = 0$ of lower energy than the huge reservoir of exciton states is the origin of the dramatic reduction in their phase scattering, since far few interactions are available to $k$=0 polaritons than to excitons.

From this dispersion, we can also note that exciton-polaritons in microcavities possess a number of advantages to the experimentalist over their counterparts in bulk semiconductors. The first is that while all polariton coupling in bulk 3D samples exists at $k$ outside the light cone (Fig. 6.4b),

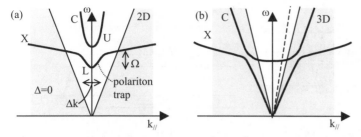

**Fig. 6.4.** In-plane dispersion relation for (**a**) 2D planar QW microcavity polaritons with strong coupling at k=0, and (**b**) 3D bulk polaritons. Only non-shaded regions have photon wavevectors accessible from outside the sample. The *dashed line* marks wavevectors at a nonzero angle of incidence

when the symmetry is broken by the planar cavity the resonant region is now easily accessible by externally impinging light. This enables us to study predictions of polariton dynamics made several decades ago [7] but now accessible directly. Previous experiments on bulk systems mapped the polariton dispersion and observed 'hyper-Raman' scattering of small magnitude which required extremely high optical intensities [8]. Bulk semiconductors only convert polaritons to light in a 'dead-layer' close to the sample surface (whose character is still under discussion [9]), while in 2D the semiconductor microcavities can be engineered for highly-specific light-matter interactions. Finally, no comparable trap exists within the 3D polariton dispersion, although it does possesses many discrete states near the resonance with varying light-matter mixtures [10].

**Wavefunction Annealing.** Our discussion of polariton dispersion relations makes a big assumption, that the in-plane momentum of each quasiparticle is a well-defined quantum number. This turns out to be true only because of the strong coupling of photon and exciton which forms the polariton trap. To see this, two extra perturbations must be considered to our simple picture of excitons: the Coulomb interaction between charged carriers, and disorder from imperfect crystalline growth. For low-energy excitons lying in local troughs of the potential energy, the disorder implies that the basis states for the center of mass motion of the excitons are not plane waves, but localised states instead. However, as long as the light-matter coupling is strong enough, it can pull out k=0 components of each localised wavefunction to make a polariton with a much larger lateral extent than the localised states. In order words, different localised excitons couple to each other by exchanging photons which diffract laterally in the planar cavity. A very similar process can occur for vertically stacked quantum wells [11]. This photon-induced delocalisation produces polaritons which are very different from their component excitons through a sort of wavefunction *annealing*, with a lateral extent no longer limited by the local potential fluctuations or alloy disorder inherent

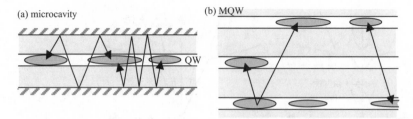

**Fig. 6.5.** Schematic wavefunction 'annealing' for polaritons arising from photon-mediated interactions of localised excitons spaced (**a**) laterally in a 2D planar microcavity, (**b**) vertically within different multiple quantum wells (MQWs)

in the semiconductor epitaxy (Fig. 6.5). In planar microcavities, the lateral coherence length $l_c = \sqrt{8\lambda_x dF} \sim 50\,\mu m$ where the finesse $F \sim 1000$ [12]. Such states are thus macroscopic quantum objects (and can be resolved with the naked eye!) however they may also be much smaller (Sect. 4).

The second perturbation which should be included is the Coulomb interaction between the electron and holes that make up excitons, which also couples together quasiparticles at different $k_{\parallel}$. The true ground state description must be a many-body wavefunction (as discussed by many theoretical groups [4,13–17]), with some similarity to a metallic or superconducting ground state, and further complicated by the non-equilibrium situation often set up in practice. In the weak coupling regime in which the mutual repulsion of $|\hat{X}\rangle$ and $|\hat{C}\rangle$ is less than the exciton linewidth, there is no polariton trap and the excitons are virtually degenerate. In this situation in which the Coulomb interaction can couple together a huge number of states exceeding the optical spectroscopy window ($k_{\parallel} > \omega/c$), it is hard to make a case for the exciton momentum being a good quantum number. Only when the polaritons form in a trap in which they are energetically below all other electronic states in the sample, is the Coulomb scattering suppressed and we can plausibly refer to quantised in-plane momentum. Experimentally this is confirmed by the observation of local populations at distinct $k_{\parallel}$ for microcavities, but a virtually instantaneous spread to all $k_{\parallel}$ when quantum well excitons are excited with a specific momentum.

## 6.2.2   Polariton Dynamics and Pair Scattering

With an understanding of the dispersion relation for polaritons in these planar structures, we can return to a question which motivated much of our research: *To what extent do polaritons represent a simple linear superposition of half an exciton and half a photon, and to what extent are their properties indicative of a new particle?* The answer to this depends on the properties of interest, since linear spectroscopy of semiconductor microcavities solely measures the mixed linear dispersion relation [4,5]. However almost any other optical property involves extra interactions of polaritons, such as the relaxation of polaritons to produce photoluminescence or nonlinear optical absorption and emission. In this section we summarize why polariton interactions are so different to the well-known properties of excitons. Such differences are important since the interaction of electrons, holes, and light is the basis for the engineering of semiconductor lasers, amplifiers and optoelectronic switches.

Excitons interact most strongly with longitudinal optical (LO) phonons which in GaAs have an energy $\omega_{LO} = 36$ meV, and at high density excitons interact strongly with each other. A heated population of excitons also cools by acoustic phonon emission on rather longer timescales ($\sim 100$ ps). Simple considerations of energy-momentum conservation allow us to see that all three processes are very different for polaritons (Fig. 6.6). Polaritons within the trap have a number of pair scattering interactions which will be discussed later. However it is difficult for polaritons in the trap to escape by absorbing acoustic phonons at these temperatures because this does not provide enough energy to get them out in a single event. Only if the sample is hot enough so that there are enough of the energetic LO phonons, or enough hot excitons available, do polaritons get ionized out of their trap [6,18]. These restricted scatterings produce the particular stability of polaritons compared to excitons, even if there is a large density of polaritons excited in the trap.

Besides the energetic separation of polaritons and excitons produced by the strong coupling, one of the other key properties introduced is the change in *shape* of the dispersion relation. Both the exciton and cavity photon mode dispersions are quadratic at small $k$, which precludes any direct non-trivial pair scattering since both energy and momentum cannot be conserved. Only through the existence of unbound states of the electron and hole can Coulomb scattering between QW excitons occur, as they can take up the extra momentum. The distorted $S$-shape of the lower polariton dispersion is responsible for turning on a whole new array of interactions which can be observed experimentally.

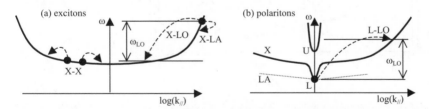

**Fig. 6.6.** In-plane dispersion relation and interactions of (**a**) excitons, and (**b**) polaritons with QW excitons. Optic phonon interactions (X-LO) involve large energy changes as compared with acoustic phonon scattering (X-LA). Exciton-exciton scattering (X-X) is allowed over a broad energy range. The *dotted line* marks the acoustic phonon dispersion, which shows the lack of available final states for polaritons in the trap absorb heat

## 6.3  Experiments

### 6.3.1  Experimental Geometry

Since the discovery of strong coupling in semiconductor microcavities [19], most experiments have concentrated on absorption or photoluminescence with non-resonant continuous-wave photoexcitation. As discussed above, this allows accurate mapping of the in-plane dispersion relation, but produces a confused picture as regards the interaction of polaritons on their dispersion curve. One key advance has been to examine the dynamics of the polaritons using pulsed excitation. This allows us to locate the products of scattering at different points along the dispersion curve, by spectroscopically probing the sample at different angles of incidence. It also allows a study of how scattering controls phase relaxation. Here we describe our experiments devised to carry out this program of research. A number of groups have now performed experiments which substantially confirm the picture described below, and also extend it further [20–25].

To ensure laser pulses arrive at the sample at the same time regardless of their angle of incidence, a femtosecond goniometer must be constructed (Fig. 6.7). Experimentally we find maximum temporal shifts when the angle of incidence is changed on the order of 100 fs (irrelevant for results presented here with 3 ps pulses), caused by imperfect realignment and systematic deviations through the optical windows of the cryostat and the sample substrate.

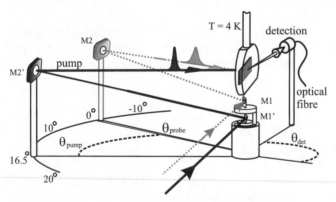

**Fig. 6.7.** Femtosecond goniometer: Each laser beam is first directed to a $\frac{1}{2}''$ mirror (M1) which is placed directly underneath the sample, and is able to rotate about a vertical axis. The beam emerges from this mirror along the direction intended as the angle of incidence and arrives at a mirror (M2) at the same height as the sample, on the end of an arm which can rotate about the same vertical axis through the sample. This final mirror then directs the beam to the sample through a 20 cm focal length lens. To change the angle of incidence, the arm is set to the correct angle of incidence and M1 rotated to redirect the beam back onto M2. The total path length to the sample is preserved and thus the arrival time is unchanged

In pump-probe experiments, one arm is kept fixed while the other is rotated allowing simple realignment on the same sample location. The goniometer-cryostat configuration is so constructed as to allow a very wide angular field of view ($\pm 70°$) on both the front and back of the sample, allowing simultaneous measurement of reflection and transmission.

Because spectroscopic measurements are needed along the dispersion curve, the wavelength and spectral bandwidth of the pulses must be carefully controlled to match that of the polaritons. The main problem is to produce synchronized pulses of different colour. This is solved by using a short-pulse Ti:sapphire laser with a wide spectral bandwidth (10 nm) from which the desired pulses can be excised, as developed for coherent control spectroscopies [26]. The basic method is to spatially separate the different colours in the pulse using a diffraction grating and bring them to a focus at a modulator plane. Here a computer-controlled liquid-crystal spatial light modulator allows control of both the amplitude and phase of each colour transmitted, before the pulse is reassembled at a matched diffraction grating. Sophisticated software allows calibration and selection of any desired pulse shape or sequence, containing any spectral content from the original laser pulse. Such an arrangement in both pump and probe beamlines allows independent tuning in wavelength, pulse-length and pulse-shape. Finally we both spectrally- and temporally-resolve the combined emission/probe transmission at any angle, using rapid-scan time-delay stages, high-frequency lock-in detection and optical fibre to convey the light to a monochromator and cooled CCD system.

The combination of these techniques rapidly produces large data sets containing signatures of the interaction processes. In addition, a number of other important degrees of freedom are explored including the power of pump and probe pulses, and their polarization state. Both these are computer-controlled using liquid-crystal waveplates and Glan-laser polarizers. The electric-field polarization in the sample depends on the angle of incidence, and care must be taken to compensate for Fresnel losses and phase shifts in the final approach arm to the sample. The purity of circular polarization is confirmed by the suppression by $> 30$ dB of interactions from opposite circularly-polarised pump and probe beams.

### 6.3.2  Microcavity Sample

Most of our measurements have been taken on high quality InGaAs-based microcavity wafers which show strong coupling below $T = 100$ K. These samples are by no means of the narrowest linewidth [27], but prove convenient because it is easy to match the spectral pulse-widths to their polariton linewidths. Many alternative designs of semiconductor planar heterostructure will produce strong coupling, but little difference in the dynamics is expected compared with the scheme presented here, as recently confirmed [20,28–30]. The cavity mirrors are fabricated from distributed Bragg reflectors (DBRs) of alternating transparent layers of different alloys of $Al_y Ga_{1-y}As$, to produce

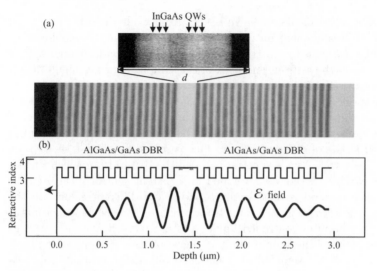

**Fig. 6.8.** (a) Microcavity sample cross-section: Scanning electron micrograph, with magnified QW region. (b) Refractive index profile of heterostructure layers, with the schematic electric field distribution at the lower polariton resonant wavelength

a monolithic microcavity. The microcavity wafer grown here consists of two pairs of three $100\,\text{Å}$ $In_{0.06}Ga_{0.94}As$ QWs in $100\,\text{Å}$ GaAs barriers, sandwiched between 17 (20) pairs of DBRs GaAs/$Al_{0.18}Ga_{0.82}As$ on top (bottom). The optical cavity length is $\sim 3\lambda_x/2\sqrt{\varepsilon}$ and varies across the sample wafer allowing access to both positive and negative detunings, $\Delta$. Careful angle-resolved spectroscopy has shown to be well fit by a transfer-matrix linear-propagation theory which uses the refractive index of each layer [31]. A slight splitting ($\sim 0.1\,\text{meV}$) exists between TE and TM field polarizations due to the planar geometry however we will see that these splittings are overwhelmed by the spin selection rules of the QW excitons.

The lowest energy excitons in strained InGaAs quantum wells are derived from the heavy-hole valance bands, which have a spin $j_{hh} = \pm\frac{3}{2}$. Together with the electron spins $j_e = \pm\frac{1}{2}$ this produces two degenerate optical transitions of opposite circular helicity $j_x = \pm 1$. Excitation with right/left-circularly polarized light creates spin-up/down excitons respectively, and subsequent spin scattering can be dynamically tracked using circularly-polarised time-delayed probe pulses (see Chap. 1). As we shall show in our results, near-resonant excitation of microcavities permits virtually no spin scattering. Spin is a good quantum number for polaritons, which implies that mixing with TE/TM photons and $j_x = \pm 2$ 'dark' excitons, is not strong.

### 6.3.3   Parametric Scattering

We now discuss pair-scattering in more detail as it turns out to dominate many of the nonlinear experiments. Two situations are of most interest, in which:

- polaritons injected at particular $k_\parallel$ can 'parametrically' scatter, or
- two excitons pair-scatter to produce a polariton in the trap

**Pulsed Interactions.** Parametric scattering (historically the first to be clearly identified experimentally [32]), can produce enormous optical nonlinearities of potential for ultrafast switching. On the other hand, exciton pair scattering (which was first to be identified theoretically [33]) is still under active investigation, with additional processes dominating those originally suggested [6,34].

In parametric scattering of polaritons, a laser beam incident at a particular angle injects a large population of polaritons at specific $k_\parallel$. Within the trap region these polaritons cannot easily lose energy, and only a small fraction make it to the polariton ground state. However at a particular 'magic' angle, $\boldsymbol{k} = \boldsymbol{k}_m$, two polaritons can parametrically scatter so that one drops down to $\boldsymbol{k} = 0$ and the other is kicked up to $\boldsymbol{k} = 2\boldsymbol{k}_m$. The magic angle is given by the condition for energy and momentum conservation of the polariton dispersion, $2E(\boldsymbol{k}_m) = E(\boldsymbol{0}) + E(2\boldsymbol{k}_m)$ which leads to $\theta_m \simeq \sin^{-1}\{\sqrt{3\varepsilon\Omega\omega_x}\} \simeq 16.5°$. This process can be easily checked by probing the sample at normal incidence ($k = 0$) while scanning the angle and energy of a pump pulse along the polariton dispersion (Fig. 6.9). A clear resonance is seen at the predicted angle, whose effect is to produce substantial *gain* at the polariton ground state. This gain is rather unexpected and different to that observed for equivalent experiments in semiconductor quantum wells.

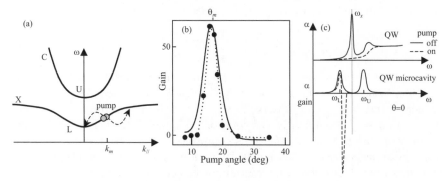

**Fig. 6.9.** (a) Parametric scattering between two pump photons at the magic angle. (b) Measured gain vs pump angle of incidence (c) linear and nonlinear absorption spectra for high power pumping of a QW and a microcavity containing the same QWs

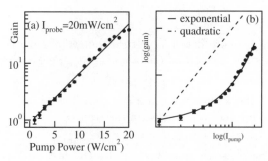

**Fig. 6.10.** Power dependence of the probe gain on (**a**) semi-log and (**b**) log plot, showing the exponential dependence of gain on pump power

Normally as an electronic state becomes occupied, its absorption reduces because the ground state empties and the excited state fills, blocking further absorption (Fig. 6.9c). In semiconductors this is termed phase-space-filling, and is accompanied by comparably-large effects from exciton screening and exchange Coulomb terms. The saturation from phase space filling arises from the fermionic statistics of the semiconductor excitations. Something radically different occurs for highly-populated polariton states.

By varying the pump power, just how different this polariton ground state is, becomes clear (Fig. 6.10). The exponential dependence of gain on pump power is a clear signature of a stimulated process. Occupation of the polariton ground state $N_0$ *increases* the scattering rate of polaritons into that state, $\frac{dN_0}{dt} \propto (1 + N_0)N_{k_m}$, hence the process accelarates until all available pump polaritons have been trapped at $k=0$ or have decayed from the sample. This conclusion is backed up by a microscopic theoretical model which reproduces all the experimental results [35,36]. The interaction driving the stimulated parametric scattering is once again the Coulomb interaction, $V$, between the electron and hole components of the excitons. This mixes polaritons at different places on the lower branch, $|\hat{L}(k)\rangle$, through an *anti-Hermitian* coupling,

$$\begin{bmatrix} \hbar\tilde{\omega}_L(0) & VN_{k_m} \\ -VN_{k_m} & \hbar\tilde{\omega}_L(2k_m) \end{bmatrix} \begin{bmatrix} |\hat{L}(0)\rangle \\ |\hat{L}^\dagger(2k_m)\rangle \end{bmatrix} = E \begin{bmatrix} |\hat{L}(0)\rangle \\ |\hat{L}^\dagger(2k_m)\rangle \end{bmatrix}. \tag{6.2}$$

The crucial negative sign in the off-diagonal mixing terms leads to an *attraction* between the energies of the two initially uncoupled polariton components, and to the gain observed [36]. Although QW excitons (nominally bosons) show predominantly the fermionic character of their constituents, the interacting polaritons behave as composite bosons. By mixing the exciton wavefunction together with a cavity photon, a new dispersion is produced which can use the Coulomb interaction for bosonic scattering at densities below those which access the underlying fermionic statistics of the electrons [37].

Both the polaritons in the trap ($k = 0$) and their high energy partners ($k = 2k_m$) have been observed in emission (Fig. 6.11). As the pump power

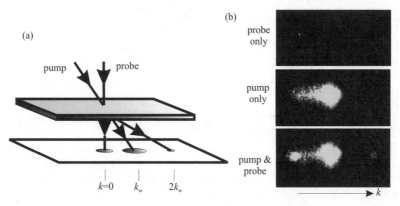

**Fig. 6.11.** Images of the emission from the microcavity when excited by pump at $k_m$ and probe at $k = 0$, showing clearly the parametric scattering process

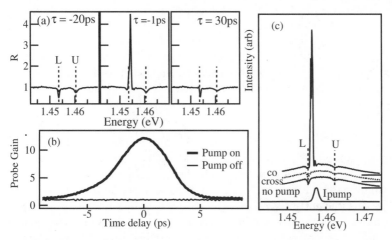

**Fig. 6.12.** (a) Reflection spectra measured by probe pulse at normal incidence arriving at indicated time delays relative to the pump pulse (injected at the magic angle). Gain is seen from polaritons in the trap ($L$). (b) Time-dependence of the reflectivity with the pump on/off. Reflectivites over 100% imply gain. (c) Probe spectra for co-circularly and cross-circularly polarised pump and probe pulses

is further increased it is possible to excite enough carriers and phonons to reach the fermionic regime at which point the strong coupling regime collapses [38]. Gains of up to $\sim 10^3$ occur at carrier densities well below that of all previous ultrafast semiconductor nonlinearities (peak powers here are below 10 kW) (Fig. 6.12). The gain in this quantum-engineered material can exceed $10^6 \, \text{cm}^{-1}$, higher than reported in any other substance, and yet still responding on ultrafast timescales. Injecting only a few aJ of energy along the sample normal produces a 30% change in the pump absorption in the sample.

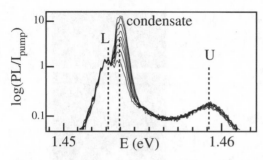

**Fig. 6.13.** Emission spectra as a function of pump intensity (normalised to the incident power) when only the pump pulse is injecting polaritons at $k = k_m$. Self-stimulated parametric scattering leads to the emergence of the new condensate line just above the lower polariton energy

These nonlinearities are thus of great interest for all-optical switching and vertical-cavity amplifiers.

Even without injecting the probe pulse, the emission in the sample normal direction is superlinear in the pump power, indicating the presence of 'self-stimulation' once a few polaritons have managed to scatter down to $k = 0$ (Fig.13). This self-stimulation shows the strong tendency for polariton trapping to occur.

**CW Interactions.** Once the parametric scattering process was confirmed in pulsed measurements, it is clear it should also be apparent in CW resonant excitation at the same magic angle. Despite the main objection that with continuous excitation the background exciton population in the sample would rise sufficiently to collapse the strong coupling regime, we observe polariton pair-scattering overwhelming all other processes. Our experiments show that once the pump power at $\theta_m$ exceeds a few tens of mW, scattering into $k = 0$ totally dominates producing a beam of light in forward and backwards directions (Fig. 6.14a). As in the case of non-resonant parametric scattering in inorganic crystals such as lithium niobate inside a cavity, the resulting emission is a manifestation of an optical parametric oscillator (OPO). Two pump polaritons parametrically scatter to produce a 'signal' ($k = 0$) polariton which resonates inside the cavity and an 'idler' ($k = 2k_m$) polariton. Typical OPOs are $> 1$ cm long due to the weak parametric scattering strength - here we have shrunk the device by a factor of $10^4$ through engineering a *resonant* parametric scattering. Our experiments clearly show the expected behaviour of the output power, angular width, and resonant linewidth narrowing (Fig. 6.14b,c)[39].

An alternative way to view this optically-pumped device depends on realising that a coherent macroscopic population of polaritons occupies the trap, each with an identical phase that is arbitrarily set by a spontaneous symmetry breaking. This is the description of a *polariton condensate*, which possesses

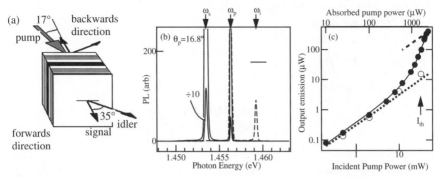

**Fig. 6.14.** (a) Schematic configuration of the microcavity OPO. (b) Emission spectra of signal and idler at $k = 0, 2k_m$ for $CW$ pumping at $k = k_m$. (c) Pump power dependence of the signal emission, showing clear threshold behaviour

coherence not just in the photon field (as in a laser) but also in the exciton fields. Such a state was predicted decades ago for bulk polaritons [10], and resembles a Bose-Einstein atomic condensate which is formed by evaporative-cooling of atoms until they are cold enough to condense. Here, the first few polaritons that scatter into the trap set the arbitrary phase for the further polaritons that they cooperatively stimulate to join them. Thus the condensate exists at a different energy and momentum to the reservoir of polaritons which efficiently feeds it, and is thus an example of a *non-equilibrium* condensate. It also differs from a laser in that an energy gap appears in the density of states [40]. Random variations in pump phase contribute only to the sum of the phase of the trapped populations since $\phi_s + \phi_i = 2\phi_p$ . In contrast to superconductors which must be observed electrically, the signature of a polariton condensate is a strong beam of emerging light. The useful nature of the polariton condensate lies in its half-matter composition, which allows the manipulation of its phase. The phase of the polaritons is identical and their phase scattering is suppressed compared to excitons. Interferometric measurements of the polariton condensate linewidth show it to be < 500 MHz (the resolution limit of our apparatus) and noise measurements suggest phase stability of several nanooseconds. By comparison to the Schalow Townes limit for lasers, the expected phase scattering $\propto \Gamma_0/N_0$ where the estimated occupation $N_0 \sim 10^4$ and the cavity decay rate $\Gamma_0 \sim (5\,\mathrm{ps})^{-1}$. This agrees well with our measurements and explains how the phase lifetime of single excitons can be extended 10,000-fold by protecting them in large coherent polariton condensates.

**Polariton Condensation.** This brings us to the second scattering process of interest, which is between pairs of excitons. It is surprising that energetic excitons produced by the idler and which remain in the sample do not affect the parametric scattering. Our explanation is that these remaining excitons

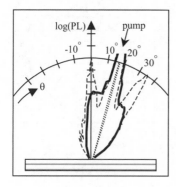

**Fig. 6.15.** Asymmetrical angular distribution of time-integrated PL (*solid*) and stimulated PL (*dashed*) when pumped by a pulse at the magic angle of 17° (*dotted*). Quantum well heterostructures show symmetrical emission

are too cold to give rise to much ionization of polaritons from the trap, and are relatively benign at low temperatures. For non-resonant excitation or at higher lattice temperatures, they can participate in exciton-polariton pair scattering releasing polaritons from the trap. Even so, recent predictions show the possibility of electrically-pumped microcavities working at room temperature [6].

To collect polaritons into the trap requires excitons to lose energy to optic- or acoustic-phonon emission or through mutual Coulomb interactions (Fig. 6.6). Only an exciton with excess energy which is $\omega_{LO}$ above the polariton ground state can emit an LO phonon of specific wavevector to drop into the trap. Such processes have indeed been seen, but do not dominate in general since relatively few excitons are able to access this relaxation channel, and the scattering rate is small because the wavefunctions of the states are so different. Acoustic phonon emission into the trap is also extremely unlikely because this cannnot remove enough energy in a single scattering event [38]. The emission of light (through decay of the cavity photon component of the polariton) is thus predicted to be dominated by the non-equilibrium dynamics of polariton interactions, rather than by thermalisation with the phonon bath as is found for a population of excitons. Measurements of luminescence have shown this to indeed be the case [41]. The photoluminescence is predicted to depend sensitively on the photon energy and momentum used to excite the sample. The non-thermalised emission can even be asymmetric with respect to the sample normal, even though this is a planar sample geometry with a cylindrical symmetry (Fig. 6.15) [12].

For two excitons at $\boldsymbol{k}_{1,2}$ in the exciton reservoir to exchange energy and momentum, dropping one of the pair into the polariton trap, requires two specific conditions (Fig. 6.16): $\boldsymbol{k}_1 + \boldsymbol{k}_2 = \boldsymbol{k}_3$ and $\omega_1 + \omega_2 = \omega_L(0) + \omega_3$ with $\omega_i = \omega_X + \frac{\hbar |k_i|^2}{2M_x}$. This condition can be rewritten as a constraint on the original momenta

$$\frac{\hbar^2}{2M_x} \boldsymbol{k}_1.\boldsymbol{k}_2 = \hbar \Omega \qquad (6.3)$$

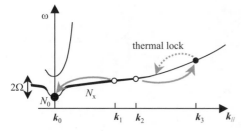

**Fig. 6.16.** Stimulated scattering process of exciton pairs from a thermal exciton population of $N_x$ into the $k = 0$ condensate and higher energy exciton at $k_3$. The high energy exciton then relaxes back to thermal equilibrium preventing re-ionization of condensate polaritons

which implies that the geometric mean of the exciton energies must equal the polariton trap depth. Only excitons which reach these high momenta can scatter into the trap, and only excitons which have nearly twice this momentum can ionize trapped polaritons. This sets the temperature and Fermi energies that control the density of the condensate $(k_B T, E_f > \hbar\Omega)$, and suggests that it is indeed possible to electrically-pump polariton condensates providing that the excitons thermalise rapidly, and the trap is sufficiently deep [6,45]. Thus although our discussion of condensates involves optically-pumped samples at low temperatures, this is by no means a fundamental limitation for condensate technologies. Currently a number of groups are exploring GaN- and ZnSe-based semiconductor microcavities which are predicted to operate at room temperature. The extension of the vertical-cavity semiconductor laser (VCSEL) to the strong coupling domain no longer seems so remote.

## 6.4   Condensate Dynamics

Given that a polariton condensate can exist in semiconductor microcavities, with a phase lifetime which can potentially extend to $\mu$s or even ms, we can now discuss the manipulation of this state. Since our knowledge of these condensates is at a nascent stage, this section is more specifically an informed speculation. However the promise in each of these areas is considerable for quantum coherent manipulation and technologies.

### 6.4.1   Polariton Interferometry

As with lasers, superconductors, atomic- and all other condensates, a clear possibility of interfering polariton condensates exists. Existing device implementations currently in use include Sagnac interferometers, superconducting interference devices (SQIDs), and atomic fountain gravitational detectors (Fig. 6.17). In each case, two condensates of fixed relative phase are created, either by a beamsplitter (light), Josephson junction (Cooper pairs) or

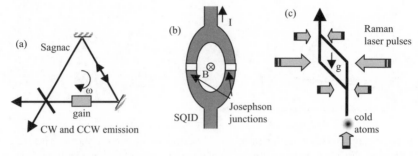

**Fig. 6.17.** Condensate interferometers: (**a**) Laser Sagnac interferometer detects angular accelarations by interfering counterpropagating laser beams (**b**) SQID Josephson junction superconducting devices detect magnetic fields in units of flux quanta trapped in the ring, and (**c**) atomic fountain gravitational accelarometers use laser beams to manipulate atoms

by an optical standing wave beamsplitter (atoms). Interference occurs when the condensates are allowed to combine at a detector which can view the photon/current/atom fringes using a beamsplitter/ammeter/ionized particle counter (respectively). In each case, the phase difference accrued by the condensates in their different passage to the detector is measured, and this provides useful information about movement (Sagnac), magnetic fields (SQID) and gravity/frequency (atomic fountain). The extreme sensitivity of each of these interferometers comes from the phase stability of each condensate and the accuracy in counting many fringes which can reduce errors down to 1 part in $10^{12}$. The excitonic component of polaritons is sensitive to magnetic and non-uniform electric fields, as well as gravity and temperature. Hence a number of all-solid-state polariton interferometers can be conceived.

One possibility is to create polaritons at one location on a sample, spatially transport the polaritons through two physically-separated paths, and recombine them in a way that detects the local fringe pattern. Sermage et al. [42] have shown lateral propagation of polaritons travelling with the same in-plane momentum as originally injected, while our work shows that polaritons can diffuse several mm to regions of the sample which let photons escape (Fig. 6.18). Lateral transport of polaritons thus requires only that the lifetime of polaritons be sufficiently long to move them around. The main problem is that the flat dispersion around the bottom of the trap insists on *stationary* condensates, whereas a moving condensate is required. Our recent work on multiple condensates suggests that this is not a fundamental problem and condensates away from $k = 0$ are possible [36]. The energy loss rate of the condensate is sufficiently fast that the device would need to be continuously (electrically or optically) pumped along its length, with the macroscopic phase set by the condensate component right up to the very edge of this layout. Possibly the greatest technological difficulty is growing microcavity samples in which the cavity thickness is sufficiently uniform that

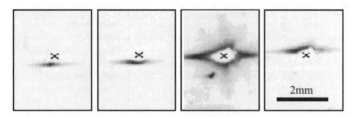

**Fig. 6.18.** Near field images of the sample emission when pumped with a pulsed pump at angle $\theta_m$, which is focused at the point marked with an X. Frames from left to right correspond to increasing detuning of cavity and exciton $\Delta = -7, -3, 0, 3\,\text{meV}$ and show diffusion of polaritons in the microcavity by $\sim 200\,\mu\text{m}$

the resonant energy varies by less than the cavity linewidth across the interferometer length, however this is already the case for certain areas of the wafers. Such an interferometer laid out on a chip would be able to detect the effect of electric and magnetic fields on the internal motion of the exciton in a similar way to the internal phase sensitivity to gravity of atoms in an atomic fountain [43]. In comparison to laser interferometers, the phase coherent exciton component gives access to new environmental fields.

An alternative device model is based on temporal rather than spatial interferometry. We have so far avoided discussion of the spin of the polariton condensates, but it is this feature which has perhaps most potential. A circularly-polarized CW pump produces condensate emission which is completely co-circularly polarized. This fits with our model that parametric scattering is spin-preserving [Fig. 6.12c] [32]. When a linearly-polarized pump is used instead, the condensate emission polarization fluctuates on short timescales and is effectively unpolarized. We can see how this would occur by noting that linear-polarizations correspond to an equal superposition of right and left circular polarizations. Thus two condensates can form, one for each exciton spin, which posses independent macroscopic condensate phases, $\psi_{\uparrow,\downarrow}$ (Fig. 6.19). As their phase difference diffuses, so the resulting linearly-polarized emission randomly rotates with an orientation , $\theta_{\text{poln}} = \psi_\uparrow - \psi_\downarrow$. It is this condition of two co-located spin condensates that is so interesting, because it can be sensitive to local magnetic fields. The g-factor of excitons is rather small ($g_x \sim -0.5$ in InGaAs quantum wells of this width) so the splitting of excitons in weak magnetic fields is much less than the exciton linewidth and hard to measure. When a polariton condensate is used, the situation is rather different since the two spin condensates now exist at slightly different energies $\hbar\Delta\omega = 2g_x\mu_B B$ which can be further apart than their small linewidths due to the long condensate phase scattering rate. Detection of the emission would reveal sinusoidal beats as the two condensates evolve in and out of phase - simple measurement of this MHz beat frequency provides the local magnetic field. Estimates of the current phase lifetime lead to magnetic field sensitivity below $100\,\mu\text{T}$, compared to typically $1\,\text{T}$ fields normally ap-

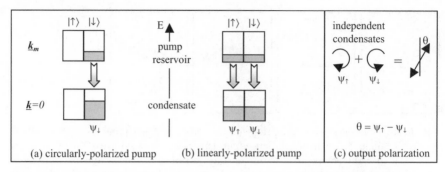

**Fig. 6.19.** Spin-dependent polariton occupations with (**a**) circularly-polarized pump producing circularly-polarized condensate (**b**) Linearly-polarized pump producing two oppositely circularly-polarized condensates of different phase $\psi$ resulting in (**c**) varying linear polarization $\theta$

plied to split excitons. Ring geometries (see below) may be used to improve this sensitivity.

Experiments are currently in progress to demonstrate the feasibility of polariton spin interferometry. In atomic laser systems this scheme is rather complicated because such spin-split two-mode lasers tend to compete for optical gain from the single pump reservoir. For semiconductor microcavities a separate reservoir exists for each condensate and, in the absence of phase locking of the condensates (for example produced by strain in the sample), the two condensates are stable. Many additional questions can also be posed such as the influence of nuclear spins in the sample (Chap. 5.6).

### 6.4.2  Macroscopic Quantum States

Although many theories of semiconductors are based on the idea of delocalised plane wave electronic states, no semiconductor system yet exists in which a particular wavefunction can be addressed from different spatial locations (Fig. 6.20). A system of this kind is needed for matter-wave tests of quantum measurement such as EPR-type correlation and entanglement. Recently a number of groups have been using near-field microscopy on samples with quantum well disorder produced by interrupting the growth, to try to identify excitonic wavefunctions which have a length-scale > 100 nm and thus (in principle) simultaneously addressable at different locations. Similar ideas have prompted local electrical probing of the phase of mesoscopic superconducting islands. The complexity and difficulty of such experiments leads us to suggest an alternative route employing polariton condensates.

Condensates have a size set by the finesse of the microcavity, the parametric scattering rate and the polariton condensate density [36,39]. In the current realization, both far-field and near-field imaging show that the 50 μm diameter of the condensate is independent of the pump focus diameter, and

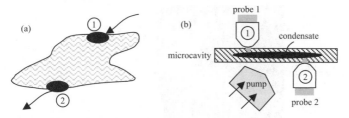

**Fig. 6.20.** (a) Macroscopic coherent state allows spatially-separated disturbance and measurement at particular times. (b) Realisation for polariton condensate probed from different sides of the sample

dependent only on the cavity finesse. Using microscope objectives placed on either side of the sample it is possible to access much smaller regions of the condensate and probe their entanglement and dynamic response. For instance, if a weak probe pulse of specific phase is injected into one part of the condensate, how long does it take for this perturbation to propagate across the macroscopic quantum state? Can the individual polaritons in a condensate be located? Answering such normally-inaccessible questions is helpful for building robust intuitions to develop new quantum technologies.

As a first test of such ideas we apply the analogy between polaritons and Cooper pairs to measure Josephson coupling between two polariton condensates. Two CW pump beams excite the sample from opposite sides at the same magic angle of 16°, with the same in-plane momentum. Independent control of each pump polarization and focal spot allows condensates of variable separation and spin to be excited. We find evidence of spin-dependent interactions in the near-field emission for spin parallel and antiparallel condensates (Fig. 6.21). The total emission (at normal incidence) increases as the condensates overlap. However the data is complicated by further spin-dependent interactions to be discussed elsewhere [44]. These first experiments show the potential for spatially-dependent condensate interactions.

**Fig. 6.21.** Two condensates pumped by circularly-polarised CW beams of equal intensity incident from the front and back of the sample at $\theta_m$. The near-field images show the difference between spin-parallel and spin-antiparallel condensates in close proximity

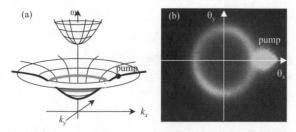

**Fig. 6.22.** (a) Different excitation conditions can enable optically-pumped lasing in higher order Airy modes, producing (b) ring emission seen imaged on a screen next to the sample

A final example of wavefunctions sensitive to spatial coherence are the higher order Airy modes of the microcavity. We first note that although the transverse modes of the empty microcavity are Airy modes, they are substantially altered by the strong coupling regime. The polariton ground state corresponds to the $l = 0$ mode and it is possible to excite higher order transverse modes in the cavity, as occasionally seen in vertical-cavity surface-emitting lasers (VCSELs). However instead of the $TEM_{10,01,11}$ modes recorded from such electrically-pumped mesas, it is possible to excite lasing ring modes in semiconductor microcavities. We have observed such behaviour (Fig. 6.22) for near-resonant selective pumping, also discussed in detail elsewhere [45]. Ring modes can be used to carry angular momentum with each Airy mode labelled by the total angular momentum $l$ (integral) and spin projection $m_l = 0, \pm 1, ..., \pm l$ [46]. Laser beams with these spatial profiles form micro-rings when brought to a focus, however these have not yet been exploited in the study of semiconductors. This is because although there is optical coherence between points on opposite sides of the ring, there is no electronic coherence - the electronic wavefunctions excited are separately localised and are not quantum-mechanically correlated. When considering the interaction of such a laser beam with a polariton condensate this is no longer true, and it should be possible to impart a specific angular momentum to the excited condensate. In analogy with magnetic vortices trapped in mesoscopic superconducting disks, we predict a panoply of electric field vortices that can exist in the polariton condensate. Such states should be sensitive to the sign of a magnetic field threading the condensate. Similarly it should be possible to use angularly-polarized pump and probe beams (together with spin polarization) to create novel coherent states in these systems. The phase stability of such states should be particularly high and suggest the possibility of creating spatially-entangled wavefunctions.

### 6.4.3 Quantum-Correlated Pairs

The parametric process that creates pairs of signal and idler polaritons is responsible for additional quantum-correlations, which exist between the photons emitted from these modes in different directions [47]. Some of the most sensitive tests of Bell's inequalities have used parametric downconversion from pump photons at $2\omega$ to produce quantum correlated signal and idler at sideband frequencies $\omega \pm \epsilon$ [48,49]. Quantum cryptography typically relies on such a process to distribute quantum ciphers, however the extreme inefficiency of this conversion often mandates high intensity pulsed sources to drive single-photon pulses.

The resonant parametric scattering process in microcavities is $10^4$ times more efficient, but a number of problems need to be addressed before exploitation is possible. The first is that signal polaritons at $k = 0$ are much more likely to be reradiated than the idler polaritons at $k = 2k_m$ which can scatter into much higher $k$ exciton states [12]. One solution is to use a different configuration of laser beams, with the pump incident at the bottom of the polariton trap at $k = 0$ and quantum correlations measured between signal and idler photons at $k = \pm q$. A second problem is that the quantum correlations are much reduced by the choice that polaritons have to randomly decay into either forward or backward directions. By suitably engineering the DBR mirrors in the heterostructure it is possible to suppress optical emission from the top of the sample, and restore the strong quantum correlations. A number of groups are currently actively pursuing this arrangement [50,51].

### 6.4.4   Conclusions

The possibilities listed so far by no means exhaust the range of quantum technologies that can be accessed through the polariton condensate. Many other prospective states are yet to be explored in this rich system such as spin waves and other collective excitations, quantum bistability and spatial solitons, pseudo-spin in double microcavities, and spin transport using inhomogeneous electric fields. Mapping potentially interesting investigations will pave the way for a new generation of solid-state quantum technologies in strongly-coupled light-matter systems. Much can be learned from the basic principle of using the coupling of fully-engineered wavefunctions to create quasiparticles with new properties. In the present realization this provides a way for Coulomb interactions to introduce a new long-lived coherent state in semiconductors. Such ideas are applicable to a wide variety of other systems including organic chromophores such as J-aggregates, liquid crystals, and even nanostructured metals.

### Acknowledgements

Many people have directly contibuted to this work including Pavlos Savvidis, Maurice Skolnick, Mark Stevenson, Pavlos Lagoudakis, Caterina Netti, Dolores Martin-Fernandez, David Whittaker, Alexei Kavokin and Guillaume Malpuech. I am indebted and grateful to their constant enthusiasm and skill.

# References

1. N.H. Bonadeo et al., Science **282** 1473 (1998)
2. F. Schmidt-Kaler G. Rempe and H. Walter, Phys. Rev. Lett. **64**, 2783 (1990)
3. M.O. Scully and M.S. Zubairy,*Quantum Optics* (CUP, Cambridge 1997)
4. G. Khitrova et al.., Rev. Mod. Phys. **71**, 1591 (1999)
5. M.S. Skolnick, T.A. Fisher and D.M. Whittaker, Semicond. Sci. Tech **13**, 645 (1998)
6. G. Malpuech, A. Kavokin, A. DiCarlo and J.J. Baumberg, Phys. Rev. B **66** (2002)
7. F. Yura and E. Hanamura, Phys. Rev. B **50**, 15457 (1994)
8. B. Honerlage, R. Levy, J.B. Grun, C. Klingshirn, and K. Bohnert, Phys. Rep. **124**, 161 (1985); S. Savasta and R. Girlanda, Phys. Rev B **59**, 15409 (1999)
9. J.J.Hopfield and D.G. Thomas, Phys. Rev. **132**, 563 (1963); F. Evangelisti, F.U. Fishbach and A. Frova, Phys. Rev. B **9**, 1516 (1974)
10. L.V. Keldysh, 'Macroscopic coherent states of excitons in semiconductors'. In *Bose-Einstein Condensation.* ed. by A. Griffin, D.W. Snoke, S. Stringari (CUP, Cambridge 1995) pp. 246–281
11. J.J. Baumberg, A.P. Heberle, A.V. Kavokin, M.R. Vladimirova, and K. Köhler, Phys. Rev. Lett. **80**, 3567 (1998)
12. P.G. Savvidis, J.J. Baumberg et al.., Phys. Rev. B **62**, R13278 (2000)
13. S. Schmitt-Rink, D. S. Chemla, and D. A. B. Miller, Adv. in Phys. **38**, 89 (1989)
14. S. Schmitt-Rink and D. S. Chemla, Phys. Rev. Lett. **57** 2752, (1986)
15. S. Schmitt-Rink, D. S. Chemla, and H. Haug, Phys. Rev. B **37**, 941 (1988)
16. H. Haug and S. W. Koch, *Quantum Theory of the Optical and Electronic Properties of Semiconductors* (World Scientific, Singapore 1993)
17. F. Jahnke, M. Kira and S.W. Koch, Z. Phys. B Cond. Mat **104**, 559 (1997)
18. A.I. Tartakowskii et al., Phys.Rev. B **62**, R2283 (2000)
19. A. Ishikawa C. Weisbuch, M. Nishioka and Y. Arakawa, Phys. Rev. Lett. **69**, 3314 (1992)
20. P. Senellart and J. Bloch, Phys. Rev. Lett. **82**, 1233 (1999)
21. R. Houdré, C. Weisbuch, R.P. Stanley, U. Oesterle, and M. Ilegems, Phys. Rev. Lett. **85**, 2793 (2000)
22. Le Si Dang et al., Phys. Rev. Lett. **81**, 3920 (1998)
23. R. Huang, F. Tassone, Y. Yamamoto, Phys. Rev. B**61**, R7854 (2000)
24. F. Tassone and Y. Yamamoto, Phys. Rev. B**59**, 10830 (1999)
25. M.Kuwata-Gonokami et al., Phys. Rev. Lett. **79**, 1341 (1997)
26. J.J. Baumberg, 'Coherent Control and Switching'. In *Semiconductor Quantum Optoelectronics.* ed by A. Miller, M. Ebrahimzadeh and D.M. Finlayson(IOP, Bristol, 1999) pp. 100–117
27. F. Quochi et al., Phys. Rev. Lett. **80**, 4733 (1998)
28. J. Erland et al., phys. stat. sol. (b)**221**, 115 (2000)
29. G. Dasbach et al., Phys. Rev. B **62**, 13076 (2000)
30. F. Boeuf et al., Phys. Rev. B **62** R2279 (2000)
31. D. Baxter, *Semiconductor Microcavities* PhD thesis (Sheffield University, Department of Physics, 1998)
32. P.G. Savvidis, J.J. Baumberg et al., Phys. Rev. Lett. **84**, 1547 (2000)
33. A. Imamoglu et al., Phys. Rev. A **53** 1996, (1996)

34. B. Sermage P. Senellart, J. Bloch and J.Y. Marzin, Phys. Rev. B **62**, R16263 (2000)
35. C. Ciuti et al., Phys. Rev. B **62**, R4825 (2000)
36. P.G. Savvidis, C. Ciuti, J.J. Baumberg et al., Phys. Rev. B **64**, 75311 (2001)
37. O.W. Greenberg and R.C. Hilborn, Phys. Rev. Lett **83**, 4460 (1999)
38. R.M. Stevenson et al., Phys.Rev.Lett. **85**, 3680 (2000)
39. J.J. Baumberg et al., Phys. Rev. B**62**, R16247 (2000)
40. P.R. Eastham and P.B. Littlewood, arXiv,cond-mat/0102009 (2001)
41. U. Oesterle R. Houdre R.P. Stanley, S. Pau and M. Ilegems, Phys. Rev. B **55**, R4867 (1997)
42. T. Freixanet et al., Phys. Rev. B **61** 7233 (2000)
43. M. Kasevich and S. Chu, Phys. Rev. Lett. **67**, 181 (1991)
44. P. Lagoudakis et al., Phys. Rev. Phys. Rev. B, **66** (2002)
45. P.G. Savvidis et al., Phys. Rev. B, **65**, 073309 (2002)
46. N.B. Simpson, K. Dholakia, L. Allen and M.J. Padgett, Opt. Lett. **22** 52 (1997)
47. A.M. Fox, J.J. Baumberg et al., Phys. Rev. Lett. **74**, 1728 (1994)
48. J.G Rarity and P.R.Tapster, Phil. Trans. Roy. Soc. Lond. A **355**, 2267 (1997) and references therein
49. Z.Y. Ou and L. Mandel, Phys. Rev. Lett. **61**, 50 (1988)
50. S.Savasta S and R. Girlanda, Phys. Rev. B **59**, 15409 (1999)
51. G. Messin et al., J. Phys. Cond. Matt. **11**, 6069 (1999)

# 7 Spins for Quantum Information Processing

David P. DiVincenzo

## 7.1 Introduction

*Spintronics* and *Quantum Computation* are both newly minted terms, and concepts, in physics. Depending on your point of view, they are almost exactly the same, or they are completely different. They both may be defined as a concept for using discrete, quantized degrees of freedom in a physical device to perform information-processing functions. But from this point of overlap, these two concepts are both more and less than the other. Spintronics concerns itself, of course, only with spins, spins in the solid state; quantum computing encompasses almost every possible quantum phenomenon in nature. But within its narrower setting, spintronics has a broader agenda, to facilitate and improve all possible forms of information processing using spin-based devices. Quantum computing actually has a narrower agenda: to devise and implement quantum-coherent strategies for computation and communication.

In this chapter I will explain a bit about the agenda of quantum computing, and explore its points of contact with spintronics. It seems clear that as these subjects move from theory to practice, they will interact repeatedly and profitably; I will give some indication here of how they might develop in tandem.

Spins are not the exclusive domain of solid-state physics; in quantum computing, the control of spin states is extensively studied in several areas of atomic physics and quantum optics, as well as in nuclear magnetic resonance spectroscopy. As I will mention later, I think that research in spintronics can profit from the traditions and techniques that have been developed in these other areas.

Before I start discussing how we use spins in quantum computing, I will briefly spell out the 5+2 requirements for the implementation of quantum computation that I have presented elsewhere. Here I simply quote these requirements *verbatim* from previous discussions:

### 7.1.1   The Requirements

1. *A scalable physical system with well characterized qubits*
2. *The ability to initialize the state of the qubits to a simple fiducial state, such as* $|000...\rangle$
3. *Long relevant decoherence times, much longer than the gate operation time*
4. *A "universal" set of quantum gates*
5. *A qubit-specific measurement capability*
6. *The ability to interconvert stationary and flying qubits*
7. *The ability faithfully to transmit flying qubits between specified locations*

The first five are the core requirements of quantum computation; the final two are the requirements for using quantum computation in conjunction with quantum communication. We make this distinction because there are many different ways in which quantum coherence effects can be brought into play in information processing. The best known and most dramatic quantum speedups are in computation, in the factoring of prime numbers and the breaking of other public-key crypto systems, for example. For these, just the first five items are necessary. But there are others types of tasks for which quantum effects are advantageous. Perhaps the first example of quantum information processing was quantum cryptography, in which the object is not to compute but to keep some data secret which is being communicated over a network. There are other applications that require some combination of communication and computation: in the use of entanglement to win multiparty games, in the ability to "share" a secret, in the computation of a function whose input data is held by many parties, as in appointment scheduling. In all these, it is necessary to move a qubit (that is, to move a particle holding a two-level quantum system in a definite state) from place to place; requirements 6 and 7 address this.

Spins are not generally considered as the preferred carrier of this moveable ("flying") qubit; most discussions assume that the mobile particle will be a photon. But, especially for short-distance applications, the spin of an electron travelling elastically through a quantum wire might also serve as a movable qubit, and the chapter by Burkard and Loss explores the details of this possibility. Also, as the discussion by Yamamoto et al. will show, there are schemes for converting electron spin qubits to photons, so a beginning has been made on a laboratory technique involving spins that addresses requirement 6.

But let me return to the main point of quantum computing, since the lion's share of the discussion of spin realizations have focussed on requirements 1–5. An isolated spin is a natural choice for a qubit: it is intrinsically a two-level system, unlike some other proposed solid state implementations (superconducting qubits, for example) in which the two-level structure has to be created by engineering. Perhaps as a consequence of this, the coherence times of spins in solids is quite long compared with many other degrees of freedom in the solid. One can say that hundreds of "Rabi flops" of the spin qubit have been seen in time-domain experiments such as those of Awschalom and coworkers (see Chapter 5); no such phenomenon has yet been seen in any other solid-state qubit.

Proposed solid state spin qubits come in three varieties, two of which are extensively reviewed in this volume. Burkard and Loss discuss the first proposal, made by Loss and DiVincenzo, to use the spin of a one-electron or few-electron quantum dot. Basic experiments, such as those done by Kouwenhoven [1], have indicated that the spin of a quantum dot can be controlled and characterized. A closely related second proposal, by Kane [2], considers as a qubit the spin of an isolated donor impurity in a semiconductor, such as P in Si. The basic physics of these two proposals is very similar: in both, the spin is held by an electron in a weakly localized orbital. The weakness of the localization permits the shape of the orbital to be changed by external electric gating, and it is this, as these chapters describe, that permits the control of quantum computing operations using these qubits. In another proposal, Yablonovich is pursuing a hybrid of these two different proposals, trying to incorporate the best of both [3].

A third proposal, not represented in this volume, involves the use of spin via an exciton; the idea is that the presence or absence of an exciton serves as the levels of a qubit. Experiments by Steel and co-workers[4] have indicated that this is a promising direction: an all-optical approach to manipulating excitonic qubits can be foreseen. Excitonic coherence times are not nearly so long as spin coherence times, but all-optical control has the prospect of being much faster than electronic control.

Before leaving this little discussion of the basics of spin qubits, I should mention that several authors, and Kane in particular, have suggested a rather different type of spin qubit that is rather further removed from spintronics. This is the nuclear-spin qubit. The nuclear-spin qubit has of course been very prominent in early discussions of quantum computing, in the area of bulk NMR quantum computing. But if we are to have the prospect of truly scalable and large-scale quantum computing, we probably would have to find ways of using the nuclear spin in a solid-state device. Kane has proposed a way of doing this, using the fact that the P nucleus has a spin-1/2 quantum number, and so can be used in conjunction with the electron-spin qubit in his model. Nuclear spins are frankly a nuisance in most electron-spin qubit proposals, as they provide a source of decoherence for the electron spins. Bringing them under control by making them act as qubits could ameliorate this problem. Nuclear-spin qubits may turn out to be excellent for memory, since they typically have very long decoherence times compared even with electron spins; in using them for logic, though, we will always face the disadvantage that the time required to switch a nuclear spin is always much longer than the switching time for an electron spin.

## 7.2   Timeline

Given this general assessment of the state of affairs with spintronic quantum computers, I would like to stick my neck out and offer a speculative timeline for the development of quantum computation. This will give me chance to offer my view on how the five requirements above will actually come into play as research continues along these lines. It will also provide the opportunity to offer up some quotable bloopers for the future when this timeline turns out to be complete rubbish.

As Fig. 7.1 indicates, I see there as being three main enablers in the initiation of quantum computing research. The simplest is the basic lab instrumentation that I lump under the heading of cryo. and magnets. This is to say that this research will be conducted in a rather well equipped low-temperature physics lab, with l-He and probably dilution refrigerators, multitesla magnets, and all the cleanliness and care that is associated with these. The level of infrastructure of this kind needed for quantum computing is probably no more than what we have now, there may be details like stability and homogeneity of the magnet, or quietness of the electromagnetic environment, that will need further development.

The second enabler I have indicated is the electronics or "probes." This could have been lumped in with the other laboratory infrastructure, but I think that unlike the other instrumentation, this will have to undergo significant development to meet the requirements of quantum computing. I have separated this technology into two categories, involving the delivery of electromagnetic energy in the frequency domain and the time domain. The fre-

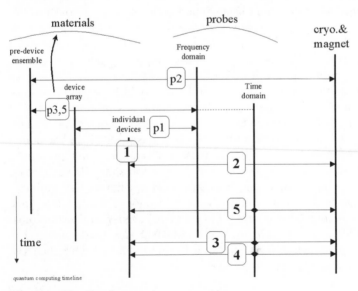

**Fig. 7.1.** Timeline for quantum computing.

quency band of interest, in the 1–100 GHz range, is routinely available in the physics laboratory, although its delivery in a clean way to a nanostructure at low temperatures is an art known only to a few. Delivering temporal electrical pulses formed from this same frequency band is an even more exotic task, and I think that some really new developments in this area will be needed for quantum computing. As I will discuss in a moment, the most interesting tests of quantum computing will require the time-domain capability.

The least predictable enabler of all, I think, will be materials and device preparation. To take one example, for the quantum-dot proposal of Loss and DiVincenzo, the ultimate desired device would have thousands or millions of quantum dots, with tailored physical properties such as strain-induced band splitting, effective mass, and g factor, each nestled into a complex enclosure of metal electrodes and magnetic and non-magnetic barriers, highly stable and free of impuririties, and with a specific isotopic composition. Even if it were clear exactly how it will function, it is a dream to imagine such a structure being put together anytime in the near future. I envision a development of this dream structure proceeding roughly in three parts:

1. First, we will consider existing materials and structures some of whose properties approximate those of the hypothetical quantum computer device. A good example of this is the semiconductor samples whose free-carrier spin dynamics have been studied by Awschalom and coworkers. This will be some approximation to a collection of qubits, so I refer to this in the figure as a "pre-device ensemble."
2. Next I speculate that we will have a "device array," which will consist of a controlled array of structures, each of which is a fair approximation to the qubit of interest; an example of this would be an array of one-electron quantum dots, missing some or most of the control apparatus.
3. Finally will come some version (doubtless initially on a small scale) of the full quantum computer structure.

The interaction of these three enablers, at all stages of their development, will permit us to begin exercising the five requirements of quantum computation. As I indicate in the figure, I think there will be a (fairly long) period in which various of the requirements will be subject to a kind of preliminary certification. One can say that this preliminary period has already begun: so, by placing a pre-device array in a magnet at low temerature, one can obtain high spin-polarization, a very simple piece of physics which serves as a preliminary achievement of requirement 2 (indicated as "p2" in the figure).

A very decisive preliminary certification could be done with the same capabilities. Measurements of the homogeneous linewidths in ensemble structures, which can be done with frequency domain techniques with some small time-domain capabilities (by spin-echo, say), can give valuable information about the coherence time of the spins. While the coherence times of the qubits made from these spins might not be the same, it would be rash to assume that they would be longer. This "p3" stage (which is also "p5", since it involves also the process of measuring the spins) can be quite decisive in that it could rule out an approach altogether, by showing that decoherence times are likely never to be long enough in the chosen materials. Thus the "back to the drawing board" arrow in the figure.

A preliminary achievement of requirement 1 can be declared quite early, but it can be an empty achievement: simply noting that electrons have a spin suffices to "demonstrate" that there are lots of qubits in the world. I would reserve this "p1" milestone for the point at which an array is first assembled is shown, by spectroscopic means, say, to have well defined two level systems.

But a real achievement of the requirements must, in my opinion, await the milestone of constructing the first real qubit device, even one with just a couple of qubits. The "1" and "2" milestones would follow almost by definition. I think that the next one likely to be achieved is "5", putting these devices together with the right time-domain probe will hopefully get us to the ability to measure specific qubits; Kane's work is, I would say, focussed on this achievement. It may yet be some time more, involving further development of precision time-domain control, that a real decoherence time is measured ("3") and real gate operations are done ("4").

It is humbling to consider that liquid-state NMR experiments and, in a different way, atomic-physics systems like ion traps, have already made it though a large part of the certification process of the figure, while solid-state systems have only begun. But the NMR work, while it has demonstrated, at a quite sophisticated level, the operation of many quantum gates and even the execution of quantum algorithms, has done so without the "cryo." enabler mentioned in the figure. It is this, primarily, that will prevent any straightforward extension of the liquid-state NMR work from becoming large-scale quantum computation. The ion trap work is well into the small-prototype stage with real qubits, although real general-purpose quantum gates have not quite been achieved. It is not a given that a solid-state quantum computer is inevitable; it could be that the "vacuum-tube" quantum computer of ion trap technology will be the way to go. But it is far too early to tell.

## 7.3  Final Thoughts

As I said at the beginning, quantum computing is less (while at the same time being more) than spintronics. But I think that the effort to make a quantum computer out of a spin-based electronic device will have definite spin offs that would aid the larger pursuit of spintronics. Quantum computing will drive a major metrological effort: magnetometry at the single-spin level is necessary for quantum information processing. The achievement of this would be a fantastic advance in the art of magnetometry, and might open up a variety of other sensing applications. Quantum computing will also require very high speed and very high precision electronic control of magnetic properties;

these specifications are surely of much more general interest, but quantum computing may have a key role in driving advances in this area. Quantum computing will require the invention of novel magnetic/semiconductor hybrid device structures, which also fits well within the overall goal of spintronics. Finally, quantum computing coupled to communication could stimulate the development of hybrid magnetic/photonic devices, and it is easy to imagine a broader technological arena in which such inventions would be useful. I am confident that quantum computing will be a powerful intellectual driver for a variety of exciting technical advances well into the future.

# References

1. T. H. Oosterkamp et al., Phys. Rev. Lett. **80**, 4951 (1998).
2. B. E. Kane, Nature **393** 133 (1998).
3. R. Vrijen et al., Phys. Rev. A **62**, 012306 (2000).
4. N.H. Bonadeo *et al.*,Science **282** 1473 (1998).

# 8  Electron Spins in Quantum Dots as Qubits for Quantum Information Processing

Guido Burkard and Daniel Loss

Coherent manipulation, filtering, and measurement of electronic spin in quantum dots and other nanostructures are new technologies which have promising applications both in conventional and in quantum information processing and transmission. We review the spintronics proposal for quantum computing, in which electron spins in quantum confined structures play the role of the quantum bits (qubits), and discuss the essential requirements for such an implementation. We describe several realizations of one- and two-qubit quantum gates and of state preparation and measurement, based on an all-electrical scheme to control the dynamics of spin. We discuss recently proposed schemes for using a single quantum dot as a spin filter and spin read-out device, and show how the decoherence time can be measured in a transport set-up. We address the issue of spin decoherence due to non-uniform hyperfine interactions with nuclei and show that for electrons confined to dots the spin decay is non-exponential. Finally, we discuss methods for producing and detecting the spin-entanglement of electronic EPR pairs, being an important resource for quantum communication.

## 8.1  Introduction

The idea to use the spin as an additional degree of freedom in electronics [1,2] has recently received strong support from experiments [3–5] showing unusually long spin dephasing times [3] in semiconductors (approaching microseconds), the injection of spin-polarized currents into semiconductor material [4,5], ultrafast coherent spin manipulation [6], as well as phase-coherent spin transport over distances of up to $100\,\mu$m [3]. This offers possibilities for finding new mechanisms for information processing and transport, both in conventional devices [1,2] as well as in quantum confined structures used for quantum computing [7]. In conventional devices, the electron spin has the potential to improve the performance, e.g. in spin-transistors (based on spin-currents and spin injection), non-volatile memories, memories based on single spins, etc. The spin of the electron is a most natural candidate for the quantum bit (qubit) [7,8] because every spin 1/2 encodes *exactly* one qubit. Treating the spin-orbit interaction and the coupling to the environment as small perturbations this means that one possible source of errors in quantum computation – leakage into "undefined" parts of the Hilbert space – has already been avoided.

Quantum computing [9,10] and quantum communication [10,11] require a radically new approach to the design of the necessary hardware. It was shown earlier [7] that the spin qubits, when located in quantum-confined structures such as semiconductor quantum dots, atoms, or molecules, satisfy all requirements necessary for a scalable quantum computer. In addition to that, being attached to a mobile electron, a spin-qubit can be transported along conducting wires [10]. An interesting application of this possibility appears to be the creation (say, in coupled quantum dots or near a superconductor-normal interface) and transport of spin-entangled electrons which act as Einstein-Podolsky-Rosen (EPR) pairs [10]. Such EPR pairs represent the fundamental resources for quantum communication [11].

In this chapter we will review the current status of our theoretical efforts toward the goal of implementing quantum computation and quantum communication with electron spins in quantum-confined nanostructures.

### 8.1.1 Quantum Computing

The interest in quantum computing is founded in the hope to outperform classical computers through new quantum algorithms which make use of the quantum computer's abilities to exist in a quantum superpositions of its "binary" basis states $|0\cdots00\rangle$, $|0\cdots01\rangle$, $|0\cdots10\rangle$,..., and to perform quantum time evolutions $U|\Psi_{\text{in}}\rangle = |\Psi_{\text{out}}\rangle$ for computation. Each qubit in a quantum computer is allowed to be in any state of a quantum two-level system (an example thereof being a spin $1/2$). For practical purposes it is very useful to know that all possible quantum algorithms can be implemented by concatenating one- and two-qubit gates [12], therefore it is sufficient when we discuss these two types of operations below in Sects. 8.3 and 8.4. It is proven that for some classes of problems speedups are impossible, however there are problems for which very powerful quantum algorithms have been discovered [13,14] which outperform all known classical algorithms. Furthermore, there is now a growing list of other quantum tasks [10,11] such as cryptography, error correcting, teleportation, etc. that have indicated even more the desirability of experimental implementations of quantum hardware.

The number of proposed physical implementations of qubits and quantum gates is growing rapidly. Some examples are: Trapped ions [15], cavity QED [16], nuclear magnetic resonance [17], superconducting devices [18–21], and our proposal [7] for using the electron spin in quantum-confined nanostructures, which is based on the Heisenberg exchange interaction between spins being controlled via electrical gates. There are a number of subsequent spin-qubit schemes in solid state also based on the exchange coupling [22–26]. In particular, Kane [23] discussed a modification of the spin scheme [7] by considering the nuclear spin of P atoms in Si as qubit. However, the basic concept in this proposal [23] – manipulation of spin and exchange via electrical gates – follows very closely the one introduced in [7].

### 8.1.2   Quantum Communication

The production of entangled pairs of particles for various tasks in quantum communication has already been shown and used in many experiments in quantum optics (see e.g. [27]), using parametric downconversion. A disadvantage of these sources is that entangled particles can only be produced in a stochastic manner and at a very low rate. For most applications in quantum communications it would be desirable to have a *deterministic* source which in addition emits EPR pairs at a rate as high as possible. Coupled quantum dots provide a natural source of such deterministic entanglement between qubits of localized but also of delocalized electrons [10,7]. This means that with a pair of tunnel-coupled quantum dots (acting as a quantum gate) it is possible to create a singlet state (one realization of an EPR pair) out of two uncorrelated electrons and subsequently separate (by electronic transport) the two electrons spatially while maintaining the entanglement of their spins. Another possibility is the use of s-wave superconductor-normal junctions as a source of spin-entangled electrons, exploiting the fact that the Cooper pairs are in a singlet (and therefore entangled) state. The implementation of such devices would allow the study of a new class of quantum phenomena in electronic nanostructures [10] such as the entanglement and non-locality of electronic EPR pairs, tests of Bell inequalities, quantum teleportation [28], and quantum cryptography [29], which promises secure transmission of information. Coupled quantum dots could also be used as an efficient and deterministic source for entangled photons obtained from the recombination of biexciton states in these structures [30]. These issues will be discussed in Sect. 8.7.

### 8.1.3   Quantum Dots

In the work which we discuss here, quantum dots play a central role and thus some general remarks about them are in order. Quantum dots are structures where charge carriers are confined in all three spatial dimensions. For quantum dots which are formed by confinement of the carriers in a metallic or semiconducting material, this means that the dot size has to be of the order of the Fermi wavelength in the host material, in semiconductor quantum dots typically between $10\,\text{nm}$ and $1\,\mu\text{m}$ [31]. The confinement is usually achieved by electrical gating of a two-dimensional electron gas (2DEG), possibly combined with etching techniques, see Fig. 8.1. Precise control of the number of electrons in the conduction band of a quantum dot (starting from zero) has been achieved in GaAs heterostructures [32]. Under the application of an external magnetic field the electronic spectrum of typical quantum dots can vary strongly [31,32], because the magnetic length pertaining to typical laboratory fields $B \approx 1\,\text{T}$ is comparable to typical dot sizes. In coupled quantum dots Coulomb blockade effects [33], tunneling between neighboring dots [31,33], and magnetization [34] have been observed as well as the formation of a delocalized single-particle state [35].

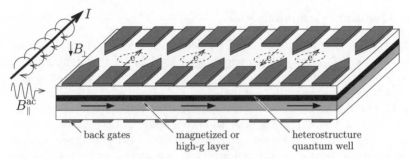

**Fig. 8.1.** Quantum dot (QD) array which can be controlled by electrical gating. The electrodes (dark gray) confine electrons, thus forming the QDs (circles). Electrons can be moved into the magnetized or high-$g$ layer by changing the gate voltage, allowing for spatially varying Zeeman splittings. Alternatively, local magnetic fields can be achieved by a current-carrying wire (indicated on the left of the QD array). Since the electron in each QD is subject to a distinct Zeeman splitting, the spins can be addressed individually, e.g. through ESR pulses of an additional in-plane magnetic ac field with the corresponding Larmor frequency $\omega_L$. This mechanism can be used for single-spin rotations and the initialization step (Sects. 8.2.7 and 8.4). The exchange coupling between the QDs can be controlled by lowering the tunnel barrier between the dots, see Sect. 8.3.1. The two rightmost QDs are drawn schematically as tunnel-coupled. Note the all-electrical control of the spin dynamics.

## 8.2 Requirements for Quantum Computing

### 8.2.1 Coherence

In theory, any quantum two-level system could represent a qubit. In practice, however, quantum systems are never truly isolated and therefore the problem of decoherence and the transition to classical behavior has to be addressed. For quantum computing, only systems with a weak coupling to their environment and therefore with slow dephasing are interesting. One has to distinguish between the relaxation time $T_1$ and the (transverse) decoherence time $T_2$. Usually $T_2 \leq T_1$, thus $T_2$ is the more relevant quantity for quantum computing. In the study of mesoscopic systems, much attention has been devoted to characterizing and understanding the decoherence of electrons in small structures. However, most of what has been probed (say in weak localization studies or the Aharonov-Bohm effect) is the *orbital* (i.e. charge) coherence of electron states, i.e. the preservation of the relative phase of superpositions of spatial states of the electron (e.g., in the upper and lower arm of an Aharonov-Bohm ring). The coherence times seen in these experiments are almost completely irrelevant to the *spin* coherence times which are important in the quantum computer proposal described here. Charge dephasing may have some effect on spin dephasing if there is strong spin-orbit coupling, but our intention is that conditions and materials should be chosen such that these effects are small.

Under these circumstances the spin coherence time – the time over which the phase of a superposition of spin-up and spin-down states is well-defined – can be completely different from the charge coherence time (a few nanoseconds), and in fact it is known from experiment that they can be orders of magnitude longer (see below). This was one of our main motivations for proposing spin [7] rather than charge as the qubit in these structures. For a detailed description of the magneto-optical experiments employed to measure the spin coherence time in bulk GaAs and a 2DEG [3], and in CdSe quantum dots [36], we refer the reader to Chap. 5 of this book (for theory see also Ref. [37]). Since none of the experiments have been done on an actual quantum computing structure as we envision it (see below), the existing results cannot be viewed as conclusive but certainly point into the right direction. However, we have proposed to measure $T_2$ of a single spin in such a device directly via a transport experiment by applying electron spin resonance (ESR) techniques [38]. The stationary current exhibits a resonance whose line width is determined by the single-spin decoherence time $T_2$ (see Sect. 8.5.6).

In general, theory can only give guidance about the mechanisms and dependencies to be looked for, but it seems rather difficult to make reliable quantitative predictions of the decoherence times, ultimately they must be extracted from experiment. In this sense, in the following Section 8.2.2, we report on recent calculations identifying the dominant spin-orbit effects in GaAs quantum dots leading to unusually low phonon-assisted spin-flip rates, which suggest also long spin-decoherence times, although more work is required here.

Yet another source of spin decoherence are spin flips caused by the nuclear spins of the host material [39]. Their rate can be suppressed by a factor $(B_n^*/B)^2/N$, where $B_n^* = AI/g\mu_B$ is the maximal magnitude of the effective nuclear field (Overhauser field), $N$ the number of nuclear spins in the vicinity of the electron, and $A$ the hyperfine coupling constant. For GaAs, the nuclear spin of both Ga and As is $I = 3/2$. In the suppression factor above, $B$ denotes either the external field, or, in the absence of an external field, the Overhauser field $B = pB_n^*$ due to the nuclear spin polarization $p$, which can be obtained e.g. by optical pumping [40] or by spin-polarized currents at the edge of a 2DEG [41]. In the latter case, the suppression of the spin flip rate becomes $1/p^2N$ [39].

## 8.2.2  Slow Spin Relaxation in GaAs Semiconductor Quantum Dots

We discuss next spin-flip processes due to phonon-assisted transitions between different discrete energy levels (or Zeeman sublevels) in GaAs quantum dots [42]. The analysis is based on a one-electron Hamiltonian approach derived from the Kane model [43]. It turns out that there are a number of different mechanisms originating from spin-orbit coupling which lead to such spin flip processes. The corresponding rates $\Gamma = T_1^{-1}$ can be evaluated in leading perturbation order, with and without a magnetic field, and it is seen that the most effective spin-flip mechanisms in 2D are related to the broken inversion symmetry, either in the elementary crystal cell or at the heterointerface. The relevant Hamiltonian for the electron, being linear in the 2D momentum operator $p_i$, i=x,y ([100] orientation of the 2D plane), is given by

$$H_{so} = \sum_{i,j=x,y} \beta_{ij} \sigma_i p_j. \tag{8.1}$$

The constants $\beta_{ij}$ depend on the strength of the confinement in z-direction and take values in the interval $(1 \div 3) \cdot 10^5$ cm/s for GaAs heterostructures. In the presence of the spin-orbit terms $H_{so}$ the spin-up state of the electron actually acquires a small admixture of the spin-down state. This leads to a non-vanishing matrix element of the phonon-assisted transition between two states with opposite spins. However, the spin relaxation of the electrons localized in the dots differs strongly from that of delocalized electrons. In quantum dots (in contrast to extended 2D states), one can remove the terms linear in $\beta$ from the Hamiltonian by a spin-dependent unitary transformation (the boundedness of the electron wave functions is essential to this procedure). As a result, the contributions to the spin-flip rate proportional to $\beta^2$ are absent in general, and appear only if we take into account either the admixture of higher states of the size quantization in the z-direction, i.e. the weak deviation from the true 2D motion, or higher orders in the expansion of the momentum in the plane, $\propto p^3$. Thus, such contributions acquire extra small factors: either the ratio of the lateral kinetic energy $E_{lat}$ to the distance between the quantized levels in the z-direction $E_z$, or $E_{lat}/E_g$, where the band gap in GaAs is approximately $E_g \approx 1.5$ meV. This results in unusually low spin-flip rates of electrons confined to dots. The finite Zeeman splitting in the energy spectrum also leads to contributions $\propto \beta^2$,

$$\Gamma \simeq \Gamma_0(B) \left( \frac{m\beta^2}{\hbar\omega_0} \right) \left( \frac{g\mu_B B}{\hbar\omega_0} \right)^2, \tag{8.2}$$

where $\hbar\omega_0$ is the distance between the orbital levels in the dot and $\Gamma_0(B)$ is the inelastic rate without a spin flip for the transition between neighbouring orbital levels. This rate depends on the energy transfer and for small energy transfer is determined by the interaction with piezo-phonons.

In the case of spin-flip transitions between Zeeman sublevels (when the energy transfer is small compared to $\hbar\omega_0$) the fact that the linear-in-$\beta$ spin-flip matrix element is already proportional to the Zeeman splitting leads to a spin-flip rate which, at low temperature, is proportional to the fifth power of the Zeeman splitting (three additional powers come from the phonon density of states),

$$\Gamma_z \simeq \frac{(g\mu_B B)^5}{\hbar(\hbar\omega_0)^4} \Lambda_p. \tag{8.3}$$

The dimensionless constant $\Lambda_{\rm p} \propto \beta^2$ characterizes the strength of the effective spin–piezo-phonon coupling in the heterostructure and ranges from $\approx 7 \cdot 10^{-3}$ to $\approx 6 \cdot 10^{-2}$ depending on $\beta$. To give a number, $\Gamma_z \approx 1.5 \cdot 10^3\,{\rm s}^{-1}$ for $\hbar\omega_0 = 10\,{\rm K}$ and at a magnetic field $B = 1\,{\rm T}$. Two-phonon contributions to the spin-flip transitions between Zeeman sublevels are also considered. It is shown that the two-phonon contributions prevail at sufficiently small Zeeman splittings. However, if the temperature is much smaller than $\hbar\omega_0$ the characteristic Zeeman splittings below which the two-phonon contribution to the spin-flip rate dominates are small and the corresponding spin-flip times are unusually long. For example, for $\hbar\omega_0 \approx 10\,{\rm K}$ this time is much longer than $1\,{\rm ms}$.

The spin-flip rates $1/T_1$ due to several other mechanisms (not related to the spin-orbit interaction) have also been estimated. Among them: 1) modulation of the hyperfine coupling with nuclei by lattice vibrations (note that this mechanism is not identical to the one described at the end of Sect. 8.2.1, which does not involve phonons), 2) the spin-spin interaction between the bound electron and the conduction electron in the leads, 3) the spin-current interaction, when the bound electron spin flip is caused by the fluctuating magnetic field of the conduction electrons. It is shown in the case of GaAs quantum dots that these latter spin-flip effects are negligible compared to the spin-orbit admixture mechanism.

More recently we have analyzed in more detail (see also end of previous subsection) the spin decoherence of a single electron confined to an isolated quantum dot due to the hyperfine interaction of the electron spin with the nuclei [44]. A rather rich behavior has emerged. The corresponding decoherence time $T_2$ is found to be shorter than the nuclear spin relaxation time $T_{n2}$ determined by the dipole-dipole interaction between nuclei. Therefore the problem can be considered in the absence of the dipole-dipole interaction between nuclei. The decoherence is caused by the spatial variation of the electron density, corresponding to a given discrete quantum state in the dot; this variation results in a wide distribution of the local values of the hyperfine interaction. We have found that in a weak external Zeeman field, which is smaller than a typical fluctuating nuclear magnetic field seen by the electron spin through the hyperfine interaction (on the order of 100 Gauss in a GaAs quantum dot), the perturbative treatment of the electron spin decoherence breaks down. It is found that the decay of the spin precession amplitude in time due to hyperfine interaction is not exponential, instead it is either described by a power law, $1/t^{d/2}$ (for finite Zeeman fields) or an inverse logarithm, $1/(\ln t)^{d/2}$ (for vanishing fields). The corresponding characteristic inverse decay time $1/T_2$ is roughly given by $A/\hbar N$, where $A$ is the hyperfine interaction constant, and $N$ is the number of nuclei within the dot, with $N$ typically $10^5$. This time is of the order of several $\mu$s. But we emphasize that there is no simple exponential decay anymore, and one needs to be cautious with the concept of $T_1$ and $T_2$ times. The many-body character of the problem manifests itself also in the fact that there is a continuum of excitations in the spectral density (with edge singularities), which leads, in 3d, to a resonance-like behavior at a certain Zeeman field value, $\epsilon_z = Ab/2$, where $b > 1$ is some non-universal number (depending on the dot shape), resulting in a cross-over behavior to a slower decay from $1/t^{3/2}$ to $1/t^{1/2}$, and to a vanishing of one of the Fourier components of the precession. This behavior is seen in the exactly solvable case of a fully polarized nuclear spin state. The amplitude of the precession part, finally reached as a result of the decay, is of order one, while the decaying part is $1/N$, in agreement with earlier results [39], discussed above. Furthermore it is also seen that there is a large difference between the values of $T_2$ (decoherence time for a single dot) and $T_2^\star$ (dephasing time for an ensemble of dots), i.e. $T_2^\star \ll T_2$. This result shows that it is desirable to have direct experimental access to single spin decoherence times.

### 8.2.3   Scalability

The speedup of quantum over classical computation can only be realized if a large number of qubits (on the order of $10^5$) is available. Thus it is essential to find a concept for quantum computing which is scalable to such a large number of qubits. Note that known error correction schemes (see Sect. 8.2.5) require that these qubits have to be operated on in parallel. The scaling requirement is well achievable with spin-based qubits confined to quantum dots, since the production of arrays of quantum dots [10,45] is feasible with today's technology, cf. Fig. 8.1. Certainly, the actual implementation of such arrays including all the needed circuits poses new experimental challenges, but at least we are not aware of physical restrictions which would exclude such an upscaling for spin-qubits.

### 8.2.4   Switching

The quantum gate operations in our model will be controlled through the effective Hamiltonian

$$H(t) = \sum_{i,j} J_{ij}(t)\,\mathbf{S}_i \cdot \mathbf{S}_j + \sum_i \mu_B g_i(t)\,\mathbf{B}_i(t) \cdot \mathbf{S}_i, \tag{8.4}$$

see also Sects. 8.3 and 8.4. The Hamiltonian depends on time via external control fields $v(t)$, such as gate voltages, etc., which determine the values of the coupling strengths $P(v(t))$ in the effective Hamiltonian, such as $J_{ij}(t)$ for the exchange coupling, and $g_i(t)\mu_B \mathbf{B}_i(t)$ for the (effective) Zeeman coupling. Note that in the following we assume the exchange coupling to be local, i.e. $J_{ij} \neq 0$ only for nearest neighbor qubits. We note however that in cavity-QED systems, there is also a long-range coupling of qubits as described in Ref. [46]; similarly long range coupling via a superconductor is possible [47], see also Sect. 8.7.6. Even if the exchange coupling is local, operations on non-neighboring qubits can be performed. This is done by swapping the state of two qubits with the help of the exchange interaction only, as we will show in Sect. 8.3. Thus, a qubit can be transported to a place where it can be coupled with a desired second qubit, or where single-qubit operations can be performed, or where it can be measured.

The mechanisms described in Sect. 8.3 and 8.4 below for performing gate operations with spin qubits are independent of the details of the pulse shape $P(v(t))$, where $P$ stands for the exchange coupling $J$ or the Zeeman interaction. It is only the value of the integral $\int_0^\tau P(v(t))dt$ (mod $2\pi$) which determines the quantum gate action. This is true provided that the parameters $P(v(t))$ are switched adiabatically, i.e. such that $|\dot{v}/v| \ll \delta\varepsilon/\hbar$, thus guaranteeing the validity of the effective Hamiltonian (8.4). If spin is conserved, the energy-level separation of a single dot is the energy scale $\delta\varepsilon$ on which excitations may occur. A rectangular pulse leads to excitation of higher levels, whereas an adiabatic pulse with amplitude $v_0$ is e.g. given by [48] $v(t) = v_0 \operatorname{sech}(t/\Delta t)$ where $\Delta t$ controls the width of the pulse. A switching time $\tau_s > \Delta t$ has to be used such that $v(t = \tau_s/2)/v_0$ becomes vanishingly small. We then have $|\dot{v}/v| = |\tanh(t/\Delta t)|/\Delta t \leq 1/\Delta t$, so we need $1/\Delta t \ll \delta\varepsilon/\hbar$ for adiabatic switching. The Fourier transform $v(\omega) = \Delta t v_0 \pi \operatorname{sech}(\pi\omega\Delta t)$ has the same shape as $v(t)$ but width $2/\pi\Delta t$. In particular, $v(\omega)$ decays exponentially in the frequency $\omega$, whereas it decays only with $1/\omega$ for a rectangular pulse.

The validity of the adiabatic approximation was recently checked explicitly by numerically integrating the microscopic Hamiltonian underlying the exchange coupling between two spins in the Hund-Mulliken approximation which includes double occupation of the dots [49]. The agreement between the two is found to be very good, even if $|\dot{v}/v| \approx \delta\varepsilon/\hbar$. Here, $\delta\varepsilon$ is given by the on-site repulsion $U_H$, i.e. the energy required for double occupation of a dot. The unwanted admixture of a state with double occupation of a dot in the final state is found to be tiny if the sech pulse is used and the adiabaticity criterion if fulfilled. Therefore, double occupancy poses no problem for quantum gate operation. The result of a numerical simulation is plotted in Fig. 8.2.

A single qubit operation can be performed for example using g-factor modulated materials, as described in Sect. 8.4. A spin is rotated by a relative angle of $\phi = \Delta g_{\text{eff}} \mu_B B\tau/2\hbar$ by changing the effective g-factor by $\Delta g_{\text{eff}}$ for a time $\tau$. Thus, a typical switching time for an angle $\phi = \pi/2$, a field $B = 1\,\text{T}$, and $\Delta g_{\text{eff}} \approx 1$ is $\tau_s \approx 30\,\text{ps}$. If slower operations are required, they are easily implemented by choosing a smaller $\Delta g_{\text{eff}}$, reducing the magnitude of the field $B$, or by replacing $\phi$ by $\phi + 2\pi n$ with integer $n$, thus "overrotating" the spin.

We proceed by discussing the adiabaticity criterion for the square-root-of-swap gate (Sect. 8.3) which is obtained under the condition $\int_0^{T_s} J(t)dt/\hbar = \pi/2$. We use a pulse (see Sect. 8.2.4) $J(t) = J_0 \operatorname{sech}(t/\Delta t)$. For typical num-

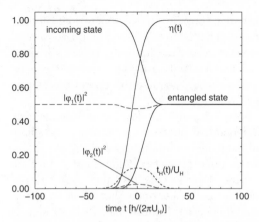

**Fig. 8.2.** Numerical simulation of a square-root-of-swap gate operation [49] (defined in Sect. 8.3), producing the entangled state $a|\uparrow\downarrow\rangle + b|\downarrow\uparrow\rangle$ (with $|a|^2 = |b|^2 = 1/2$) from an unentangled incoming state $|\uparrow\downarrow\rangle$ (cf. Sect. 8.5.3). The probability of double occupancies $|\varphi_2|^2$ (due to non-adiabatic corrections) is strongly suppressed after the tunneling pulse $t_H(t)/U_H$. The quantum mechanical weights $|a|^2$ and $|b|^2$ are plotted as thick solid lines. The amount of entanglement $\eta(t)$ is also plotted as a function of time (see Ref. [49] for a definition and discussion of $\eta$).

bers $J_0 = 80\,\mu eV \approx 1\,K$ and $\Delta t = 4\,ps$ we obtain the switching time $\tau_s \approx 30\,ps$, while the adiabaticity criterion is $\hbar/\Delta t \approx 150\,\mu eV \ll \delta\varepsilon$. Reducing the dot size further, there is no principal limitation to reach switching times in the femtosecond range. We note that such fast switching is made possible by the strong Coulomb interaction (see below). Again, the switching time can be easily increased by adding $2\pi n$ with integer $n$ to the integrated pulse $\int_0^{\tau_s} J(t)/\hbar$, i.e. by "overswapping" the two spins. This increased switching time allows a slower switching of $J(t)$ if required. We also note that the total time consumed by an algorithm can be optimized considerably by simultaneously switching different parameters of the Hamiltonian, i.e. producing parallel instead of serial pulses. As an example, we have shown that for an error-correcting algorithm using only three qubits, a speed-up of a factor of two can be achieved [48]. For algorithms handling a larger number of qubits, we expect a more drastic optimization.

### 8.2.5   Quantum Error Correction

Quantum error correction [50] is an indispensable ingredient for the physical realization of scalable quantum computation. Schemes for fault-tolerant quantum computation exist, meaning that a reliable computation (including error correction itself) can be performed using only error-prone qubits and quantum gates. The known schemes for fault-tolerant quantum computation work if the error rate does not exceed a certain threshold, usually about $10^{-4}$

(depending on the exact scheme) [51]. It we take the ratio of the dephasing time from Sect. 8.2.1, $T_2 \geq 100$ ns, and the switching times from Sect. 8.2.4, $\tau_s \approx 30$ ps, we obtain a value very close to this threshold. Once the error rate is below the threshold, an arbitrary upscaling of a quantum computer becomes feasible and is no further limited by decoherence and lacking gate precision, at least when systems with a scalable number of qubits are considered. We note that a larger number of qubits also requires a larger total number of gate operations to be performed, in order to implement the error-correction schemes. Therefore it is inevitable to perform these operations in parallel; otherwise the pursued gain in computational power is used up for error correction. Hence, one favors concepts where a localized control of the gates can be realized such that operations can be performed in parallel. However, since there are still many milestones to reach before sophisticated error-correction schemes can be applied, one should by no means disregard setups where gate operations are performed in series.

## 8.2.6  Gate Precision

A quantum computation cannot only be spoiled by decoherence, but also by the limited precision of the gates, i.e. by the limited precision with which the Hamiltonian is controlled. In order for error correcting schemes to work, the (time integrated) exchange and Zeeman interaction need to be controlled again in about one part in $10^4$ [51]. This requirement applies for all quantum computer proposals, and it emphasizes the importance of gates with fine control. After a gate operation was performed on two qubits, one should be able to turn off the coupling between these qubits very efficiently, e.g. exponentially in the external fields, such that errors resulting from the remaining coupling can be reduced efficiently (if there is still a remaining coupling this can easily result in correlated errors which would pose new problems since standard error correction schemes explicitly exclude them). The exchange coupling between two quantum dots can be indeed suppressed *exponentially*, as we will describe below in Sect. 8.3. A further possible source of errors are fluctuating charges in the environment (e.g. moving charges in the leads attached to the gates) since they can lead to unknown shifts of the electrostatic potentials raised and lowered for switching. However, it is known from experiments on single quantum dots that such charge fluctuations can be controlled on the scale of hours [52] which is sufficiently long on the time-scale set by the spin decoherence time which can be on the order of $10^{-6}$ s. Still, the ability to suppress $1/f$ noise will be very important for well-controlled switching in quantum computation. Finally, we note that uncontrolled charge switching is not nearly so great a problem for spin qubits as for charge qubits, since this switching does not couple directly to the spin degree of freedom.

### 8.2.7  Initialization

As a starting point for most quantum algorithms as well as for error correcting schemes, initialized qubits are required, i.e. qubits in a well defined state such as spin up, $|\uparrow\rangle$. Single spins can be polarized by exposing them to a large magnetic field $g\mu_B B \gg kT$ and letting them relax to the ground state. This magnetic field could be applied locally or realized by forcing the electrons (via external gates) into a magnetized layer/dot, into a layer with a different effective g- factor [7,10] or with polarized nuclear spins (Overhauser effect) [39] etc., see also Fig. 8.1 and Sect. 8.4. If a spin-polarized current can be produced, such as by spin-polarizing materials [4,5] or by spin-filtering with the help of another dot [53] (see Sect. 8.5.4), polarized electrons can be injected into an empty quantum dot.

For some algorithms, it is favorable to start with a given initial state, such as $|0110\ldots\rangle$, instead of a ground state $|0000\ldots\rangle$. This can be readily implemented with spins as qubits using standard electron spin resonance (ESR) techniques [39]: We start with a ground state $|0000\ldots\rangle$. Then we produce a Zeeman splitting by applying a static local magnetic field for these spins, which should be initialized into state $|1\rangle$. An ac magnetic field is then applied perpendicularly to the first field with a resonant frequency that matches the Larmor frequency $\omega_L = g\mu_B B/\hbar$. Due to paramagnetic resonance, this causes spin-flips in the quantum dots with the corresponding Zeeman splitting, thus producing the desired state. We note that since we do not want to affect the other spins (having a different Zeeman splitting) the amplitude of the ac field must be switched adiabatically, see Sect. 8.2.4. Of course, spin precession can also be used to perform single-spin rotations (see Sect. 8.4).

## 8.3  Coupled Quantum Dots as Quantum Gates

No computer design can do without a multi-(qu)bit gate, which allows calculations through logical combination of several (qu)bits. Since two-qubit gates are (in combination with single-qubit operations) sufficient for quantum computation [12] – they form a universal set – we now focus on a mechanism that couples pairs of spin-qubits. Such a mechanism exists in coupled quantum dots, resulting from the combined action of the Coulomb interaction and the Pauli exclusion principle. Two coupled electrons in the absence of a magnetic field have a spin-singlet ground state, while the first excited state in the presence of sufficiently strong Coulomb repulsion is a spin triplet. Higher excited states are separated from these two lowest states by an energy gap, given either by the Coulomb repulsion or the single-particle confinement. The low-energy dynamics of such a system can be described by the effective Heisenberg spin Hamiltonian

$$H_s(t) = J(t)\, \mathbf{S}_1 \cdot \mathbf{S}_2, \tag{8.5}$$

where $J(t)$ denotes the exchange coupling between the two spins $\mathbf{S}_1$ and $\mathbf{S}_2$, i.e. the energy difference between the triplet and the singlet. After a pulse of $J(t)$ with $\int_0^{\tau_s} dt J(t)/\hbar = J_0 \tau_s/\hbar = \pi$ (mod $2\pi$), the time evolution $U(t) = T \exp(i \int_0^t H_s(\tau) d\tau/\hbar)$ corresponds to the "swap" operator $U_{\text{sw}}$, whose application leads to an interchange of the states in qubit 1 and 2 [7]. While $U_{\text{sw}}$ is not sufficient for quantum computation, any of its square roots $U_{\text{sw}}^{1/2}$, say $U_{\text{sw}}^{1/2}|\phi\chi\rangle = (|\phi\chi\rangle + i|\chi\phi\rangle)/(1 + i)$, turns out to be a *universal* quantum gate. Thus, it can be used, together with single-qubit rotations, to assemble any quantum algorithm. This is shown by constructing the known universal quantum gate XOR[54] (also known as CNOT), through combination of $U_{\text{sw}}^{1/2}$ and single-qubit operations $\exp(i\pi S_i^z/2)$, applied in the sequence [7],

$$U_{\text{XOR}} = e^{i(\pi/2)S_1^z} e^{-i(\pi/2)S_2^z} U_{\text{sw}}^{1/2} e^{i\pi S_1^z} U_{\text{sw}}^{1/2}. \tag{8.6}$$

These universal gates allow us to reduce the study of general quantum computation to the study of single-spin rotations (see Sect. 8.4) and the *exchange mechanism*, in particular how $J(t)$ can be controlled experimentally. The central idea is that $J(t)$ can be switched by raising or lowering the tunneling barrier (via electrical gates) between the dots. In the following, we shall review our detailed calculations to describe such a mechanism. We note that the same principles can also be applied to other spin systems in quantum-confined structures, such as coupled atoms in a crystal, supramolecular structures, overlapping shallow donors in semiconductors [23,25] etc., using similar methods as explained below. We point out that, beyond the mechanisms described in Sect. 8.3.1 and Sect. 8.3.2, spins in quantum dots can also be coupled on a long distance scale by using a cavity-QED scheme [46] or by using superconducting leads to which the quantum dots are attached [47], see Sect. 8.7.6.

### 8.3.1  Lateral Coupling

First, we consider a system of two coupled quantum dots in a two-dimensional electron gas (2DEG), containing one (excess) electron each, as described in Sect. 8.1.3. The dots are arranged in a plane, at a sufficiently small distance $2a$, such that the electrons can tunnel between the dots (for a lowered barrier) and an exchange interaction $J$ between the two spins is produced. We model this system of coupled dots with the Hamiltonian $H = \sum_{i=1,2} h_i + C + H_Z$ $= H_{\text{orb}} + H_Z$, where the single-electron dynamics in the 2DEG ($xy$-plane) is described through

$$h_i = \frac{1}{2m}\left(\mathbf{p}_i - \frac{e}{c}\mathbf{A}(\mathbf{r}_i)\right)^2 + V(\mathbf{r}_i), \tag{8.7}$$

with $m$ being the effective mass and $V(\mathbf{r}_i)$ the confinement potential as given below. A magnetic field $\mathbf{B} = (0, 0, B)$ is applied along the $z$-axis, which couples to the electron spin through the Zeeman interaction $H_Z$ and to the charge through the vector potential $\mathbf{A}(\mathbf{r}) = \frac{B}{2}(-y, x, 0)$. In almost depleted regions, like few-electron quantum dots, the screening length $\lambda$ can be expected to be much larger than the screening length in bulk 2DEG regions (where it is 40 nm for GaAs). Thus, for small quantum dots, say $\lambda \gg 2a \approx 40$ nm, we need to consider the bare Coulomb interaction $C = e^2/\kappa|\mathbf{r}_1 - \mathbf{r}_2|$, where $\kappa$ is the static dielectric constant. The confinement and tunnel-coupling in (8.7) for laterally aligned dots is modeled by the quartic potential

$$V(x,y) = \frac{m\omega_0^2}{2}\left[\frac{1}{4a^2}\left(x^2 - a^2\right)^2 + y^2\right], \tag{8.8}$$

with the inter-dot distance $2a$ and $a_\mathrm{B} = \sqrt{\hbar/m\omega_0}$ the effective Bohr radius of the dot. Separated dots ($a \gg a_\mathrm{B}$) are thus modeled as two harmonic wells with frequency $\omega_0$. This is motivated by the experimental evidence that the low-energy spectrum of single dots is well described by a parabolic confinement potential [32].

Now we consider only the two lowest orbital eigenstates of $H_\mathrm{orb}$, leaving us with one symmetric (spin-singlet) and one antisymmetric (spin-triplet) orbital state. The spin state for the singlet is $|S\rangle = (|\uparrow\downarrow\rangle - |\downarrow\uparrow\rangle)/\sqrt{2}$, while the triplet spin states are $|T_0\rangle = (|\uparrow\downarrow\rangle + |\downarrow\uparrow\rangle)/\sqrt{2}$, $|T_+\rangle = |\uparrow\uparrow\rangle$, and $|T_-\rangle = |\downarrow\downarrow\rangle$. For temperatures with $kT \ll \hbar\omega_0$, higher-lying states are frozen out and $H_\mathrm{orb}$ can be replaced by the effective Heisenberg spin Hamiltonian (8.5). The exchange energy $J = \epsilon_\mathrm{t} - \epsilon_\mathrm{s}$ is given as the difference between the triplet and singlet energy. For calculating these energies, we use the analogy between atoms and quantum dots and make use of variational methods similar to the ones in molecular physics. Using the Heitler-London ansatz with ground-state single-dot orbitals, we find [39],

$$J = \frac{\hbar\omega_0}{\sinh\left(2d^2\left(2b - 1/b\right)\right)}\left\{\frac{3}{4b}\left(1 + bd^2\right)\right.$$

$$\left. + c\sqrt{b}\left[e^{-bd^2} I_0\left(bd^2\right) - e^{d^2(b-1/b)} I_0\left(d^2\left(b - 1/b\right)\right)\right]\right\}, \tag{8.9}$$

where we have introduced the dimensionless distance $d = a/a_\mathrm{B}$ between the dots and the magnetic compression factor $b = B/B_0 = \sqrt{1 + \omega_L^2/\omega_0^2}$ with the Larmor frequency $\omega_L = eB/2mc$. The zeroth order Bessel function is denoted by $I_0$. In (8.9), the first term comes from the confinement potential, while the terms proportional to the parameter $c = \sqrt{\pi/2}(e^2/\kappa a_\mathrm{B})/\hbar\omega_0$ result from the Coulomb interaction $C$; the exchange term is recognized by its negative sign. We are mainly interested in the weak coupling limit $|J/\hbar\omega_0| \ll 1$, where the ground-state Heitler-London ansatz is self-consistent. We plot $J$ (8.9) in Fig. 8.3 as a function of $B$ and $d$. We note that $J(B{=}0) > 0$, i.e. the singlet is the ground state, which is generally true for a two-particle system with time-reversal invariance. We observe that over a wide range of the parameters $c$ and $a$, the sign of $J(B)$ changes from positive to negative at a finite value of $B$ (for the parameters chosen in Fig. 8.3a at $B \approx 1.3\,\text{T}$). $J$ is suppressed exponentially either by compression of the electron orbitals through large magnetic fields ($b \gg 1$), or by large distances between the dots ($d \gg 1$), where in both cases the orbital overlap of the two dots is reduced. This exponential suppression, contained in the $1/\sinh$ prefactor in (8.9), is partly compensated by the exponentially growing exchange term $\propto \exp(2d^2(b - 1/b))$. In total, $J$ decays exponentially as $\exp(-2d^2 b)$ for large $b$ or $d$. Since the sign reversal of $J$ – signalling a singlet-triplet crossing – results from the long-range Coulomb interaction, it is not contained in the standard Hubbard model which takes only short-range interaction into account. In this latter model one finds $J = 4t^2/U > 0$ in the limit $t/U \ll 1$ (see Fig. 8.3). The Heitler-London result (8.9) was refined by taking higher levels and double occupancy of the dots into account (implemented in a Hund-Mulliken approach), which leads to qualitatively similar results [39], in particular concerning the singlet-triplet crossing. These results have been confirmed by numerical calculations which take more single-particle levels into account [55].

We remark again that the exponential suppression of $J$ is very desirable for minimizing gate errors, see Sect. 8.2.6. In the absence of tunneling between the dots we still might have direct Coulomb interaction left between the electrons. However, this has no effect on the spins (qubit) provided the spin-orbit coupling is sufficiently small, which is the case for s-wave electrons in symmetric GaAs structures (this would not be so for hole-doped systems since the hole has a much stronger spin-orbit coupling due to its p-wave character). The spin-orbit interaction can lead to anisotropies in the effective spin Hamiltonian (8.5), see Sect. 8.3.3. Finally, the vanishing of $J$ can be exploited for switching by applying a constant homogeneous magnetic field to an array of quantum dots to tune $J$ to zero (or close to some other desirable

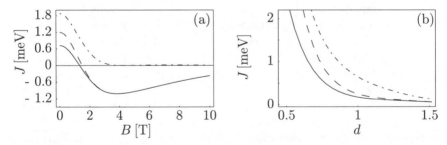

**Fig. 8.3.** The exchange coupling $J$ (full line) for GaAs quantum dots with confinement energy $\hbar\omega = 3\,\text{meV}$ and $c = 2.42$. For comparison the usual short-range Hubbard result $J = 4t^2/U$ (*dashed-dotted line*) and the extended Hubbard result [39] $J = 4t^2/U + V$ (*dashed line*) are plotted. In (**a**), $J$ is plotted as a function of the magnetic field $B$ at fixed inter-dot distance $d = a/a_\text{B} = 0.7$, while in (**b**) as a function of the inter-dot distance $d = a/a_\text{B}$ at $B = 0$.

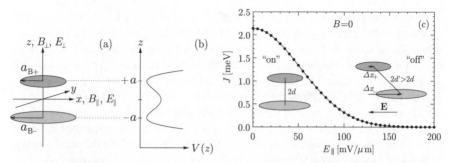

**Fig. 8.4.** (a) Two vertically coupled quantum dots with different lateral radii $a_{B+}$ and $a_{B-}$. In the text, we discuss magnetic and electric fields applied either in-plane $(B_\parallel, E_\parallel)$ or perpendicularly $(B_\perp, E_\perp)$. (b) The quartic double-well potential used for modeling the vertical confinement $V_v$, see text. (c) Switching of the spin-spin coupling between dots of different size by means of an in-plane electric field $E_\parallel$ at $B = 0$. We have chosen $\hbar\omega_z = 7\,\text{meV}$, $d = 1$, $\alpha_{0+} = 1/2$ and $\alpha_{0-} = 1/4$. For these parameters, $E_0 = \hbar\omega_z/ea_B = 0.56\,\text{mV/nm}$ and $A = (\alpha_{0+}^2 - \alpha_{0-}^2)/2\alpha_{0+}^2\alpha_{0-}^2 = 6$. The exchange coupling $J$ decreases exponentially on the scale $E_0/2A = 47\,\text{mV/}\mu\text{m}$ for the electric field. Thus, the exchange coupling is switched "on" for $E_\parallel = 0$ and "off" for $E_\parallel \approx 150\,\text{mV/}\mu\text{m}$, see text.

value). Then, for switching $J$ on and off, only a small gate pulse or a small local magnetic field is needed.

### 8.3.2  Vertical Coupling

We have also investigated the case of vertically tunnel-coupled quantum dots [56]. Such a setup of the dots has been produced in multilayer self-assembled quantum dots (SAD) [57] as well as in etched mesa heterostructures [58]. We apply the same methods as described in Sect. 8.3.1 for laterally coupled dots, but now we extend the Hamiltonian (8.7) from two to three dimensions and take a three-dimensional confinement $V = V_l + V_v$. We implement the vertical confinement $V_v$ as a quartic potential similar to (8.8), with curvature $\omega_z$ at $z = \pm a$ see Fig. 8.4b, implying an effective Bohr radius $a_B = \sqrt{\hbar/m\omega_z}$ and a dimensionless distance $d = a/a_B$. We have modeled a harmonic potential for the lateral confinement, while we have allowed for different sizes of the two dots $a_{B\pm} = \sqrt{\hbar/m\alpha_{0\pm}\omega_z}$. This results in additional switching mechanisms as explained in the next paragraph.

Since we are considering a three-dimensional setup, the exchange interaction is not only sensitive to the magnitude of the applied fields, but also to their direction. We now give a brief overview of our results [56] for in-plane $(B_\parallel, E_\parallel)$ and perpendicular $(B_\perp, E_\perp)$ fields; this setup is illustrated in Fig. 8.4a: (1) An in-plane magnetic field $B_\parallel$ suppresses $J$ exponentially; a perpendicular field in laterally coupled dots has the same effect (Sect. 8.3.1). (2) A perpendicular magnetic field $B_\perp$ reduces on the one hand the exchange coupling between identically sized dots $\alpha_{0+} = \alpha_{0-}$ only slightly. On the other hand, for different dot sizes $a_{B+} < a_{B-}$, the behavior of $J(B_\perp)$ is no longer monotonic: Increasing $B_\perp$ from zero amplifies the exchange coupling $J$ until both electronic orbitals are magnetically compressed to approximately the same size, i.e. $B \approx 2m\alpha_{0+}\omega_z c/e$. From this point on, $J$ decreases weakly, as for identically sized dots. (3) A perpendicular electric field $E_\perp$ detunes the single-dot levels, and thus reduces the exchange coupling; the very same finding was made for laterally coupled dots and an in-plane electric field [39]. (4) An in-plane electric field $E_\parallel$ and different dot sizes provide another switching mechanism for $J$. The dots are shifted parallel to the field by $\Delta x_\pm = E_\parallel/E_0\alpha_{0\pm}^2$, where $E_0 = \hbar\omega_z/ea_B$. Thus, the larger dot is shifted a greater distance $\Delta x_- > \Delta x_+$ and so the mean distance between the electrons grows as $d' = \sqrt{d^2 + A^2(E_\parallel/E_0)^2} > d$, taking $A = (\alpha_{0+}^2 - \alpha_{0-}^2)/2\alpha_{0+}^2\alpha_{0-}^2$. Since the exchange coupling $J$ is exponentially sensitive to the inter-dot distance $d'$, it is suppressed exponentially when an in-plane electric field is applied, $J \approx \exp[-2A^2(E_\parallel/E_0)^2]$, which is illustrated in Fig. 8.4c. We have thus described an exponential switching mechanism for a quantum gate relying only on a tunable electrical field, as an alternative for the magnetically driven switching.

### 8.3.3  Anisotropic Exchange

The exchange Hamiltonian (8.5) can acquire anisotropic terms due to spin-orbit coupling during tunneling between the quantum dots [59]. However, the first-order effect of the spin-orbit coupling during quantum gate operations can be eliminated by using time-symmetric pulse shapes for the coupling between the spins [60]. Moreover, the spin-orbit effects *exactly* cancel in the gate sequence on the right hand side of (8.6) required to produce the quantum XOR gate, provided that the pulse form for the spin-orbit and the exchange couplings are the same [61]. Since the XOR gate is universal together with single-qubit operations, this result ascertains that the spin-orbit coupling can be dealt with in any quantum computation. In practice, the pulse shapes for the exchange and the spin-orbit coupling cannot be chosen completely identical, but it is possible to choose two pulse shapes which are very similar; in this case it was shown that the cancellation still holds to a very good approximation, i.e. the effect of the spin-orbit coupling is still strongly suppressed [61]. The effect of the dipole coupling between adjacent spins, providing another anisotropic coupling, can also be treated as an anisotropic contribution to (8.5) and therefore cancels out in the gate sequence (8.6) for the same reasons as the spin-orbit interaction.

The spin-orbit coupling for a conduction-band electron is given by (8.1). The isotropic Heisenberg coupling with exchange energy $J$ and the anisotropic exchange between two localized spins $\mathbf{S}_1$ and $\mathbf{S}_2$ ($s = 1/2$) are combined in the Hamiltonian [60] $H(t) = J(t)(\mathbf{S}_1 \cdot \mathbf{S}_2 + \mathcal{A}(t))$ where [59],

$$\mathcal{A}(t) = \boldsymbol{\beta}(t) \cdot (\mathbf{S}_1 \times \mathbf{S}_2) + \gamma(t)(\boldsymbol{\beta}(t) \cdot \mathbf{S}_1)(\boldsymbol{\beta}(t) \cdot \mathbf{S}_2), \tag{8.10}$$

and $\beta_i = \sum_j \beta_{ij} \langle \psi_1 | i p_j | \psi_2 \rangle$ is the spin-orbit field, $|\psi_i\rangle$ the groundstate in dot $i = 1, 2$, and $\gamma \approx O(\beta^0)$. For $\mathcal{A} = 0$, the quantum XOR gate can be obtained by applying $H(t)$ twice, together with single-spin rotations, see (8.6). Also, if $\mathcal{A} = 0$, then the Hamiltonian $H(t)$ commutes with itself at different times and the time-ordered exponential

$$U(\varphi) = T \exp\left[-i \int_{-\tau_s/2}^{\tau_s/2} H(t)\,dt\right]$$

is only a function of the integrated interaction strength,

$$\varphi = \int_{-\tau_s/2}^{\tau_s/2} J(t) dt \ .$$

In particular, $U(\varphi = \pi/2) = U_{\text{sw}}^{1/2}$ is the "square-root of swap" gate.

In the case where $\mathcal{A} \neq 0$, but $\boldsymbol{\beta}$ and $\gamma$ (and thus $\mathcal{A}$) are time-independent, $H(t)$ still commutes with itself at different times and we can fix a coordinate system in which $\boldsymbol{\beta}$ points along the $z$ axis, and in which the anisotropy can be written as

$$\mathcal{A} = \beta(S_1^x S_2^y - S_1^y S_2^x) + \delta S_1^z S_2^z \ ,$$

with $\delta = \gamma \beta^2$. In the basis $\{|T_+\rangle = |\uparrow\uparrow\rangle, |S\rangle = (|\uparrow\downarrow\rangle - |\downarrow\uparrow\rangle)/\sqrt{2}, |T_0\rangle = (|\uparrow\downarrow\rangle + |\downarrow\uparrow\rangle)/\sqrt{2}, |T_-\rangle = |\downarrow\downarrow\rangle\}$ we obtain from the gate sequence (8.6) including the anisotropy (8.10),

$$U_g = \text{diag}(i e^{-i\varphi(1+\delta)}, 1, 1, -i e^{-i\varphi(1+\delta)}), \tag{8.11}$$

where $\text{diag}\,(x_1, \ldots, x_4)$ denotes the diagonal matrix with diagonal entries $x_1, \ldots, x_4$. The pulse strength $\varphi$ and the spin-orbit parameters only enter $U_g$ in the $S^z = \pm 1$ subspaces. Note that the linear term in $\boldsymbol{\beta}$ has cancelled out exactly in $U_g$. By choosing $\varphi = \pi/2(1 + \delta)$ we obtain the conditional phase flip gate $U_g = U_{CPF} = \text{diag}(1, 1, 1, -1)$, being equivalent to the XOR up to a basis change. This is the main result of this section – the anisoptropic terms $\mathcal{A} = \text{const.}$ in the spin Hamiltonian have exactly cancelled in the gate sequence for the quantum XOR.

In real systems, however, the anisotropic terms in the Hamiltonian $H$ cannot be expected to be exactly proportional to $J(t)$, i.e. $\mathcal{A}(t)$ is time-dependent. In general, both $\boldsymbol{\beta}$ and $\gamma$ depend on time. Under these circumstances, we cannot exactly eliminate the effect of the anisotropy because of the time-ordering in the definition of $U$ and since the Hamiltonian does, in general, not commute with itself at different times, $[H(t), H(t')] \neq 0$. The gate errors $\epsilon = ||U_g - U_{CPF}||^2$ due to the anisotropy in the Hamiltonian in the case where $\mathcal{A}(t)$ is only weakly time-dependent can be estimated as $\epsilon \leq \Delta^2$ where we use $\Delta\beta(t) = \beta(t) - \beta_0$ to write $\Delta = (|\varphi|\beta_0/2) \max_{|t| \leq \tau_s/2} \times |(J(t)/J_0)(\beta(t)/\beta_0 - 1)|$, where $J_0$ denotes the average exchange coupling, $J_0 = \varphi/\tau_s \neq 0$. It was shown [61] that for tunnel-coupled quantum dots, it is possible to choose a weakly time-dependent $\mathcal{A}$ by using (8.9) for the exchange coupling and the result [61] $b(d,q) \equiv |J(d,q)\beta(d,q)| = b_0\sqrt{q}d\exp(-2qd^2)$, where $b_0 = a/a_B^0$, $a_B^0 = \sqrt{\hbar/m\omega_0}$, $a$ being a constant depending on the spin-orbit parameter (for a 5 nm wide [100] GaAs quantum well $a \approx 2$ meV nm), $q = \omega/\omega_0$. Here, $\omega_0$ denotes the minimal value of the quantum dot confinement energy $\omega$. The switching process can be modeled e.g. by a time-dependent confinement strength $q(t) = \omega(t)/\omega_0 = \cosh^2(\alpha t/\tau_s)$ (here, we choose $\alpha = 4$). This pulse shape is suited for adiabatic switching [39,49] (Sect. 8.2.4) and leads to a pulsed exchange interaction $J(t) = J(d,q(t))$ and spin-orbit field $b(t) = b(d,q(t))$, where $-\tau_s/2 \le t \le \tau_s/2$. We find $\Delta \approx 7 \cdot 10^{-3}$. Therefore, the gate errors occur at a rate $\epsilon \approx 4\Delta^2 \approx 2 \cdot 10^{-4}$ which is around the currently known threshold for fault tolerant quantum computation [51] and could therefore be corrected by quantum error correction. In cases where the error $\epsilon$ is too large for quantum error correction, pulses with smaller intensity, with a long period of constant $\mathcal{A}$ between their rise and fall, can be used to further reduce $\epsilon$ at the cost of a slower gate operation.

### 8.3.4   Superexchange

If there is a sizable distance between the two quantum dots hosting the qubits before the barrier is lowered, then one has to deal with "extended" quantum dots whose energy level spacing $\delta\varepsilon$ is small (i.e. $\delta\varepsilon < k_BT$). Then, the singlet-triplet splitting becomes vanishingly small, and it would not take much energy to excite higher-lying states. Thus, the adiabatic switching time (Sect. 8.2.4) which is proportional to $\hbar/\delta\varepsilon$ becomes arbitrarily large. A better scenario for coupling two spin-qubits of this type is to make use of the superexchange mechanism [62] to obtain the Heisenberg interaction, (8.5).

Consider three aligned quantum dots where the middle dot is empty and so small that only its lowest levels will be occupied by one or two electrons in a virtual hopping process. The left and right dots can be much larger but still small enough such that the Coulomb charging energies $U_L \approx U_R$ are high enough (compared to $k_BT$) to suppress any double occupancy. We assume now that the middle dot has energy levels higher than the ground states of right and left dots, the latter two having approximately the same energy. These energies include the single particle energy (set to zero) and the Coulomb charging energy $N^2e^2/2C$, where $N$ denotes the number of electrons and $C$ the capacitance of the middle dot. The ground state energy of the middle dot is 0 when empty, $\epsilon = e^2/2C$ for one electron, and $4\epsilon$ for two electrons. We denote the tunnel coupling between the dots by $t$. There are two types of virtual processes possible which couple the spins.

First, the electron of the left (right) dot hops on the middle dot, and then the electron from the right (left) dot hops on the *same* level on the middle dot, and therefore, due to the Pauli principle, the two electrons on the middle dot form a singlet, giving the desired entanglement. Then they hop off again into the left and right dots, respectively. (Note that we need $U > k_B T$, otherwise real processes involving two electrons in the left or right dot will be allowed). This virtual process leads to an effective Heisenberg exchange interaction (8.5) with exchange constant $J = 4t^4/4\epsilon^3$, where the virtual energy denominators follow the sequence $1/\epsilon \to 1/4\epsilon \to 1/\epsilon$. Second, the left (right) electron can hop via the middle dot into the right (left) dot and form a singlet there, giving $J = 4t^4/U_R\epsilon^2$. However, this second process has vanishing weight because there are many nearby states available in the outer dots for which there is no spin correlation required by the Pauli principle. Thus most of the virtual processes, for which there are two electrons in the left (right) dot, do not produce spin correlations, and thus we can neglect the virtual processes of the second type altogether. This scenario is valid for $T \gg T_K$, where $T_K$ is the Kondo temperature (see also Sect. 8.3.5). When $T$ approaches $T_K$, many-particle effects become important and we can no longer ignore virtual processes since they induce new spin correlations.

Finally we note that it should be possible to produce ferroelectrically defined nanostructures for which superexchange is the dominant mechanism for coupling neighboring electrons.

### 8.3.5 Accessing the Exchange Interaction $J$ Between the Spins in Coupled Quantum Dots Via the Kondo Effect

An experimental characterization of a double dot (DD) can be performed using transport measurements. Here we consider two lateral quantum dots connected in series between two Femi leads with density of states $\nu$ and at chemical potentials $\mu_L$, $\mu_R$, with bias $\Delta\mu = \mu_L - \mu_R$ (see Fig. 8.5a). The DD is tuned into the Coulomb blockade regime, with one electron in each dot, by means of a common gate with gate voltage $V_g$. The dots are tunnel coupled with the amplitude $t$, and each dot is tunnel coupled to the lead on its side with the amplitude $t_{L(R)}$, defined at $t = 0$. The presence of a Fermi sea produces Kondo correlations at low temperatures, provided there is a degeneracy of the DD ground state. The Kondo correlations are particularly pronounced at a singlet-triplet transition ($J = 0$), because the four-fold degeneracy on the DD favors more spin scattering processes, and leads to an enhanced Kondo temperature, as compared to the case when the dots are detached from each other[1], or when the DD spins are locked into a spin 1 (triplet side). From a perturbative renormalization group (RG) approach [63], we derive that the Kondo energy scale at $J = 0$ is given by [64] $T_K = D_0 \exp(-\gamma/\nu J_0)$, where $\gamma \leq 0.5$ is a non-universal number, depending on the internal features of the DD, $D_0 \simeq \hbar\omega_0$ is the cutoff energy, and $J_0 = (t_L^2 + t_R^2)/E_C$. Here, $E_C = 2E_+E_-/(E_+ + E_-)$ where $E_\pm$ are the Coulomb blockade addition/extraction energies. We use the Hund-Mulliken method to model a

---

[1] At $t = 0$ the two dots undergo independent spin 1/2 Kondo effects with the enegy scales $T_{K,L(R)}^0 = D_0 \exp\left(-E_C/4\nu t_{L(R)}^2\right)$.

realistic DD with long range Coulomb interaction between the electrons in the DD, and in the presence of a magnetic field $B$, which allows us to tune the DD to a singlet-triplet degeneracy point and study the transport properties as a function of temperature, magnetic field, interdot tunnel coupling, and bias [64]. Readily the differential cotunneling conductance $G(\Delta\mu)$ through the DD shows distinct features (steps at $\Delta\mu = \pm J$) [64], allowing one to measure experimentally the exchange interaction $J$ between the spins on the DD. Such a step-like $I$-$V$ characteristic can be observed at temperatures $T$, such that $|J|/k_B \gg T \gg T_K$. The height of the step on the singlet side is 3 times larger than the height on the triplet side [64], thus providing information about the ground state of the DD. At $T \leq T_K$ one obtains Kondo peaks at $\Delta\mu = \pm J$, which can also be used to estimate the value of $J$ in a DD isolated from the leads.

An alternative way to infer the value of the exchange interaction $J$ is the temperature dependence of the linear conductance [64]. Here again the cotunneling conductance allows one to measure $J$, as the value of $T$ at which the $G(T)$ has a maximum. However, this maximum is not very well pronounced in the cotunneling regime. An efficient flow of the coupling constant, responsible for the inelastic scattering on the DD, makes the peak in the temperature dependence of $G$ better pronounced in the Kondo regime [64]. The left inset of Fig. 8.5b shows both the cotunneling contribution (dotted line) and full conductance (solid line), renormalized due to Kondo correlations.

The long range part of the Coulomb interaction reveals itself in a novel peak in the linear conductance versus $t$ at $B = 0$ [64]. The origin of the first peak on the left in Fig. 8.5b can be understood within a chain Hamiltonian with onsite (short range) Coulomb repulsion [65], and it is due to a competition between the Kondo correlations and the exchange interaction $J$. However, the second peak at larger $t$ is present only if $J$ deviates from $4t^2/U$ by a contribution coming from the long range Coulomb interaction [39] (compare the solid lines with the dot-dashed line on Fig. 8.5b). We find that the second peak is sensitive to an applied weak ($J > 0$) magnetic field, see Fig. 8.5b. We point out that the second peak can be used for studying the screening of the Coulomb interaction in a DD.

The $B$ dependence of the linear conductance shows a peak at the singlet-triplet transition, which grows when $T$ is lowered down to $T_K$ [64], see the right inset of Fig. 8.5b. Note that the energy scale for the Kondo effect on the triplet side ($J < 0$) is monotonically decreasing with increasing $|J|$ [66,67], as follows from a two-stage RG procedure working on that side. Furthermore, the strong coupling limit (not shown in Fig. 8.5b) occurs in two stages with lowering $T$ on the triplet side, resulting first in an increase and then, at a lower energy scale, in a decrease of the conductance [68]. Finally, we note that attaching leads to the DD certainly shifts its energy spectrum, and hence, modifies the value of $J$. However such changes can be accounted for as considered in Ref. [64].

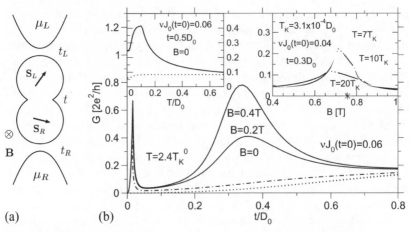

(a)          (b)

**Fig. 8.5.** (a) Double dot (DD) system with tunnel coupling $t_L$ and $t_R$ to leads with chemical potentials $\mu_L$ and $\mu_R$. The inter-dot tunneling amplitude is denoted $t$. (b) Linear conductance $G$ at different values of $B$. *Dotted lines*: cotunneling contributions. *Dot-dashed line*: $G$ vs $t$ at $B = 0.4\,\mathrm{T}$ neglecting the long range part of the Coulomb interaction. For definiteness we keep the DD in the midle of the Coulomb blockade valley by adjusting the gate voltage $V_g$ when varying $t$, and choose $t_L = t_R$. *Left inset*: $G$ vs $T$ showing a peak at $T \simeq |J|$; the dotted line is the cotunneling contribution. *Right inset*: $G$ vs $B$ at the singlet-triplet transition; the kinks in the *dotted-line regions* come from a simplified treatment of the Kondo effect crossover regions and will be smoothened in an exact treatment; the *star* denotes the value of $B$ at which the singlet-triplet transition occus at high temperatures.

## 8.4 Single-Spin Rotations

Besides at least one (non-trivial) two-qubit gate (e.g. XOR or square-root-of-swap), it is required for quantum computation to perform one-qubit operations, which in the context of spin physics are equivalent to single-spin rotations. For this, it must be feasible to expose any specific qubit to a time-varying Zeeman coupling $(g\mu_B\mathbf{S} \cdot \mathbf{B})(t)$ [39], which is controlled through the magnetic field $\mathbf{B}$ (Sect. 8.4.1) and/or the g-factor $g$ (Sect. 8.4.2). Note that since only relative rotations are relevant, it is sufficient to rotate all spins of the system at once (e.g. by an external field $B$), but with a different Larmor frequency.

In many cases it appears that localized magnetic fields are more difficult to achieve than the nearest-neighbor exchange coupling. It turns out that there are several possibilities to use exclusively the exchange mechanism to achieve both the two-qubit and the single-qubit gates. These "exchange only" techniques are discussed in Sect. 8.4.3. The mechanisms described in Sects. 8.4.2 and 8.4.3 show that for both single-spin and two-spin operations, the control over the *spin* can be achieved through *charge* (i.e. orbital wave function) manipulation [7].

Controlled coherent rotations of electron spins in semiconductors have also been performed using laser pulses with a duration on the order of 100 fs [6]. The strong "tipping" pulse with a frequency which is below the band gap of the semiconductor creates an effective magnetic field on the order of 20 T via the optical Stark effect, i.e. via the virtual excitation of carriers from the valence band into the conduction band.

### 8.4.1   Local Magnetic Coupling

Localized magnetic fields can be generated with the magnetic tip of a scanning force microscope, a magnetic disk writing head, by placing the dots above a grid of current-carrying wires, or by placing a small wire coil above the dot. Alternatively one can use electron-spin-resonance (ESR) techniques [39] (see also Sec. 8.2.7) to perform single-spin rotations, e.g. if we want to flip a certain qubit (say from $|\uparrow\rangle$ to $|\downarrow\rangle$) we apply an ac-magnetic field perpendicular to the $\uparrow$- axis that matches the Larmor frequency of that particular electron. Due to paramagnetic resonance the spin can flip.

### 8.4.2   Local g-Factor Coupling

The coupling of the qubit-spins to a layer of material with a higher (or lower) g-factor can be achieved by changing the equilibrium position of the electron by electrical gating [7,10,45] (cf. Fig. 8.1). If the electron wave function is pushed into a region with a different (effective) g-factor, a relative rotation around the direction of $\mathbf{B}$ by an angle of $\phi = (g'B' - gB)\mu_B\tau/2\hbar$ is produced. In bulk semiconductors the free-electron value of the g-factor $g_0 = 2.0023$ is modified by spin-orbit coupling. Similarly, the g-factor can be drastically enhanced by doping the semiconductor with magnetic impurities [4,5]. In confined structures such as quantum wells, wires, and dots, the g-factor is further modified and becomes sensitive to an external bias voltage [69]. We have numerically analyzed a system with a layered structure (AlGaAs-GaAs-InAlGaAs-AlGaAs), in which the effective g-factor of electrons is varied by shifting their position from one layer to another by electrical gating. We have found that in this structure the effective g-factor can be changed by about $\Delta g_{\text{eff}} \approx 1$ [45]. This effect has now been observed [70].

### 8.4.3   Quantum Computing with Exchange Interactions Only

By means of *exchange* coupling to ferromagnetic (FM) dots [7] (Fig. 8.6) or to a FM layer (Fig. 8.1), the coupling of a single spin $\mathbf{S}$ to an *effective* Zeeman field $(g\mu_B\mathbf{B})(t)$ can be obtained. Regions (layers) with increased magnetic field can be provided by a FM material while an effective magnetic field can be produced e.g. with dynamically polarized nuclear spins (Overhauser effect) [39]. Ferromagnetic dots can be made of magnetic metals (e.g. Co or Dy) or magnetic semiconductors, e.g. Mn-doped GaAs (in this case, exchange effects are equivalent to an effective g-factor description, see Sect. 8.4.2). The magnetic dots should be coupled to the spin-carrying dots through either lateral (Fig. 8.6a) or vertical [32] (Fig. 8.6b) exchange coupling. Note that for aligning the magnetic dots, only homogenous (no localized) magnetic fields are required. Preparing two sets of magnetic dots with different orientations $\mathbf{m}_1$ and $\mathbf{m}_2$ can be achieved by using two magnets with different Curie temperatures $T_{C1} > T_{C2}$ and successively cooling them at two different (weak) external fields, i.e. cooling below $T_{C1}$ at a field parallel to $\mathbf{m}_1$ (fixing $\mathbf{m}_1$ for the first set of dots), and then applying the other field parallel to $\mathbf{m}_2$ when crossing $T_{C2}$, magnetizing the second set of dots without changing the magnetization of the first set.

Finally, another possibility for exchange-only quantum computation consists of using an appropriate encoding for the qubits [71]. It turns out that the Heisenberg interaction (8.5) between the spins representing the qubits alone is sufficient to (exactly) perform any quantum computation if each qubit is encoded using (at least) three spins (instead of only one). A possible encoding for this is $|0_L\rangle = |S\rangle|\uparrow\rangle$, $|1_L\rangle = \sqrt{2/3}|T_+\rangle|\downarrow\rangle - \sqrt{1/3}|T_0\rangle|\uparrow\rangle$. It can now easily be seen how to perform one-qubit gates on the coded qubits by applying the exchange interaction between the spins of the code; the Hamiltonian $H_{12} = J_{12}\mathbf{S}_1 \cdot \mathbf{S}_2$ e.g. generates rotations about the $z$-axis for the coded state $|\psi_L\rangle = \alpha|0_L\rangle + \beta|1_L\rangle$. More generally, it can be shown that any single-qubit rotation in code space can be performed with a sequence involving at most four steps of applying the exchange $H_{ij}$ between the adjacent spins $i$ and $j$ in the case where the spins are arranged in a line (i.e. with couplings $H_{12}$ and $H_{23}$), or three steps if the arrangement is in a "triangle" i.e. $H_{13}$ is also possi-

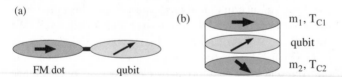

**Fig. 8.6.** Sketch of structures allowing spin manipulations via the charge, i.e. without localized external magnetic fields. The exchange interactions are controlled by applying gate voltages allowing the electron wavefunction to extend into the FM dot for some switching time. (**a**) Quantum dot hosting a qubit, represented by the spin 1/2 of its excess electron, and being coupled via the exchange interaction to a nearby ferromagnetic (FM) quantum dot. (**b**) Vertical quantum dot structure in which the middle dot hosts the qubit. In order to perform arbitrary single-qubit rotations, the spin can become exchange-coupled to two ferromagnetic dots with non-parallel magnetizations $\mathbf{m}_1$ and $\mathbf{m}_2$ and different Curie temperatures $T_{C1}$ and $T_{C2}$. The process of preparing the magnetic dots is explained in the text.

ble. In order to find the exact number of computational steps for performing XOR between two coded qubits, a numerical study was performed [71] with the result that XOR involves 19 sequential applications of a nearest-neighbor exchange interaction. If parallel (simultaneous) exchange couplings are allowed, we find that 7 or 8 steps are required, depending on the geometry of the arrangement. These results quantify the expense in number of devices (a factor of three more) and computational steps (roughly a factor of ten more) if one wants to dispense with the control of local magnetic fields using the encoding method.

## 8.5   Read-Out of a Single Spin

The final step of each (quantum) computation consists in reading out the result of the computation. For this it is sufficient to know the state of a single specified qubit, i.e. to measure if the electron spin is in the $|\uparrow\rangle$ or $|\downarrow\rangle$ state (we do not need to know the coherent superposition). However, it is very hard to detect an electron spin over its tiny (of the order of $\mu_B$) magnetic moment directly, although it is not out of experimental reach [72]. To circumvent this difficulty we have proposed several schemes [7,53] for single spin measurements, the most promising ones make use again of the Pauli principle, yielding an effective spin-to-charge conversion, which allows then to reduce the spin measurement to a charge or a current measurement. We review them now in turn.

### 8.5.1   Spontaneous Magnetization

One scheme for reading out the spin of an electron on a quantum dot is implemented by tunneling of this electron into a supercooled paramagnetic dot [7]. There the spin induces a magnetization nucleation from the paramagnetic metastable phase into a ferromagnetic domain, whose magnetization direction $(\theta, \phi)$ is along the measured spin direction and which can be measured by conventional means. Since this direction is continuous rather than only one of two values, we describe this generalized measurement in the formalism of positive-operator-valued (POV) measurements [73] as projection into the overcomplete set of spin-1/2 coherent states $|\theta, \phi\rangle = \cos(\theta/2)|\uparrow\rangle + e^{i\phi}\sin(\theta/2)|\downarrow\rangle$. Thus if we interpret a magnetization direction in the upper hemisphere as $|\uparrow\rangle$, and using $(1/2\pi)\int_{\theta \geq \pi/2} d\Omega \, |\langle \uparrow |\theta, \phi\rangle|^2 = 3/4$, we have a 75%-reliable measurement.

### 8.5.2   Measuring Spin Via Charge

Spins have the advantage of long decoherence times, but it is very hard to measure a single spin directly via its magnetic moment. However, it is known how to build electrometers with single-charge detection capabilities; resolutions down to $10^{-8}$ of one electron charge have been reported [74]. This gives rise to the idea of detecting the spin of an electron via measuring charge, i.e. voltage or current [7,53].

A straightforward concept yielding a potentially 100% reliable measurement requires a switchable "spin-filter" tunnel barrier which allows only, say, spin-up but no spin-down electrons to tunnel. When the measurement of a spin in a quantum dot is to be performed, tunneling between this dot and a second dot, connected to an electrometer, is switched on, but only spin-up electrons are allowed to pass (spin-filtering). Thus if the spin had been up, a charge would be detected in the second dot by the electrometer [7], and no charge otherwise. Again, this is a POV type of measurement (see above). Spin filtering and also spin-state measurements can be achieved by tunneling through a quantum dot [53] as we shall discuss below.

### 8.5.3   Coupled Dots as Entangler

If it is possible to encode a single spin 1/2 state $|\alpha\rangle$ into a singlet/triplet then one can measure the state of the qubit represented by $|\alpha\rangle$, provided that a measurement device capable of distinguishing singlet from triplet states is available (see e.g. Sect. 8.7.5). Moreover, this operation acts as an "entangler" for electron pairs used in quantum communication (see Sect. 8.7). Indeed, we can construct such a two-qubit operation explicitly. While quantum dot 1 is in state $|\alpha\rangle$, we prepare the state of quantum dot 2 to $|\uparrow\rangle$, perform a $U_{\mathrm{sw}}^{1/2}$ gate and finally apply a local Zeeman term, generating the time evolution $\exp\{i(\pi/2)S_1^z\}$, thus

$$\left.\begin{array}{c}|\uparrow\uparrow\rangle \\ |\downarrow\uparrow\rangle\end{array}\right\} \xrightarrow{\mathrm{e}^{i\frac{\pi}{2}S_1^z}U_{\mathrm{sw}}^{1/2}} \left\{\begin{array}{l}\mathrm{e}^{i\frac{\pi}{4}}|\uparrow\uparrow\rangle, \\ -\mathrm{i}\left(|\downarrow\uparrow\rangle-|\uparrow\downarrow\rangle\right)/\sqrt{2}.\end{array}\right. \tag{8.12}$$

In other words, this operation maps the triplet $|\uparrow\uparrow\rangle$ (and $|\downarrow\downarrow\rangle$) into itself, while $|\downarrow\uparrow\rangle$ is mapped into the singlet (and $|\uparrow\downarrow\rangle$ into the triplet ($|\uparrow\downarrow\rangle + |\downarrow\uparrow\rangle)/\sqrt{2}$), up to phase factors.

### 8.5.4   Spin Filter

A quantum dot attached to in- and outgoing current leads can be operated as a spin filter (Sect. 8.5.4), or as a read-out device, or as a spin-memory (Sect. 8.5.8) where a single spin stores the information [53]. In this proposal the spin-degeneracy is lifted with *different* Zeeman splittings in the dot and in the leads, e.g. by using materials with different effective g-factors for leads and dot. This results in Coulomb blockade peaks and spin-polarized currents which are uniquely associated with the spin state on the dot.

Using the standard tunneling Hamiltonian ($H_T$) approach in the Coulomb blockade regime [31] (where the charge on the dot is quantized), and using the master equation for the reduced density matrix of the dot the transition rates in a "golden-rule" approach are calculated up to 2nd order in $H_T$ [53]. The current in first order in $H_T$ is the sequential tunneling current $I_s$ [31], where the electron number on the dot fluctuates. The second-order contribution is the cotunneling current $I_c$ [75], involving only a virtual intermediate state with a different number of electrons on the dot (see also Sect. 8.7.5). We consider a system where the Zeeman splitting in the leads is negligible while on the dot it is given as $\Delta_z = \mu_B|gB|$, and assume a small bias $0 < \Delta\mu = \mu_1 - \mu_2 < \Delta_z$ between the leads at chemical potentials $\mu_{1,2}$ and low temperatures such that $\Delta\mu$, $kT < |E_S - E_T|$, where $E_S$, $E_T$ are the energies of singlet and triplet states. If the dot contains more that one electron in the ground state the last requirement is crucial.

If the dot is in its ground state, filled with an odd number of electrons with total spin 1/2, which we assume to be $|\uparrow\rangle$, then we find that $I_s$ is spin polarized (spin $|\downarrow\rangle$). However, the cotunneling current $I_c$ contains a spin $|\downarrow\rangle$ component which potentially reduces the efficiency of the spin-filtering effect. The ratio of spin-polarized to unpolarized current is [53]

$$I_s(\downarrow)/I_c(\uparrow) \sim \frac{\Delta_z^2}{(\gamma_1 + \gamma_2)\max\{k_B T, \Delta\mu\}}, \tag{8.13}$$

where $\gamma_l$ is the tunneling rate between lead $l$ and the dot. In the sequential tunneling regime we have $\gamma_l < k_B T, \Delta\mu$, thus, for $k_B T, \Delta\mu < \Delta_z$ the ratio (8.13) is large and the spin-filter is efficient.

A spin *pumping* effect at zero bias $\Delta\mu = 0$ but with an externally applied ESR source (see also Sect. 8.5.6 below) can also be achieved [38]. For a single quantum dot one finds that the current is proportional to $\gamma_1^\uparrow\gamma_2^\downarrow - \gamma_1^\downarrow\gamma_2^\uparrow$, where $\gamma_l^\sigma$ denotes the tunneling amplitude between dot and lead $l$ for electrons with spin $\sigma$. On the one hand it is clear that the pumping current vanishes if the barriers are not spin-selective, i.e. $\gamma_l^\uparrow = \gamma_l^\downarrow$. On the other hand, it is sufficient if one of the barriers is spin-selective, and this can be achieved by using a second quantum dot as a spin filter as described above. The combined device will then function as a spin inverter, which pumps a spin-polarized current.

### 8.5.5  Berry Phase Controlled Spin Filter

A further approach for building a spin filter is to use a quantum interference effect, where one component of the spin current is filtered out via destructive inferference while the other component remains unaffected. Such effects can be produced by using Aharonov-Bohm (AB) loops in an orientationally inhomogeneous magnetic field. We assume that the electrons move sufficiently slowly through the AB ring such that the adiabatic assumption is justified, i.e., the spin state evolves as instantaneous eigenstate of the (local) Zeeman interaction in the inhomogeneous field. Then, when the spin $\alpha$ has followed a closed contour in spin space, it has acquired a Berry phase $\Phi_\alpha^g$ (also known as geometrical phase), which is half the solid angle enclosed by this contour. For instance, for a crown-shaped field texture, when the magnetic field is tilted with an angle $\eta$ from the $z$ axis and winds once around the $z$ axis in one turn around the ring (which lies in the $xy$ plane), the Berry phase is $\Phi^g_\pm = \pm\pi(1 - \cos\eta)$. To obtain a spin filter, $\eta$ is tuned such that $\cos\eta = \frac{1}{2}$, e.g., by varying in addition a homogeneous external field in $z$ direction. Thus, the Berry phase difference of the two spin components is $\Phi^g_- - \Phi^g_+ = \pi$. Further, the Aharonov-Bohm phase $2\pi\phi/\phi_0$ can be tuned independently of $\eta$, since generally only a small modulation of the magnetic field in $z$ direction is required to change the flux $\phi$ through the ring on the order of $\phi_0 = h/e$. The spin filter effect is then obtained (in a ballistic ring), when $\phi$ is tuned such that one spin channel interferes constructively; the other spin channel is offset by a phase $\pi$ and thus will be suppressed due to destructive interference. A similar effect occurs for diffusive rings in an inhomogeneous magentic field. In the impurity-averaged magnetoconductance, coherence effects occur with a period $h/2e$ and, similarly, the Berry phase contribution is $\pm 2\pi(1 - \cos\eta)$. Then, to obtain a phase difference of $\pi$ for the two spin channels, $\eta$ must be tuned such that $\cos\eta = 1/4, 3/4$. For these "magic angles" $\eta$, a spin filtering effect is possible [76].

### 8.5.6   Detection of Single-Spin Decoherence

A somewhat different measurement of the current through a single quantum dot as the one described in Sect. 8.5.4 is proposed in Ref. [38]. The idea is to extract the decoherence time $T_2$ by probing the sequential tunneling current through a quantum dot which is attached to two leads in the presence of an applied electron spin resonance (ESR) field, producing spin-flips on the dot. We assume a situation in which the Zeeman splitting on the dot is $g\mu_B B > \Delta\mu, k_B T$, while the Zeeman splitting in the leads is different, such that the effect of the ESR field on the leads can be neglected. This can be achieved, e.g., by using materials with different $g$-factors for the dot and the leads. Then, the stationary master equation for the reduced density matrix of the quantum dot in the basis $|\uparrow\rangle$, $|\downarrow\rangle$, $|S\rangle$ (with corresponding energies $0 = E_\uparrow < E_\downarrow < E_S$) is derived. We can assume that the triplet is higher in energy and does not contribute to the sequential tunneling current. In the regime $E_S > \mu_1 > E_S - g\mu_B B > \mu_2$, the current is blocked in the absence of the ESR field due to energy conservation. Only in the presence of the ESR field there can be a sequential tunneling current; after some calculation we find for the stationary current [38]

$$I(\omega) \propto \frac{V_{\downarrow\uparrow}}{(\omega - g\mu_B B)^2 + V_{\downarrow\uparrow}^2}, \tag{8.14}$$

where the width of the resonance at $\omega = g\mu_B B$ is given by the total spin decoherence rate $V_{\downarrow\uparrow} = (W_{S\uparrow} + W_{S\downarrow})/2 + 1/T_2$. Here, $W_{S\sigma}$ denotes the rate

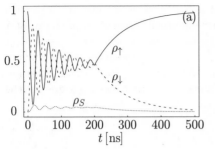
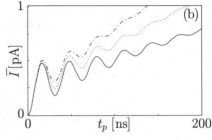

**Fig. 8.7.** Single spin Rabi oscillations, generated by ESR pulses of length $t_p$, visible in the time-averaged current $I(t_p)$ [77]. We consider the same regime as in Sect. 8.5.6 and take the amplitude of ESR field as $B_x^0 = 20$ G (and $g = 2$), and $\Delta\mu > kT$, $\gamma_1 = 2 \times 10^7\,\mathrm{s}^{-1}$, $\gamma_2 = 5\gamma_1$, $T_1 = 1\,\mu s$, and $T_2 = 150\,\mathrm{ns}$. (**a**) Evolution of the density matrix $\rho$, where a pulse of length $t_p = 200$ ns is switched on at $t = 0$, obtained by numerical integration of the master equation. (**b**) Time-averaged current $\bar{I}(t_p)$ (*solid line*) for a pulse repetition time $t_r = 500$ ns. We also show the current where $\gamma_1$ and $\gamma_2$ are increased by a factor of 1.5 (*dotted*) and 2 (*dash-dotted*).

for the transition from the state $|\sigma\rangle = |\uparrow\rangle, |\downarrow\rangle$ to the singlet $|S\rangle$ due to the tunnel coupling to the leads. Therefore, the inverse of the observed line width $1/V_{\downarrow\uparrow}$ represents a lower bound for the intrinsic single-spin decoherence time $T_2$. For finite temperatures in the linear response regime $\Delta\mu < kT$ the current has roughly the standard sequential tunneling peak shape $\cosh^{-2}[(E_S - E_\downarrow - \mu)/2k_B T]$ as a function of the gate voltage $V_{\text{gate}} \propto \mu = (\mu_1 + \mu_2)/2$, while the width of the resonance in (8.14) as a function of $\omega$ remains unaffected.

### 8.5.7   Rabi Oscillations and Pulsed ESR

The spin of a quantum dot in the presence of an ESR field shows coherent Rabi oscillations. Then, the current through the quantum dot may depend on the spin state of the dot, e.g., as seen in Sect. 8.5.6. Thus, the Rabi oscillations can be observed via the time-dependent current [38]. However, as shown in Ref. [77], it is also possible to observe the Rabi oscillations of a single spin without measuring time-resolved currents. For this, the time-averaged currents can be measured (over arbitrarily long times) while ESR pulses are applied. Then, the time-averaged current $\bar{I}(t_p)$ as a function of the pulse length $t_p$ exhibits the Rabi oscillations of the spin-state of the dot, for both polarized (Sect. 8.5.8) and unpolarized (Sect. 8.5.6 and Fig. 8.7) leads. Note that since arbitrarily long times (and thus a large number of electrons) can be used, the measurement of $\bar{I}$ is significantly simpler compared to measurements of time-dependent currents.

We assumed a rectangular envelope for the ESR pulse with length $t_p$ and repetition time $t_r$ (thus $t_p < t_r$). The time during two pulses, $t_r - t_p$, should be long enough such that the dot can relax into its ground state, which ensures that the state of the dot is identical at the beginning of each pulse. Then, $\bar{I}(t_\mathrm{p})$ can be calculated by numerical integration of the master equation and by subsequently averaging the (time-dependent) current over the interval $[0, t_\mathrm{r}]$. The result is shown in Fig. 8.7b for unpolarized leads in the same regime as discussed in Sect. 8.5.6, and the Rabi oscillations are clearly visible in $\bar{I}(t_\mathrm{p})$. We have calculated the current contributions for the interval $[t_\mathrm{p}, t_\mathrm{r}]$ analytically, and obtained

$$\bar{I}(t_\mathrm{p}) \approx \frac{e}{t_\mathrm{r}} \frac{\gamma_2^\uparrow + \gamma_2^\downarrow}{\gamma_1^\downarrow + \gamma_2^\downarrow} \left[\rho_\downarrow(t_\mathrm{p}) + \rho_\mathrm{S}(t_\mathrm{p})\right] \propto 1 - \rho_\uparrow(t_\mathrm{p}), \qquad (8.15)$$

up to the background contribution $\bar{I}_\mathrm{bg}$ for times $t < t_\mathrm{p}$, which is roughly linear in $t_\mathrm{p}$. In particular, we see from (8.15) that $\bar{I}(t_\mathrm{p})$ is determined by $\rho_\uparrow(t_\mathrm{r})$, i.e, by the probability that the dot is in spin state $|\uparrow\rangle$ at the end of the ESR pulse. Thus, current $\bar{I}$ probes the spin state of the dot at time $t_\mathrm{p}$, and therefore allows one to measure the Rabi oscillations of the spin.

### 8.5.8   Spin Read-Out

In Sect. 8.5.4 we have argued that a quantum dot can be used as a spin filter if the Zeeman splitting on the dot is much larger than in the attached wires. In the opposite case where the leads are fully spin polarized with a much smaller Zeeman splitting on the dot [53] (a situation which can be realized with magnetic semiconductors [4] or in the quantum Hall regime where spin-polarized edge states are coupled to a quantum dot [78]) the same device can be used as read-out for the spin state on the dot. If the spin polarization in both leads is $|\uparrow\rangle$, and the ground state of the dot contains an odd number of electrons with total spin 1/2, then the leads can provide and absorb only spin-up electrons. Thus, $I_s \neq 0$ only if the dot state is $|\downarrow\rangle$ (to form a singlet with the incoming electron, whereas the triplet is excluded by energy conservation). Hence, the current is much larger for the spin on the dot being in $|\downarrow\rangle$ than it is for $|\uparrow\rangle$. Again, there is a small cotunneling leakage current for the dot-state $|\uparrow\rangle$, with a ratio of the two currents given by (8.13), replacing $\Delta_z$ by $|E_\mathrm{T} - E_\mathrm{S}|$ (we assume here $\Delta_z \ll |E_\mathrm{T} - E_\mathrm{S}|$). Thus, we can probe (read out) the spin-state on the quantum dot by measuring the current which passes through the dot. Given that the sequential tunneling current is typically on the order of 0.1–1 nA [31], we can estimate the read-out frequency $I/2\pi e$ to be on the order of 0.1–1 GHz. Combining this with the initialization and read-in techniques from Sect. 8.2.7, i.e. ESR pulses to switch the spin state, we have a *spin memory* at the ultimate single-spin limit, whose relaxation time is just the spin relaxation time $T_1$ (which can be expected longer thato exceed a few $\mu$s [79]) and can be read out via the currents when they switch from high to low due to a spin flip on the dot [53].

A detailed analysis of the dynamics (master equation) confirms these rough estimates [38]. If the spin on the dot is $|\downarrow\rangle$, then the probability for no electron being transmitted after time $t$ is $P_\downarrow(t) = \exp(-Wt)(1 + Wt)$, where $W = 2I/e$ is the rate of tunneling from one of the leads to the dot. For example, after time $2e/I$ the spin state can be determined with more than 90% probability.

### 8.5.9  Optical Measurements

The Faraday rotation (see Chap. 4) originating from a pair of coupled electrons would allow us to distinguish between spin singlet and triplet [56]: In the singlet state ($S = 0$, no magnetic moment) there is no Faraday rotation, whereas in the triplet state ($S = 1$) the polarization of linearly polarized light is rotated slightly due to the presence of the magnetic moment. A single spin $|\alpha\rangle$ can be measured either directly via Faraday rotation or by first entangling it with another spin $|\uparrow\rangle$ and then applying the singlet/triplet-measurement. This entanglement is achieved by applying the gate defined in Sect. 8.5.3, resulting in either a triplet or singlet, depending on whether $|\alpha\rangle$ was $|\uparrow\rangle$ or $|\downarrow\rangle$. However, much more work is required to analyze the Faraday rotation (in particular to calculate the oscillator strength for such processes) in order to assess its efficiency for spin measurements.

## 8.6  Quantum Information Processing with Large-Spin Systems

So far we have concentrated on spin 1/2 systems. Although a bit outside the mainline of this review we briefly digress on systems with spin $S > 1/2$. We begin by remarking again that Shor and Grover demonstrated that a quantum computer can outperform any classical computer in factoring numbers [13] and in searching a database [14] by exploiting the parallelism of quantum mechanics. While Shor's algorithm requires both superposition and entanglement of a many-particle system [84], the superposition of single-particle quantum states is sufficient for Grover's algorithm [83]. Recently, Ahn et al. [85] have successfully implemented Grover's algorithm in atomic physics using Rydberg atoms. In Refs. [86] and [87] we focused on solid state systems and proposed the implementation of Grover's algorithm in large-spin systems, such as molecular magnets [88] (with e.g. spin $S = 10$) and nuclear spins in GaAs [89] (with spin $I = 3/2$), respectively. We have shown that molecular magnets and nuclear spins can be used to build dense and efficient memory devices based on the Grover algorithm. Due to fast electron spin resonance (ESR) pulses that decode and read out a number between 0 and $10^5$ stored in, say, a $Mn_{12}$ crystal, access times as short as $10^{-10}$ s can be attained.

Suppose we want to find a phone number in a phone book consisting of $N = 2^n$ entries. Usually it takes $N/2$ queries on average to be successful. Even if the $N$ entries were encoded binary-like, a classical computer would need approximately $\log 2N$ queries to find the desired phone number [14]. Remarkably, the computational parallelism provided by the superposition and interference of quantum states enables the Grover algorithm to reduce the search to one single query [14]. We have shown that this query can be implemented in terms of a unitary transformation applied to a single spin of a molecular magnet [86] (or more recently to a nuclear spin in GaAs [87]). The ensemble nature of such crystals provides a natural amplification of the magnetic moment of a single spin. However, for the Grover algorithm to work, it is necessary to find ways for generating arbitrary superpositions of spin eigenstates. For spins larger than $1/2$ this turns out to be a non-trivial task since spin excitations induced by magnetic dipole transitions in conventional ESR can occur only in discrete steps of one $\hbar$, i.e. single steps by two or more $\hbar$'s are excluded by selection rules. To circumvent such physical limitations we proposed a novel scenario which, in principle, allows the controlled generation of arbitrary spin superpositions through the use of multifrequency coherent magnetic radiation being in the microwave and radiofrequency range. In particular, we demonstrated by means of the S-matrix and time-dependent high-order perturbation theory that using advanced ESR techniques it is possible to coherently populate and manipulate many spin states simultaneously by applying one single pulse of a magnetic ac field containing an appropriate number of matched frequencies. This ac field creates a nonlinear response of the magnet via multiphoton absorption processes involving particular sequences of $\sigma$ and $\pi$ photons [86], or only $\sigma$ photons [87] which allows the encoding and, similarly, the decoding of states. Finally, the subsequent readout of the decoded quantum state can be achieved by means of pulsed ESR techniques thereby exploiting the non-equidistance of energy levels which is due to spin anisotropies of molecular magnets or due to quadrupolar splitting of nuclear spin states. These high spin excitation effects are interesting in their own right, and it would be interesting to see experimentally if states of spin systems with $s > 1/2$ can be coherently populated.

## 8.7    Quantum Communication

Two particles (qubits) are entangled if their wavefunction cannot be expressed as a tensor product of two single-particle wavefunctions (a similar definition exists for mixed states). Many tasks in quantum communication require maximally entangled states of two qubits (EPR pairs) such as the spin singlet [11]. The triplet $|T_0\rangle$ (see Sect. 8.3.1) is another maximally entangled state, while the other two triplets $|T_\pm\rangle$ are not entangled. The quantum gate mechanism described in Sect. 8.5.3 is one possibility for producing such entangled states for electrons (we call in general such a device an *entangler*, for which a number of realizations are conceivable). Here we first present another method of producing entangled electrons making use of Andreev tunneling at a superconductor-normal interface. Then, we go on by discussing the effect of interactions when entangled particles are transported in wires which are described by a normal Fermi liquid. Finally, three experimental setups by which the entanglement of electrons can be detected via their charge in transport and noise measurements in mesoscopic nanostructures are discussed [10,47,80,81]. This investigation touches on fundamental issues such as the non-locality of quantum mechanics, especially for massive particles, and genuine two-particle Aharonov-Bohm effects which are fascinating topics in their own right. The main idea here is to exploit the unique relation between the symmetry of the orbital state and the spin state (for two electrons) which makes it possible to detect the spin state again via the charge (orbital) degrees of freedom of the electrons.

We emphasize here that entanglement *per se* is rather the rule than the exception in condensed matter systems; every ground state of a many-electron system is entangled simply by the antisymmetry requirement for the wave function. However, it is crucial to have separate control over each specified particle which belongs to an entangled many-particle state. A measure of entanglement which excludes pure antisymmetrization was defined in Refs. [49,82].

In quantum optics, violations of Bell inequalities and quantum teleportation with photons have been investigated [27,90], while so far no corresponding experiments for electrons let alone in a solid-state environment are reported.

### 8.7.1  Andreev Entangler

In s-wave superconductors (SC), the electrons form Cooper pairs with singlet spin-wavefunctions [91] and can thus act as a natural source of spin-entangled electrons. A proposed setup is shown in Fig. 8.8. We assume that the SC is held at the chemical potential $\mu_S$, and is weakly coupled by tunnel barriers to two separate quantum dots $D_1$ and $D_2$ which are themselves weakly coupled to Fermi liquid leads $L_1$ and $L_2$, respectively, both held at the same chemical potential $\mu_1 = \mu_2$. The tunneling amplitudes between SC and dots, and dots and leads, are denoted by $T_{SD}$ and $T_{DL}$, respectively (for simplicity we assume them to be equal for both dots and leads).

The tunnel-coupling of a SC to a normal region allows for coherent transport of two electrons of opposite spin due to Andreev tunneling, while single-electron tunneling is suppressed [92]. In the present setup, we envision a situation where the two electrons are forced to tunnel coherently into *separate* leads rather than both into the same lead, which can be enforced by two intermediate quantum dots in the Coulomb blockade regime [31] so that the state with double occupation of one dot is strongly suppressed, and thus tunneling into separate dots and subsequently separate leads is preferred.

In order to obtain transport of entangled electrons from the SC via the dots to the leads, a bias voltage $\Delta\mu = \mu_S - \mu_l > 0$ is applied. The chemical potentials $\epsilon_1$ and $\epsilon_2$ of the two quantum dots can be tuned by external gate voltages [31] such that the coherent tunneling of two electrons into different leads is at resonance, described by a two-particle Breit-Wigner resonance peaked at $\epsilon_1 + \epsilon_2 = 2\mu_S$. The resonance for the coherent tunneling of two electrons into the same lead is suppressed by the on-site Coulomb repulsion $U$ of a quantum dot.

Here, we briefly introduce the relevant parameters describing the proposed device and specify their regime of interest. We work at the resonances $\epsilon_n = \mu_S$ since then the total current and the desired suppression for tunneling into the same lead is maximized. The injection of the two electrons into separate leads at the *same orbital energy* is then achieved; this turns out to be crucial for the detection of entanglement which is described in Sect. 8.7.4. It is most efficient to work in the regime where the dot levels $\epsilon_n$ have vanishing occupation probability. For this purpose we require that the dot-lead coupling is much stronger than the SC-dot coupling, i.e. $|T_{SD}| \ll |T_{DL}|$, so that electrons which enter the quantum dots from the SC will leave the quantum dots to the leads much faster than new electrons can be provided from the SC. The stationary occupation due to the coupling to the leads is indeed exponentially small if $\Delta\mu > k_B T$, $T$ being the temperature and $k_B$ the Boltzmann constant. In this asymmetric barrier case, the resonant dot levels $\epsilon_n$ can be occupied only during a virtual process.

The quantum dots are allowed to contain an arbitrary but even number of electrons, $N_D$ = even, with total spin zero in the ground state (i.e. antiferromagnetic filling of the dots). An odd number $N_D$ must be excluded since a simple spin-flip on the quantum dot would be possible in the transport process and as a result the desired spin-entanglement would be lost. Moreover, we have to make sure that also spin flip processes of the following kind are

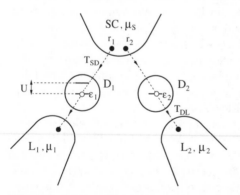

**Fig. 8.8.** Andreev Entangler. Two entangled electrons initially forming a Cooper pair hop with amplitude $T_{SD}$ from two points $\mathbf{r}_1$, $\mathbf{r}_2$ of the superconductor, SC, (distance $\delta r = |\mathbf{r}_1 - \mathbf{r}_2|$) to two dots $D_{1,2}$ by means of Andreev tunneling. The dots are coupled to normal leads $L_{1,2}$ with tunneling amplitude $T_{DL}$. In order to maximize the efficiency of the device, we require asymmetric barriers, $|T_{SD}|/|T_{DL}| \ll 1$. The chemical potentials of the SC and leads are $\mu_l$ and $\mu_S$, respectively.

excluded. Consider an electron that tunnels from the SC into a given dot. In principle, it is possible (e.g. in a sequential tunneling process [31]) that another electron with the opposite spin leaves the dot and tunnels into the lead, and, again, the desired entanglement would be lost. However, such spin flip processes will be excluded if the energy level spacings of the dots, $\delta\epsilon$, (assumed to be similar for both dots) exceeds both, temperature $k_B T$ and bias voltage $\Delta\mu$. A serious mechanism for the loss of entanglement is given by electron hole-pair excitations out of the Fermi sea of the leads during the resonant tunneling events. However, one can show that such many-particle contributions are suppressed if the resonance width $\gamma_l = 2\pi\nu_l |T_{DL}|^2$ is smaller than $\Delta\mu$ (for $\epsilon_n \simeq \mu_S$), where $\nu_l$ is the density of states (DOS) per spin of the leads at the chemical potential $\mu_l$.

Finally, an additional energy scale that enters the consideration is the superconducting energy gap $\Delta$, which is half the minimal energy it costs to break up a Cooper pair into two quasiparticles. The time delay between subsequent coherent tunneling events of the two electrons of a Cooper pair is determined by $\Delta^{-1}$. In order to exclude single-electron tunneling where the creation of a quasiparticle in the SC is a final excited state we require that $\Delta \gg \Delta\mu, k_B T$. Summarizing all above inequalities, we can specify the following regime of interest for entanglement production [93]

$$\Delta, U, \delta\epsilon > \Delta\mu > \gamma_l, k_B T \text{ , and } \gamma_l > \gamma_S . \tag{8.16}$$

In this regime, we have calculated and compared the stationary charge current of two entangled electrons for two competing transport channels, first for the desired transport of the two entangled electrons each into *different* leads (current $I_1$) and second for the unwanted transport of both electrons into the *same* lead (current $I_2$). We have calculated the currents $I_1$, $I_2$ by making use of a $T$-matrix approach which is well-adapted for describing Breit-Wigner resonances. The final result for the ratio of the two currents is [93]

$$\frac{I_1}{I_2} = \frac{2\mathcal{E}^2}{\gamma^2} \left[ \frac{\sin(k_F \delta r)}{k_F \delta r} \right]^2 e^{-2\delta r/\pi\xi}, \qquad \frac{1}{\mathcal{E}} = \frac{1}{\pi\Delta} + \frac{1}{U}, \tag{8.17}$$

where $\gamma = \gamma_1 + \gamma_2$. The current $I_1$ becomes exponentially suppressed with increasing distance $\delta r = |\mathbf{r}_1 - \mathbf{r}_2|$ between the tunneling points on the SC, the scale given by the superconducting coherence length $\xi$. This does not pose severe restrictions for a conventional s-wave material with $\xi$ typically being on the order of $\mu$m. In the important case $0 \leq \delta r \sim \xi$ the suppression is only polynomial $\propto 1/(k_F \delta r)^2$, with $k_F$ being the Fermi wavevector in the SC. On the other hand we see that the effect of the quantum dots consists in the suppression factor $(\gamma/\mathcal{E})^2$ for tunneling into the same lead. Thus, in addition to (8.16) we have to impose the condition $k_F \delta r < \mathcal{E}/\gamma$, which can be satisfied for small dots with $\mathcal{E}/\gamma \approx 100$ and $k_F^{-1} \approx 1$ Å. We would like to stress that the suppression (rather than only the absolute current) is maximized by working around the resonance $\epsilon_n \simeq \mu_S = 0$. We remark that incoherent transport (sequential tunneling) is negligible as long as the scattering rate $\Gamma_\varphi$ is much smaller than $\gamma_l$ since $(I_{\text{seq}}/I_{\text{coh}}) \simeq (\Gamma_\varphi/\gamma_l)$ [94]. We note that the efficiency of the desired injection of the two electrons into different leads can even be enhanced by using lower dimensional SC. In two dimensions (2D) we find that $I_1 \propto 1/k_F\delta r$ for large $k_F\delta r$ [95], and in one dimension (1D) there is no suppression of the current and only an oscillatory behavior in $k_F\delta r$. An actual experimental realization of a 2D SC is e.g. realized by using a SC on top of a two-dimensional electron gas (2DEG) [96], where superconducting correlations are induced via the proximity effect in the 2DEG, or in 1D it is claimed that superconductivity can also be present in ropes of single-wall nanotubes [97].

## 8.7.2   Andreev Entangler with Luttinger Liquid Leads

We discuss a further implementation of an entangler for electron spins. We again consider an s-wave SC as discussed in Sect. 8.7.1 which is tunnel-coupled to the center (bulk) of two spatially separated one-dimensional wires 1 and 2 with Luttinger liquid (LL) properties such as nanotubes [98]. The interaction between different leads is negligible. The two leads are assumed to be infinitely extended and are described by conventional LL-theory [95]. In the limit of vanishing backscattering the low energy excitations of these LL are long-wavelength charge and spin density oscillations propagating with velocities $u_\rho = v_F/K_\rho$ for the charge and $u_\sigma = v_F$ for the spin respectively [99], where $v_F$ is the Fermi velocity and $K_\rho < 1$ due to interaction. Transfer of electrons from the SC to the LL-leads is described by the tunneling Hamiltonian $H_T = t_0 \sum_{ns} \psi^\dagger_{ns} \Psi_s(\mathbf{r}_n) + \text{H.c.}$ The operator $\Psi_s(\mathbf{r}_n)$ annihilates an electron with spin $s$ at the point $\mathbf{r}_n$ on the SC nearest to the LL-lead $n = 1, 2$, and $\psi^\dagger_{ns}$ creates it again with same spin and amplitude $t_0$ at the point $x_n$ in LL $n$. By applying a bias $\mu = \mu_S - \mu_l$ between the SC, with chemical potential $\mu_S$, and the leads, held at the same chemical potential $\mu_l$, a stationary current of pairwise spin-entangled electrons can flow from the SC to the leads. We calculate the current for the two competing processes of tunneling into different leads ($I_1$) and into the same lead ($I_2$) in lowest order in $H_T$ again within a T-matrix approach. For the desired process of tunneling into different leads we obtain, in leading order in $\mu/\Delta$ and at zero temperature [95]

$$I_1 = \frac{I_1^0}{\Gamma(2\gamma_\rho + 2)} \frac{v_F}{u_\rho} \left[ \frac{2\Lambda\mu}{u_\rho} \right]^{2\gamma_\rho}, \quad I_1^0 = 4\pi e\gamma^2 \mu \frac{\sin^2(k_F\delta r)}{(k_F\delta r)^2} e^{-2\delta r/\pi\xi}, \quad (8.18)$$

where $\Gamma$ is the Gamma function, $\Lambda$ a short distance cut-off on the order of the lattice spacing in the LL, $\gamma = 2\pi\nu_S\nu_l|t_0|^2$ the probability per spin to tunnel from the SC to the LL-leads, $\nu_S$ and $\nu_l$ the energy DOS per spin for the superconductor and the LL-leads at the chemical potentials $\mu_S$ and $\mu_l$, respectively, and $\delta r$ the separation between the tunneling points on the SC.

The current $I_1$ has its characteristic non-linear form $I_1 \propto \mu^{2\gamma_\rho + 1}$, with an interaction dependent exponent $\gamma_\rho = (K_\rho + K_\rho^{-1})/4 - 1/2 > 0$, which is the exponent for tunneling into the bulk of a single LL, i.e. $\rho(\varepsilon) \sim |\varepsilon|^{\gamma_\rho}$, where $\rho(\varepsilon)$ is the single-particle DOS [99]. The non-interacting limit $\gamma_\rho = 0$ is given by $I_1^0$. Again, the coherence length $\xi$ of the Cooper pairs should exceed $\delta r$ in order to obtain a finite measurable current. The suppression of the current by $1/(k_F \delta r)^2$ can be considerably reduced by going over to lower dimensional SCs [95], see Sect. 8.7.1. Now we compare this result with the unwanted process when the two electrons tunnel into the same lead and having $\delta r = 0$. We find after some calculation that the current $I_2$ for tunneling into the same lead (1 or 2) is suppressed if $\mu < \Delta$ with the result, in leading order in $\mu/\Delta$,

$$ I_2 = I_1 \sum_{b = \pm 1} A_b \left( \frac{2\mu}{\Delta} \right)^{2\gamma_{\rho b}}, \tag{8.19} $$

where $A_b$ is an interaction dependent constant of order one [95], and where $\gamma_{\rho+} = \gamma_\rho$, and $\gamma_{\rho-} = \gamma_{\rho+} + (1 - K_\rho)/2 > \gamma_{\rho+}$. We remark that in (8.19) the current $I_1$ is to be taken at $\delta r = 0$. The non-interacting limit $I_2 = I_1 = I_1^0$ is rediscovered by putting $\gamma_\rho = \gamma_{\rho b} = 0$, and $u_\rho = v_F$. The result for $I_2$ shows that the unwanted injection of two electrons into the same lead is suppressed compared to $I_1$ by a factor of $(2\mu/\Delta)^{2\gamma_{\rho+}}$, where $\gamma_{\rho+} = \gamma_\rho$, if both electrons are injected into the *same* branch (left or right movers), or by $(2\mu/\Delta)^{2\gamma_{\rho-}}$ if the two electrons travel in *different* directions. Since $\gamma_{\rho-} > \gamma_{\rho+}$, it is more likely that the two electrons move in the same direction than in opposite directions. The suppression of the current $I_2$ by $1/\Delta$ shows very nicely the two-particle correlation effect in the LL, when the electrons tunnel into the same lead, which is similar to the Coulomb blockade effect found in the previous case of tunneling into quantum dots. The larger $\Delta$, the shorter is the delay time between the arrivals of the two partner electrons of a Cooper pair, and, in turn, the more the second electron tunneling into the same lead will feel the existence of the first one which is already present in the LL. By increasing the bias $\mu$ the electrons can tunnel faster through the barrier since there are more channels available into which the electron can tunnel, and therefore the effect of the gap $\Delta$ is less pronounced. Also note that this correlation effect disappears when interactions are absent (i.e. when $\gamma_\rho = \gamma_{\rho b} = 0$) in the LL. Actual experimental systems which show LL-behavior are e.g. metallic carbon nanotubes with similar exponents as derived here [100,101].

### 8.7.3 Entangled Electrons in a Fermi Sea

When talking about the injection of entangled electrons into a Fermi sea, one must keep in mind that there is always Coulomb interaction present with all other electrons in the leads. Therefore we have analyzed the effect of interaction on entanglement [10,81]. Specifically, when we add an electron in state $q$ to a Fermi sea (lead), the quasiparticle weight of that state will be renormalized by $0 \leq z_q \leq 1$ (see below), i.e. some weight $1 - z_q$ to find the electron in the original state $q$ will be distributed among all the other electrons [10,81]. This rearrangement of the Fermi system due to the Coulomb interaction happens very quickly, on a time scale given by the inverse plasma frequency. We are now interested in quantifying this renormalization. More precisely, when a triplet/singlet electron pair ($t$ and $s$ for short) is injected from an entangler into two leads 1 and 2, we obtain the state

$$|\psi_{\mathbf{nn}'}^{t/s}\rangle = \frac{1}{\sqrt{2}} (a_{\mathbf{n}\uparrow}^{\dagger} a_{\mathbf{n}'\downarrow}^{\dagger} \pm a_{\mathbf{n}\downarrow}^{\dagger} a_{\mathbf{n}'\uparrow}^{\dagger}) |\psi_0\rangle, \qquad (8.20)$$

with the filled Fermi sea $|\psi_0\rangle$, $\mathbf{n} = (\mathbf{q}, l)$, $\mathbf{q}$ the momentum of an electron, and $l$ the lead number. The operator $a_{\mathbf{n}\sigma}^{\dagger}$ creates an electron in state $\mathbf{n}$ with spin $\sigma$. The propagation of the triplet or singlet, interacting with all other electrons in the Fermi sea, can be described by the 2-particle Green's function $G^{t/s}(\mathbf{12}, \mathbf{34}; t) = \langle \psi_{\mathbf{12}}^{t/s}, t | \psi_{\mathbf{34}}^{t/s} \rangle$. If a triplet (singlet) is prepared at $t = 0$, we define the *fidelity of transmission*[2] $P(t) = |G^{t/s}(\mathbf{12}, \mathbf{12}; t)|^2$ as the probability for finding a triplet (singlet) after time $t$. Assuming sufficiently separated leads with negligible mutual interaction, we find [10,81] $P = z_F^4$ for times much smaller than the quasiparticle lifetime (but larger than the inverse plasma frequency). For a spin-independent Hamiltonian with bare Coulomb interaction only and within RPA [102], the quasiparticle weight for a 2DEG is given by [10,81] $z_F = 1 - r_s (1/2 + 1/\pi)$, in leading order of the interaction parameter $r_s = 1/k_F a_B$, where $a_B = \epsilon_0 \hbar^2 / m e^2$ is the Bohr radius and $k_F$ the Fermi wavevector. In a typical GaAs 2DEG we have $a_B = 10.3$ nm and $r_s = 0.614$, and thus we obtain $z_F = 0.665$, therefore the fidelity of transmission of the injected singlet will only be around $P \approx 0.2$. However, for large electron density (small $r_s$), $P$ is closer to unity. Note the fidelity of the ("postselected") singlet pairs which can successfully be removed from the Fermi sea, is equal to 1, provided that (as assumed here) the spin-scattering effects are negligible. That this is indeed the case in GaAs 2DEGs is supported by experiments where the electron spin has been transported phase-coherently over distances of up to 100 $\mu$m [3].

### 8.7.4 Noise of Entangled Electrons

It has been known [103,104] for quite some time that bosons such as photons show "bunching" behavior when measuring the correlations between particles ("noise") in an incoming particle current. More recently, the opposite behavior for fermions, "antibunching", was expected theoretically [105–107] and found experimentally [108], in particular for electrons. As pointed out

---

[2] Here, we use the standard terminology from the literature on quantum communication by defining the *fidelity* as the probability $P$ and not as the modulus of the amplitude $\sqrt{P}$, as in Ref. [81].

recently [10] the noise of electrons in current-carrying wires is sensitive only to the symmetry of the *orbital* part of its wavefunction, at least if no spin-scattering processes are present. Thus, if we now consider a two-electron state, we expect antibunching for the triplet states, since they have an antisymmetric orbital wave function, whereas the orbital wave function associated with the spin singlet state is symmetric, and so we expect bunching. This leads to an observable decrease or increase in noise for electrons, depending on their common spin state, as we shall discuss next [81].

We assume that an entangler (a possible implementation is discussed in Sect. 8.7.1) generates pairs of entangled electrons which are then injected into lead 1 and 2, one electron each, as shown in Fig. 8.9. A beam splitter is inserted in order to create two-particle interference effects in the sense that there is an equal probability amplitude for incoming electrons (from lead 1 or 2) to leave into lead 3 or 4. The quantity of interest is then the noise, i.e. the current-current correlations, measured in leads 3 and/or 4.

The probability of recovering a singlet or triplet after injecting it into an interacting Fermi sea is reduced by a factor of $z_F^4$ (see Sect. 8.7.3), which depends on the carrier density. Except for this renormalization, the entanglement of the singlet or triplet is not affected by the interacting electrons in the filled Fermi sea. If $z_F^4$ is close to one we can now calculate transport quantities using the standard scattering theory for non-interacting quasiparticles in a Fermi liquid [105]. We consider the entangled incident states $|\pm\rangle \equiv |\psi_{12}^{t/s}\rangle$ with one electron per lead and the quantum numbers $\mathbf{n} = (\varepsilon_n, n)$, where $\varepsilon_n$ is the electron energy. Considering a multiterminal conductor with density of states $\nu$, we assume that the leads consist of only one quantum channel; the generalization to more channels is straightforward.

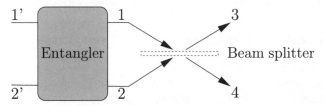

**Fig. 8.9.** The proposed setup for measuring noise of entangled electrons. The entangler takes uncorrelated electrons from the Fermi leads $1'$ and $2'$. Pairs of entangled electrons (singlet or triplet) are produced in the entangler and then injected into the leads 1 and 2, one electron per lead. The current of these two leads are then mixed with a beam splitter (to induce scattering interference) and the resulting noise is measured in lead 3 and 4: no noise (antibunching) for triplets, whereas we get enhanced noise (bunching) for singlets (i.e. EPR pairs).

We find for the spectral densities of the fluctuations $\delta I_\alpha = I_\alpha - \langle I_\alpha \rangle$ between the leads $\alpha$ and $\beta$ in the absence of backscattering and at zero frequency [81], $S_{\alpha\beta}(\omega) = \lim_{T\to\infty} (h\nu/T) \int_0^T dt\, e^{i\omega t} \mathrm{Re}\langle \pm | \delta I_\alpha(t) \delta I_\beta(0) | \pm \rangle$,

$$S_{33} = S_{44} = -S_{34} = 2\frac{e^2}{h\nu} T (1 - T)(1 \mp \delta_{\varepsilon_1\varepsilon_2}). \tag{8.21}$$

Here, the minus (plus) sign refers to the spin triplet (singlet), $T$ is the transmission coefficient of the beam splitter, and $\nu$ denotes the density of states in the leads. If two electrons with the same energies, $\varepsilon_1 = \varepsilon_2$, in the singlet state are injected into the leads 1 and 2, the shot noise is enhanced by a factor of two compared to the value for uncorrelated particles [105,109], $2e^2 T(1 - T)/h\nu$. This amplification of the noise arises from *bunching* of the electrons due to their symmetric orbital wavefunction, such that the electrons preferably appear in the same outgoing leads. If the electron pairs are injected as a triplet, an *antibunching* effect appears, completely suppressing the noise, i.e. $S(\omega = 0) = 0$. We stress that the sign of cross-correlations does not carry any signature of statistics, e.g. here the different signs of $S_{34}$ and $S_{33} = S_{44}$ (8.21) merely reflect current conservation and absence of backscattering. Since bunching appears only for a state with a symmetric orbital wave function, which cannot occur for unentangled states, measuring noise enhancement in the outgoing arms of the beamsplitter provides unique evidence for entanglement [81]. A crucial requirement is $\varepsilon_1 = \varepsilon_2$, which can be satisfied for finite-size systems of length $L$ (including the beam splitter), with discrete spectrum $\varepsilon$ such that $\Delta\varepsilon > k_B T > \gamma$ or $L < \hbar v_F / k_B T \approx 2\,\mu m$ at $1\,K$. Here, $\gamma$ is the level width of $\varepsilon$ caused by the coupling to the current reservoirs. Finally, we need also to assume that inelastic scatterings which could violate the requirement $\varepsilon_1 = \varepsilon_2$ can be neglected, for this we require that $\ell_{\mathrm{inelastic}} > L$ which can be achieved at sufficiently small temperatures.

## 8.7.5   Double-Dot with Normal Leads

Now we consider a setup by which the entanglement of two electrons in a double-dot can be measured through current and noise [80]. The double-dot is supposed to be weakly coupled, with tunneling amplitude $\mathcal{T}$, to in- and outgoing leads at chemical potentials $\mu_{1,2}$. The dots are put in parallel in contrast to the standard series connection. We work in the Coulomb blockade regime [31] where the charge on the dots is quantized and in the cotunneling regime [75,110], with $U > |\mu_1 \pm \mu_2| > J > k_B T, 2\pi\nu\mathcal{T}^2$, where $U$ is the single-dot charging energy, $\nu$ the lead density of states, and $J$ the exchange coupling (see Sect. 8.3). The cotunneling current involves a coherent virtual process where an electron tunnels from a dot to, say, lead 2 and then a second electron tunnels from lead 1 to this dot. Assuming $|\mu_1 - \mu_2| > J$, elastic as well as inelastic cotunneling occurs. Further, $\mathcal{T}$ is assumed to be sufficiently small so that the double-dot will return to its equilibrium state before the next electron passes through. Since an electron can either pass through the upper or lower dot, a closed loop is formed by these two paths, and in the presence of a magnetic flux the upper and the lower paths collect a phase difference given by the Aharonov-Bohm phase $\phi = ABe/\hbar$ (with $A$ being the loop area), thus leading to interference effects. If the two electrons on the double-dot are in the *singlet state*, then the tunneling current acquires an additional phase of $\pi$ (see below) leading to a sign reversal of the coherent contribution compared to that for triplets. Explicitly, we find for the cotunneling current [80]

$$I = e\pi\nu^2 \mathcal{T}^4 \frac{\mu_1 - \mu_2}{\mu_1\mu_2}\left(2 \pm \cos\phi\right), \tag{8.22}$$

and for the shot noise power $S(0) = -e|I|$, where the upper sign refers to the triplet states in the double-dot and the lower sign to the singlet state.

We have evaluated the noise also for finite frequencies [80], and found that again $S(\omega) \propto (2 \pm \cos\phi)$, and, moreover, that the odd part of $S(\omega)$ leads to slowly decaying oscillations of the noise in real time, $S(t) \propto \sin(\mu t)/\mu t$, $\mu = (\mu_1 + \mu_2)/2$, which can be ascribed to a charge imbalance on the double dot during an uncertainty time $\mu^{-1}$.

We finally note that the three triplets can be further distinguished by an orientationally inhomogeneous magnetic field which results in a spin-Berry phase [111,80] that leads to left, right or no phase-shift in the Aharonov-Bohm oscillations of the current (noise).

### 8.7.6  Double-Dot with Superconducting Leads

We have considered a further scenario of double-dots (DD) [47], where the dots are aligned in parallel between the leads, as in Sect. 8.7.5, but now no direct coupling is assumed between them. However, they are coupled with a tunneling amplitude $\mathcal{T}$ to two superconducting leads. The s-wave superconductor (SC) favors an entangled singlet-state on the dots (like in a Cooper pair) and further provides a mechanism for detecting the spin state via the Josephson current. It turns out that in leading order $\propto \mathcal{T}^4$ the spin coupling is again described by a Heisenberg Hamiltonian [47]

$$H_{\text{eff}} \approx J\left(1 + \cos\varphi\right)\left(\mathbf{S}_a \cdot \mathbf{S}_b - \frac{1}{4}\right), \tag{8.23}$$

where $J \approx 2\mathcal{T}^2/\epsilon$, and the energy of the dot is $\epsilon$ below the lead Fermi energy. Here, $\varphi$ is the average phase difference across the SC-DD-SC junction. We can modify the exchange coupling between the spins by tuning the external control parameters $\mathcal{T}$ and $\varphi$. Thus, we have presented here another implementation of a two-qubit quantum gate (see Sect. 8.3) or an "entangler" for EPR transport (see Sect. 8.7.4). Furthermore, the spin state on the dot can be probed if the superconducting leads are joined with one additional (ordinary) Josephson junction with coupling $J'$ and phase difference $\theta$ into a SQUID-ring. The supercurrent $I_S$ through this ring is given by [47]

$$I_S/I_J = \begin{cases} \sin(\theta - 2\pi f) + (J'/J)\sin\theta \,, & \text{singlet}, \\ (J'/J)\sin\theta \,, & \text{triplets}, \end{cases} \qquad (8.24)$$

where $I_J = 2eJ/\hbar$. The spin of the DD is probed by measuring the spin- and flux-dependent critical current $I_c = \max_\theta\{|I_S|\}$; this is done by biasing the system with a dc current $I$ until a finite voltage $V$ appears for $|I| > I_c$ [47]. The SC can also be used to effect a long-distance coupling between spins residing in dots separated by $\delta r$. The exchange coupling is found to be

$$J(\delta r) = J(0)\left[\frac{\sin(k_F\delta r)}{k_F\delta r}\right]^2 e^{-2\delta r/\xi}. \qquad (8.25)$$

### 8.7.7  Biexcitons in Coupled Quantum Dots as a Source of Entangled Photons and Electrons

Despite their wide use for implementations of quantum communication schemes [11], spontaneous parametric down-conversion sources for entangled photons have two main shortcomings, their low efficiency and the stochastic (non-deterministic) production of EPR pairs. It is therefore interesting to study the possibility of producing entangled *photons* using semiconductor quantum dots (QDs) for which one or both of these problems could be overcome. The use of electron-hole recombination in a single QD [112] and in coupled quantum dots [30] has been suggested. Here, we review the latter idea, the production of polarization-entangled photons (or, alternatively, spin-entangled electrons) using biexcitonic states in *two tunnel-coupled* QDs. Since the polarization-entangled photons are emitted from two different QDs, their spatial separation, which is required for quantum communication, is easier to achieve than for photon pairs emitted from the same QD. The low-energy biexcitonic states in coupled QDs in the presence of external magnetic and electric fields and the optical properties of the low-energy states have been studied [30], concentrating on the spin configuration of the calculated states – being related to the orbital wavefunction via the Fermi statistics which is implemented in a Heitler-London (HL) ansatz for the coupled excitons – as done previously for coupled conduction-band electrons [39] (see Sect. 8.3.1). Both the tunneling of electrons and holes between the coupled QDs and the long range Coulomb interaction are taken into account.

Each biexciton consists of two bound excitons. Excitons are formed as a bound state of a conduction-band electron and a valence-band hole in a semiconductor. Starting from a strong confinement ansatz, $a_X \gg a_e, a_h$, where $a_X$ is the bulk exciton radius and $a_e, a_h$ the electron and hole effective Bohr radii in the QD, i.e. from independent electrons and holes (two of each

species), and then using the HL approximation to include the Coulomb interaction and the tunneling, the low-energy (spin-resolved) biexciton spectrum
in which the electrons and holes each form either a spin singlet or triplet has
been obtained. As in the case of two electrons in coupled QDs (Sect. 8.3.1),
the biexciton in two coupled QDs is modeled by a Hamiltonian $H$ similar to
the one given by (8.7) but for four interacting particles, two electrons and
two holes. The low-energy physics of QDs filled with few particles can be
obtained under the assumtion of approximately 2D parabolic confinement.
Furthermore, one can assume the simultaneous confinement of electrons and
holes which can be realized e.g. in QDs formed by thickness fluctuations in
a quantum well [113] or by self-assembled QDs [114]. The valence band is
assumed to be split into well-separated heavy and light hole bands and only
heavy-hole excitations are considered. From the Fock-Darwin ground states
$|D\rangle_\alpha$ in the QD $D = 1, 2$ [39,30] a *variational* HL ansatz

$$|I\rangle^\alpha = (|12\rangle_\alpha + (-1)^I|21\rangle_\alpha)/\sqrt{2(1 + (-1)^I|_\alpha\langle 1|2\rangle_\alpha|^2)}, \qquad (8.26)$$

is made, where $|s\rangle^\alpha \equiv |I = 0\rangle^\alpha$ is a spin singlet and $|t\rangle^\alpha \equiv |I = 1\rangle^\alpha$ a
spin triplet, and where the notation $|DD'\rangle_\alpha = |D\rangle_\alpha \otimes |D'\rangle_\alpha$ for two-particle
states has been used. Now, the four biexciton states $|IJ\rangle = |I\rangle^e \otimes |J\rangle^h$ can be
constructed, where $I = 0$ (1) for the electron singlet (triplet) and $J = 0$ (1)
for the hole singlet (triplet). The electron-hole exchange for the GaAs QDs
considered here is reported to be only on the order of tens of $\mu$eV [115] and
can therefore be neglected. It is found [30] that at $B = 0$ the state $|ss\rangle$ is
the ground state. In this state, both the electron spins and the hole spins are
entangled and the recombination of the biexciton leads to entangled photon
polarization (see below). The entanglement of the ground state biexciton
is robust if the temperature is lower than the splitting to the next higher
biexciton state in the double QD.

The oscillator strength of the biexciton-exciton and the exciton-vacuum
transitions have also been calculated [30]. The oscillator strength $f$ is a measure for the coupling of exciton states to the electromagnetic field and is
proportional to the optical transition rates. It is found that the effect of an
electric field is to spatially separate the electron and the hole, which leads to a
reduction of $f$. In principle, this effect can be used to control the time when
the excitons recombine. Transformation of a HL biexciton state $|IJ\rangle$ into
the basis of two coupled excitons yields a superposition of dark ($S_z = \pm 2$)
and bright ($S_z = \pm 1$) exciton states. The polarization state of the emitted
photons is (assuming that they enclose an azimuthal angle of $\Delta\varphi = 0, \pi$)

$$|\chi_{\text{photon}}\rangle \propto |+1, \theta_1\rangle|-1, \theta_2\rangle + (-1)^{I+J}|-1, \theta_1\rangle|+1, \theta_2\rangle, \qquad (8.27)$$

where $|\sigma, \theta\rangle \propto m_{\sigma,+1}(\theta)|\sigma_+\rangle + m_{\sigma,-1}(\theta)|\sigma_-\rangle$ is the state of the emitted photon
from the recombination of an electron with spin $S_z = \sigma/2 = \pm 1/2$ and a
heavy hole with spin $S_z = 3\sigma/2$ in a direction which encloses the angle $\theta$ with

the normal to the plane of the 2D electron and hole motion, and $m_{\sigma\lambda}(\theta) = (\cos(\theta) - \sigma\lambda)/2$. The states of right and left circular polarization are denoted $|\sigma_{\pm}\rangle$. The entanglement in the state (8.27) for $I = J$ can be quantified by the von Neumann entropy $E = \log_2(1 + x_1 x_2) - x_1 x_2 \log_2(x_1 x_2)/(1 + x_1 x_2)$, with $x_i = \cos^2(\theta_i)$. Note that only the emission of both photons perpendicular to the plane ($\theta_1 = \theta_2 = 0$) results in maximal entanglement ($E = 1$) since only in this case $|+1, \theta_i\rangle$ is orthogonal to $|-1, \theta_i\rangle$. In particular, the two photons are not entangled ($E = 0$) if at least one of them is emitted in-plane ($\theta_i = \pi/2$).

Reversing the scheme outlined above, spin-entangled electrons can be produced by optical absorption in the case where the relaxation to the ground state, e.g. $|ss\rangle$, is faster than the recombination of the involved excitons (as seen, e.g. in experiments under low excitation density [116]. After each QD has been filled with an exciton, the recombination can efficiently be "switched off" using an electric field, as discussed above.

## 8.8 Conclusions

To conclude, we have described a spin-based quantum computer concept for quantum-confined nanostructures, in particular quantum dots. Theoretical proposals for manipulation, coupling and detection of spins in such structures have been presented and analyzed. We have discussed the requirements for initialization, read-in, gate operations, read-out, coherence, switching times and precision and their actual realization in "spintronics" devices. All of those elements can be taken together as a concept for a scalable, all-electrically controlled quantum computer based on spin-qubits.

Furthermore, we have made a link between mesoscopic transport phenomena and quantum communication via the production, detection and transport of electronic EPR pairs. A variety of experimental setups which would probe novel spin-based phenomena in open and closed mesoscopic nanostructures have been proposed and discussed. In addition to their potential applications, spin-coherence and non-local quantum correlations are of fundamental interest in their own right and deserve further investigation.

### Acknowledgments

We thank C. Bruder, M.-S. Choi, D. P. DiVincenzo, and E. V. Sukhorukov for useful discussions. We are grateful to H.-A. Engel, V. N. Golovach, O. Gywat, A. V. Khaetskii, M. N. Leuenberger, and P. Recher for their contributions to this chapter. This work has been supported by the NCCR Nanoscience, Swiss NSF, DARPA, and ARO.

# References

1. G. A. Prinz, Phys. Today **45**(4), 58 (1995); Science **282**, 1660 (1998).
2. S. A. Wolf, D. D. Awschalom, R. A. Buhrman, J. M. Daughton, S. von Molnár, M. L. Roukes, A. Y. Chtchelkanova, and D. M. Treger, Science **294**, 1488 (2001).
3. J. M. Kikkawa, I. P. Smorchkova, N. Samarth, and D. D. Awschalom, Science **277**, 1284 (1997); J. M. Kikkawa and D. D. Awschalom, Phys. Rev. Lett. **80**, 4313 (1998); D. D. Awschalom and J. M. Kikkawa, Phys. Today **52**(6), 33 (1999).
4. R. Fiederling et al., Nature **402**, 787 (1999).
5. Y. Ohno et al., Nature **402**, 790 (1999).
6. J. A. Gupta, R. Knobel, N. Samarth, and D. D. Awschalom, Science **292**, 2458 (2001).
7. D. Loss and D. P. DiVincenzo, Phys. Rev. A **57**, 120 (1998); cond-mat/9701055.
8. G. Burkard, H.-A. Engel, and D. Loss, Fortschr. Phys. **48**, 965 (2000).
9. A. Steane, Rep. Prog. Phys. **61**, 117 (1998).
10. D. P. DiVincenzo, D. Loss, J. Magn. Magn. Mater. **200**, 202 (1999); cond-mat/9901137.
11. C. H. Bennett and D. P. DiVincenzo, Nature **404**, 247 (2000).
12. D. P. DiVincenzo, Phys. Rev. A **51**, 1015 (1995).
13. P. W. Shor, in *Proc. 35th Symposium on the Foundations of Computer Science*, (IEEE Computer Society Press), 124 (1994).
14. L. K. Grover, Phys. Rev. Lett. **79**, 325 (1997); *ibid* **79**, 4709 (1997); **80**, 4329 (1998).
15. J. I. Cirac and P. Zoller, Phys. Rev. Lett. **74**, 4091 (1995); C. Monroe *et al.*, *ibid.* **75**, 4714 (1995).
16. Q. A. Turchette et al., Phys. Rev. Lett. **75**, 4710 (1995).
17. D. Cory, A. Fahmy, and T. Havel, Proc. Nat. Acad. Sci. U.S.A. **94**, 1634 (1997); N. A. Gershenfeld and I. L. Chuang, Science **275**, 350 (1997).
18. A. Shnirman, G. Schön, and Z. Hermon, Phys. Rev. Lett. **79**, 2371 (1997); Y. Makhlin, G. Schön, and A. Shnirman, Phys. Rev. Lett. **85**, 4578 (2000).
19. D. V. Averin, Solid State Commun. **105**, 659 (1998).
20. L. B. Ioffe et al., Nature **398**, 679 (1999).
21. T. P. Orlando et al., Phys. Rev. B **60**, 15398 (1999).
22. V. Privman, I. D. Vagner, and G. Kventsel, Phys. Lett. A **239**, 141 (1998).
23. B. E. Kane, Nature **393**, 133 (1998).
24. C. H. W. Barnes, J. M. Shilton, A. M. Robinson, Phys. Rev. B **62**, 8410 (2000).
25. R. Vrijen et al., Phys. Rev. A **62**, 012306 (2000).
26. J. Levy, Phys. Rev. A **64**, 052306 (2001).
27. D. Bouwmeester et al., Nature **390**, 575 (1997); D. Boschi, S. Branca, F. De Martini, L. Hardy, S. Popescu, Phys. Rev. Lett. **80**, 1121 (1998).
28. C. H. Bennett et al., Phys. Rev. Lett. **70**, 1895 (1993).
29. A. K. Ekert, Phys. Rev. Lett. **67**, 661 (1991).
30. O. Gywat, G. Burkard, and D. Loss, cond-mat/0109223.

31. L. P. Kouwenhoven, G. Schön, and L. L. Sohn, *Mesoscopic Electron Transport*, NATO ASI Series E, Vol. 345, Kluwer Academic Publishers (1997).
32. S. Tarucha, D. G. Austing, T. Honda, R. J. van der Hage, and L. P. Kouwenhoven, Phys. Rev. Lett. **77**, 3613 (1996).
33. F. R. Waugh et al., Phys. Rev. Lett. **75**, 705 (1995); C. Livermore et al., Science **274**, 1332 (1996).
34. T. H. Oosterkamp et al., Phys. Rev. Lett. **80**, 4951 (1998).
35. R. H. Blick et al., Phys. Rev. Lett. **80**, 4032 (1998); *ibid.* **81**, 689 (1998); T. H. Oosterkamp et al., Nature **395**, 873 (1998); I. J. Maasilta and V. J. Goldman, Phys. Rev. Lett. **84**, 1776 (2000).
36. J. A. Gupta, D. D. Awschalom, X. Peng, and A. P. Alivisatos, Phys. Rev. B **59**, R10421 (1999).
37. M. Flatte and J. Byers, Phys. Rev. Lett. **84**, 4220 (2000).
38. H.-A. Engel and D. Loss, Phys. Rev. Lett. **86**, 4648 (2001).
39. G. Burkard, D. Loss, and D. P. DiVincenzo, Phys. Rev. B **59**, 2070 (1999).
40. M. Dobers, K. v. Klitzing , J. Schneider, G. Weimann, and K. Ploog, Phys. Rev. Lett. **61**, 1650 (1988).
41. D. C. Dixon, K. R. Wald, P. L. McEuen, and M. R. Melloch, Phys. Rev. B **56**, 4743 (1997).
42. A. V. Khaetskii and Y. V. Nazarov, Phys. Rev. B **61**, 12639 (2000); Phys. Rev. B **64**, 125316 (2001); A. V. Khaetskii, Physica E **10**, 27 (2001).
43. V. F. Gantmakher and Y. B. Levinson, *Carrier scattering in metals and semiconductors*, North-Holland, Amsterdam, Chapter 1 (1987).
44. A. V. Khaetskii, D. Loss, and L. I. Glazman, cond-mat/0201303.
45. D. P. DiVincenzo, G. Burkard, D. Loss, and E. Sukhorukov, in *Quantum Mesoscopic Phenomena and Mesoscopic Devices in Microelectronics*, eds. I.O. Kulik and R. Ellialtoglu (NATO ASI), p. 399 (2000); see cond-mat/99112445.
46. A. Imamoğlu, D. D. Awschalom, G. Burkard, D. P. DiVincenzo, D. Loss, M. Sherwin, and A. Small, Phys. Rev. Lett. **83**, 4204 (1999).
47. M.-S. Choi, C. Bruder, and D. Loss, Phys. Rev. B **62**, 13569 (2000).
48. G. Burkard, D. Loss, D. P. DiVincenzo, and J. A. Smolin, Phys. Rev. B **60**, 11404 (1999).
49. J. Schliemann, D. Loss, and A. H. MacDonald, Phys. Rev. B **63**, 085311 (2001).
50. P. W. Shor, Phys. Rev. A **52**, R2493 (1995); A. M. Steane, Phys. Rev. Lett. **77**, 793 (1996); D. P. DiVincenzo and P. W. Shor, *ibid.* **77**, 3260 (1996); E. Knill and R. Laflamme, Phys. Rev. A **55**, 900 (1997); D. Gottesman, *ibid.* **54**, 1862 (1996).
51. J. Preskill, Proc. R. Soc. London Ser. A **454**, 385 (1998); J. Preskill, in *Introduction to Quantum Computation and Information*, edited by H.-K. Lo, S. Popescu, and T. Spiller (World Scientific, Singapore, 1998), pp. 213–269.
52. L. Kouwenhoven and C. Marcus, private communication.
53. P. Recher, E.V. Sukhorukov, and D. Loss, Phys. Rev. Lett. **85**, 1962 (2000).
54. A. Barenco et al., Phys. Rev. A **52**, 3457 (1995).
55. X. Hu and S. Das Sarma, Phys. Rev. A **61**, 062301 (2000).
56. G. Burkard, G. Seelig, and D. Loss, Phys. Rev. B **62**, 2581 (2000).
57. R. J. Luyken et al., Physica E **2**, 704 (1998).
58. D. G. Austing et al., Physica B **249-251**, 206 (1998).
59. K. V. Kavokin, Phys. Rev. B **64**, 075305 (2001).

60. N. E. Bonesteel, D. Stepanenko, and D. P. DiVincenzo, Phys. Rev. Lett. **87**, 207901 (2001).
61. G. Burkard and D. Loss, Phys. Rev. Lett. **88**, 047903 (2002).
62. P. Recher, J. Levy, and D. Loss, in *Macroscopic Quantum Coherence and Computing*, eds. D. Averin, P. Silvestrini, (Plenum Press, NY, 2000); cond-mat/0009270.
63. P. W. Anderson, J. Phys. C **3**, 2436 (1970).
64. V. N. Golovach and D. Loss, cond-mat/0109155.
65. W. Izumida and O. Sakai, Phys. Rev. B **62**, 10260 (2000), and references therein.
66. M. Eto, Y. Nazarov, Phys. Rev. B **64**, 085322 (2001).
67. M. Pustilnik and L. I. Glazman, Phys. Rev. B **64**, 045328 (2001).
68. M. Pustilnik and L. I. Glazman, Phys. Rev. Lett. **87**, 216601 (2001).
69. E. L. Ivchenko, A. A. Kiselev, and M. Willander, Solid State Comm. **102**, 375 (1997).
70. G. Salis, Y. Kato, K. Ensslin, D. C. Driscoll, A. C. Gossard, and D. D. Awschalom, Nature **414**, 619 (2001).
71. D. P. DiVincenzo, D. Bacon, J. Kempe, G. Burkard, and K.B. Whaley, Nature **408**, 339 (2000).
72. P. C. Hammel, Z. Zhang, G. J. Moore, and M. L. Roukes, Journ. of Low Temp. Phys. **101**, 59 (1995); J. A. Sidles, J. L. Garbini, K. L. Bruland, D. Rugar, O. Zueger, S. Hoen, and C. S. Yannoni, Rev. Mod. Phys. **67**, 249 (1995).
73. A. Peres, *Quantum Theory: Concepts and Methods* (Kluwer, Dordrecht, 1993).
74. M. Devoret, D. Estève, and Ch. Urbina, Nature (London) **360**, 547 (1992).
75. D. V. Averin and Yu. V. Nazarov, in *Single Charge Tunneling*, eds. H. Grabert and M. H. Devoret, NATO ASI Series B, Vol. 294, Plenum Press, New York, 1992.
76. H.-A. Engel and D. Loss, Phys. Rev. B **62**, 10238 (2000).
77. H.-A. Engel and D. Loss, to be published in Phys. Rev. B, April 2002, cond-mat/0109470.
78. M. Ciorga et al., Phys. Rev. B **61**, R16315 (2000).
79. T. Fujisawa, Y. Tokura, and Y. Hirayama, Phys. Rev. B **63**, R081304 (2001).
80. D. Loss and E.V. Sukhorukov, Phys. Rev. Lett. **84**, 1035 (2000).
81. G. Burkard, D. Loss, and E.V. Sukhorukov, Phys. Rev. B **61**, R16303 (2000).
82. J. Schliemann, J. I. Cirac, M. Kus, M. Lewenstein, and D. Loss, Phys. Rev. A **64**, 022303 (2001); quant-ph/0012094.
83. S. Lloyd, Phys. Rev. A **61**, R010301 (1999).
84. A. K. Ekert and R. Jozsa, Rev. Mod. Phys. **68**, 733 (1996).
85. J. Ahn, T. C. Weinacht, and P. H. Bucksbaum, Science **287**, 463 (2000).
86. M. N. Leuenberger and D. Loss, Nature **410**, 789 (2001).
87. M. N. Leuenberger, D. Loss, M. Poggio, and D. D. Awschalom, to be published.
88. M. N. Leuenberger and D. Loss, Phys. Rev. B **61**, 1286 (2000); *ibid.* **61**, 12200 (2000).
89. G. Salis et al., Phys. Rev. Lett. **86**, 2677 (2001). G. Salis, D. D. Awschalom, Y. Ohno, H. Ohno, Phys. Rev. B **64**, 195304 (2001).
90. A. Aspect, J. Dalibard, and G. Roger, Phys. Rev. Lett. **49**, 1804 (1982); W. Tittel, J. Brendel, H. Zbinden, and N. Gisin, Phys. Rev. Lett. **81**, 3563 (1998).

91. J. R. Schrieffer, *Theory of Superconductivity* (Benjamin/Cummings, NY, 1964).
92. F. W. J. Hekking, L. I. Glazman, K. A. Matveev, and R. I. Shekhter, Phys. Rev. Lett. **70**, 4138 (1993).
93. P. Recher, E. V. Sukhorukov, and D. Loss, Phys. Rev. B **63**, 165314 (2001).
94. S. Datta, *Electronic Transport In Mesoscopic Systems* (Cambridge University Press 1995), p. 260.
95. P. Recher and D. Loss, Journal of Superconductivity and Novel Magnetism **15**, 49 (2002); P. Recher and D. Loss, cond-mat/0112298; for similar results, see also C. Bena et al., cond-mat/0202102.
96. A.F. Volkov, P.H.C. Magnée, B.J. van Wees, and T.M. Klapwijk, Physica C **242**, 261 (1995).
97. M. Kociak et al., Phys. Rev. Lett. **86**, 2416 (2001).
98. M. Bockrath et al., *Nature* **397**, 598 (1999).
99. H.J. Schulz, Phys. Rev. Lett. **64**, 2831 (1990).
100. R. Egger and A. Gogolin, Phys. Rev. Lett. **79**, 5082 (1997).
101. C. Kane, L. Balents, and M.P. Fisher, Phys. Rev. Lett. **79**, 5086 (1997).
102. G. D. Mahan, *Many Particle Physics*, 2nd Ed. (Plenum, New York, 1993).
103. R. Loudon, Phys. Rev. A **58**, 4904 (1998).
104. R. Hanbury Brown and R. Q. Twiss, Nature (London) **177**, 27 (1956).
105. M. Büttiker, Phys. Rev. Lett. **65**, 2901 (1990); Phys. Rev. B **46**, 12485 (1992).
106. T. Martin and R. Landauer, Phys. Rev. B **45**, 1742 (1992).
107. E.V. Sukhorukov and D. Loss, Phys. Rev. B **59**, 13054 (1999).
108. R. C. Liu, B. Odom, Y. Yamamoto, and S. Tarucha, Nature **391**, 263 (1998); M. Henny et al., Science **284**, 296 (1999); W.D. Oliver et al., *ibid.*, 299 (1999).
109. V. A. Khlus, Zh. Eksp. Teor. Fiz. **93**, 2179 (1987).
110. J. König, H. Schoeller, and G. Schön, Phys. Rev. Lett. **78**, 4482 (1997).
111. D. Loss and P. Goldbart, Phys. Rev. B **45**, 13544 (1992).
112. O. Benson et al., Phys. Rev. Lett. **84**, 2513 (2000); E. Moreau et al., Phys. Rev. Lett. **87**, 183601 (2001).
113. G. Chen et al., Science **289**, 1906 (2000).
114. R. J. Luyken *et al.*, Physica E **2**, 704 (1998); T. Lundstrom et al., Science **286**, 2312 (1999).
115. D. Gammon et al., Science **273**, 87 (1996).
116. B. Ohnesorge et al., Phys. Rev. B **54**, 11532 (1996); E. Dekel et al., Phys. Rev. B **61**, 11009 (2000).

# 9 Regulated Single Photons and Entangled Photons From a Quantum Dot Microcavity

Yoshihisa Yamamoto, Matthew Pelton, Charles Santori, Glenn S. Solomon, Oliver Benson, Jelena Vuckovic, and Axel Scherer

## 9.1 Introduction

Quantum cryptography has emerged as a significant field of study over the last fifteen years, because it offers the promise of private communication whose security is assured by the laws of quantum mechanics. Most implementations of quantum cryptography so far have used a protocol introduced by Bennet and Brassard, generally known as BB84 [1]. The message can be encoded on the polarization state of single photons, with a random choice between two non-orthogonal polarization bases when the photons are sent and received. Since an eavesdropper does not know what bases have been chosen, any measurement she makes will impose a detectable back-action on the states of the transmitted photons. Using error correction and privacy amplification, the communicating parties can distill the transmitted message into a secure key, about which the eavesdropper knows arbitrarily little.

However, sources of single photons have not been generally available. Experimental implementations of BB84 have used pulses from lasers or light-emitting diodes, attenuated to the point where the average photon number per pulse is significantly less than one [2]. However, the number of photons in these pulses is described by Poissonian statistics, so that there is always a possibility of more than one photon being sent in a given pulse. Such pulses are vulnerable to a photon-splitting attack, where the eavesdropper removes one photon from the pulse, leaving the remaining photons undisturbed [3]. This increases the required privacy amplification, reducing the length of the secure string. In other words, for communication over a given channel, the number of secure bits per pulse will be greater for a true single-photon source than for Poissonian light [4].

A pulse stream with reduced multi-photon probability compared to the Poissonian case is said to be *antibunched*, and can only be described only quantum mechanically. Mathematically, such non-classical photon statistics can be described using the second-order correlation function $g^{(2)}(\tau)$, defined as follows:

$$g^{(2)}(\tau) = \frac{\langle \hat{a}^\dagger(t)\hat{a}^\dagger(t+\tau)\hat{a}(t+\tau)\hat{a}(t)\rangle}{\langle \hat{a}^\dagger \hat{a}\rangle^2}, \tag{9.1}$$

where $\hat{a}^\dagger$ and $\hat{a}$ are the photon creation and annihilation operators, respectively. A pulsed source with Poissonian statistics will have a $g^{(2)}(\tau)$ function

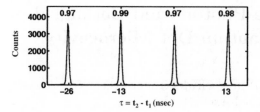

**Fig. 9.1.** Measured photon-photon correlation function for pulses from a mode-locked Ti:Sapphire laser with a 13 ns repetition period. The numbers above the peaks are the normalized areas.

consisting of a series of peaks with unit area, when normalized by the pulse repetition period. This can be seen in Fig. 9.1, which shows a measured second-order correlation function for attenuated pulses from a mode-locked Ti:Al$_2$O$_3$ laser. (See below for an explanation of the measurement method.) This reflects the fact that the probability of detecting a photon in a given pulse is independent of whether a photon has already been detected. For an ideal single-photon source, the central peak at $\tau = 0$ is absent, indicating that, once a single photon has been detected in a pulse, another one will never be detected. An antibunched source will have a zero-delay peak with an area between zero and one. This area gives an upper bound on the probability $P(n_j \geq 2)$ that two or more photons are present in the same pulse:

$$2P(n_j \geq 2)/\langle n \rangle^2 \leq \frac{1}{T}\int_{-\epsilon}^{\epsilon} g^{(2)}(\tau)d\tau. \tag{9.2}$$

where $\langle n \rangle$ is the mean photon number per pulse, $T$ is the pulse repetition period.

Recently, several systems have been investigated for production of single photons. Continuous streams of antibunched photons have been produced using single atoms and ions [5], single molecules [6], and color centers in diamond crystals [7]. Pulses of antibunched photons have been produced by exciting single molecules using a laser pulse [8] or using adiabatic following [9], or by controlled injection of single carriers into a quantum well [10]. Progress has also been made towards a single-photon source using single atoms strongly coupled to an optical cavity [11]. The common factor in all of these sources is that the photons are produced by a single emitter, be it a molecule, atom, ion, color center, or an electron-hole pair in a quantum-well. The essential principle of operation is always the same: a single emitter can only emit one photon at a time.

Excitons in quantum dots are also promising as single-photon sources, since they too behave as single emitters. It has been shown that the fluorescence from a single quantum dot exhibits antibunching [12]. We have achieved triggered generation of antibunched photons from a single quantum dot by exciting with a pulsed laser and spectrally filtering the emission [13].

The normalized area of the $g^{(2)}(0)$ peak can be as low as 0.03. This system for generating single photons, which has also been reported by other groups [14] is stable over long periods of time, and is compatible with mature semiconductor technologies. This allows for the possibility of injecting carriers into the dot electrically instead of optically, producing arrays of sources, and integrating into larger structures.

For example, the quantum dot can be fabricated within a microscopic optical cavity in order to increase the efficiency of the single-photon source. If an emitter is located inside a cavity with a small mode volume and a high finesse, its spontaneous-emission properties are changed. For example, if the emitter is on resonance with a cavity mode, the spontaneous emission rate into this mode is enhanced [15]. This so-called Purcell effect has been seen in atoms for over fifteen years [16]. More recently, ensembles of quantum dots have been coupled to three-dimensionally confined modes in semiconductor microcavities, and changes in their spontaneous-emission lifetimes have been seen [17]. We have succeeded in enhancing the spontaneous emission rate from a single quantum dot into a single mode of an optical microcavity [18]. Because of the enhanced coupling into a single cavity mode, the spontaneous emission becomes directional. A large fraction of the photons can thus be coupled into downstream optical components.

Finally, the single quantum dot can serve as a source of other non-classical radiation states. For example, there may be the possibility to create triggered pairs of polarization-entangled photons [19]. These photon pairs could be used in quantum key distribution systems that use different schemes, such as the Ekert protocol [20] or Bennett, Brassard and Mermin protocol [21].

Generation of single photons and entangled photon-pairs is also a key for quantum computation and network based on photonic qubits. Optical quantum computation based on nonlinear Fredkin gates [22–24] and linear optical elements [25] requires the synchronized parallel production of many single photons.

## 9.2   Single InAs/GaAs Quantum Dots

Semiconductor quantum dots are small regions of a low-bandgap semiconductor inside a crystal of a larger-bandgap semiconductor. The bandgap difference acts as a potential barrier for carriers, confining them inside the dot. Quantum dots are small enough that the carriers form standing waves inside the confinement region, and can only occupy discrete energy levels. Interest in quantum dots was stimulated by the discovery that structures with the required characteristics formed spontaneously during epitaxial growth of lattice-mismatched materials [26]. For example, when InAs is deposited on GaAs, a strained planar layer, known as a wetting layer, initially forms. The strain energy that builds up in this layer is eventually partially relieved by the formation of nanometer-scale islands on the surface. The islands form without defects, and can subsequently be covered with a capping layer of GaAs. Transitions between confined conduction-band and valence-band states involve the absorption or emission of photons at near-optical frequencies. These structures have been extensively studied, both for their intrinsic physical interest and for possible applications in optoelectronic devices [27].

We grew InAs/GaAs quantum dots by molecular beam epitaxy (MBE). In MBE, high-quality heterostructures are obtained by epitaxial deposition in ultra-high vacuum [28]. Elemental sources (In, Ga, and As) are heated to produce molecular beams, which impinge on a heated substrate. Atomic layer-by-layer deposition is achieved by using low growth rates, and shutters in front of the sources are used to control growth time. By controlling the growth rate, the substrate temperature, the ratio of As to In impinging on the surface, and the amount of material deposited, it is possible to control the size and density of the quantum dots [29]. The samples used in our experiments were grown under conditions that give relatively sparse dots, with a surface density of 11–75 $\mu m^{-2}$. Figure 9.2 shows an atomic-force microscope image of dots similar to the dots used in our experiments, except for the absence of a GaAs capping layer.

Single dots are isolated by etching mesas in the MBE-grown sample. The mesas are fabricated by electron-beam lithography and dry etching [30]. Figure 9.3 outlines the steps used. An electron-beam resist is spin-coated onto the surface of the sample. In our case, we use a two-layer resist consisting of poly-methyl methacrylate (PMMA) of different molecular weights. The sample is then introduced into a scanning electron microscope. An electron beam is moved across the surface in order to expose a pattern of circles in the resist. The exposed sample is developed in a solvent mixture, which dissolves away the portions of the resist that were exposed to the electron beam. A thin layer of gold is then deposited on the surface using electron-beam evaporation. Next, the remaining PMMA is dissolved in acetone, which removes both the resist and the gold above it, leaving behind a pattern (etch mask) of metal circles. The sample is then introduced into an electron-cyclotron resonance plasma etcher. This etcher is a vacuum chamber which contains a mixture of Ar, $Cl_2$, and $BCl_3$ gases at low pressure. Microwaves at the elec-

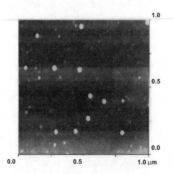

**Fig. 9.2.** Atomic-force-microscope image of sparse self-assembled InAs quantum dots grown on GaAs.

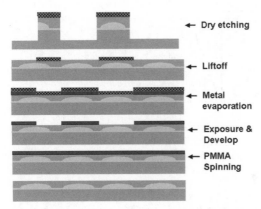

Dry etching

Liftoff

Metal evaporation

Exposure & Develop

PMMA Spinning

**Fig. 9.3.** Schematic illustration of the process used to fabricate microposts in a sample containing InAs/GaAs quantum dots. The process steps are shown sequentially from bottom to top. Starting with an MBE-grown sample, an electron-beam-sensitive resist is spun on top. Electron-beam lithography is used to draw small holes in the resist, and metal is then evaporated on top. The resist is dissolved in a solvent, lifting off the portions of metal above it. A dry etching process is then used to remove the parts of the sample not protected by the metal mask.

tron cyclotron frequency ionize the gas, and permanent magnets confine the resulting plasma. A radio-frequency field is applied to the sample, creating an effective bias in the plasma above the surface. Ions are thus accelerated towards the sample, where they react with and remove GaAs and InAs. This etch process is highly directional, leading to small mesas below the metal circles. The mesas used for isolating single dots are about 120 nm tall and 200 nm wide, and are spaced 50 μm apart.

The single quantum dots are probed by photoluminescence (PL). Fig. 9.4 shows the experimental apparatus used. The sample is held at 4 K in a continuous-flow liquid-helium cryostat, and is held close to the cryostat win-

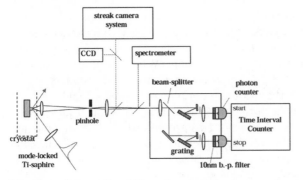

**Fig. 9.4.** Schematic of the setup used to measure luminescence from single quantum dots.

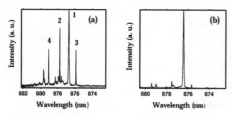

**Fig. 9.5.** Photoluminescence spectra from a single InAs/GaAs self-assembled quantum dot. For (**a**), the pump laser had energy above the GaAs bandgap. For (**b**), the pump energy was resonant with a higher-order transition in the dot, so that excitons are created only in the dot.

dow. Light from a tunable, mode-locked Ti:Sapphire laser is focused onto a post from a steep angle, down to an 18 $\mu$m effective spot diameter. The laser can be tuned to have a photon energy larger than the GaAs bandgap. In this case, a large number of electron-hole pairs are created in the GaAs matrix surrounding the quantum dot. The carriers diffuse towards the dot, where they are rapidly trapped, and quickly relax to the lowest confined states [31]. Alternatively, the laser energy can be tuned so that it is equal to the transition energy between higher-lying confined states in the dot. This eliminates the necessity for carriers to diffuse to the dot and be captured into the confined states, and ensures charge neutrality in the dot.

Light emitted from the dot is collected with an aspheric lens that has a numerical aperture of 0.5, and is focused onto a pinhole that effectively selects a 5 $\mu$m region of the sample for collection. The light is then sent to a charge-coupled device (CCD) camera, a spectrometer, a streak camera, or a Hanbury Brown-Twiss configuration (described below) for measuring the second-order correlation function. The CCD camera allows us to monitor the sample through the collection lens, making alignment possible. The spectrometer has a resolution of 0.05 nm and a cooled CCD array on the output. Figure 9.5 shows PL spectra from the quantum dot used to generate single photons. With continuous-wave (CW) excitation above the GaAs bandgap, the emission spectrum displays several lines [32]. We believe that these lines all come from a single dot, because other mesas show nearly identical emission pattern (peak heights, spacings and widths), except for an overall wavelength shift, suggesting that this pattern is not random. When the laser is tuned to an absorption resonance at 857.5 nm of an excited but QD confined state, thus creating excitons directly inside the dot, emission peaks 3 and 4 almost disappear. We therefore believe that they represent emission from charged states of the dot [33]. We identify peak 1 as ground-excitonic state emission after the capture of a single exciton, and peak 2 as "biexcitonic" emission after the capture of two excitons. A biexcitonic energy shift of 1.7 meV is due to electrostatic interactions among carriers.

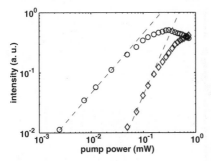

**Fig. 9.6.** Luminescence intensity from a single quantum dot as a function of incident pump power. *Circles* represent the integrated intensity under the single-exciton line, while *diamonds* represent the biexciton line.

Assignment of the different peaks is supported by the dependence of the emission line intensities on pump power, as shown in Fig. 9.6. We can see linear growth of peak 1 and quadratic growth of peak 2 in the weak-pump limit, as expected for excitons and biexcitons, respectively. Further support for the peak identification comes from time-dependent spectra, as collected by the streak camera [34]. The camera produced two-dimensional images of intensity *vs.* wavelength and time after exciting with a laser pulse. Integration times were about 5 minutes, corresponding to about 20 billion pulses. Time resolution as determined by the spectrometer is about 25 ps. The images were corrected for background counts, non-uniform sensitivity, and a small number of cosmic ray events. By integrating intensity within frequency windows corresponding to the peaks shown in Fig. 9.5, time-dependent intensities are obtained for the different lines. The results are shown in Fig. 9.7 for different pump powers.

Under weak excitation, the single-exciton line (line 1) appears quickly after the excitation pulse, and then decays exponentially. This decay time has been measured accurately under resonant excitation to be 0.47 nsec. Under higher excitation power, however, line 1 reaches its maximum only after a long delay. Most of the emission immediately after the excitation pulse now comes from line 2. A simple explanation for this behavior is that, since the laser pulse now initially creates several exciton-hole pairs on average, some time is required before the population of the dot reduces to one electron-hole pair, and only then can the single-exciton emission occur.

In an even high excitation power, the bi-excitonic emission is delayed and the third peak (tri-excitonic emission at a slightly longer wavelength than the bi-exciton line) appears immediately after the excitation pulse.

A statistical and quantitative model can describe these results. Photons from the laser excitation pulse are assumed to be absorbed independently from each other, so that the number of created excitons follows a Poisson distribution. We also assume that excitons decay independently from each

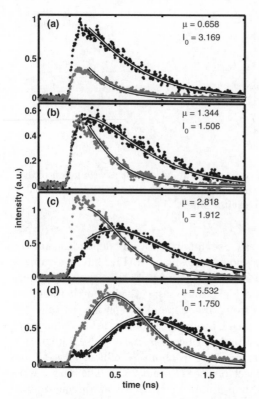

**Fig. 9.7.** Time-dependence of luminescence intensities of the single-exciton line 1 (*black*) and the biexciton line 2 (*gray*) after pulsed excitation above the GaAs bandgap. *Hollow lines* show model fit results, with the parameter values for the fits shown in the figures. The pump powers are (**a**) 27 μW, (**b**) 54 μW, (**c**) 108 μW, and (**d**) 432 μW.

other. This last assumption is not obvious, but we shall see that it matches the data very well. These assumptions result in a Poisson-distributed exciton population for all times after the excitation pulse, with an exponentially decaying mean:

$$P_n(t) = (\mu e^{-\gamma t})^n \exp(-\mu e^{-\gamma t})/n! \,, \tag{9.3}$$

where $P_n(t)$ is the probability that $n$ excitons exist in the dot at time $t$, $\mu$ is the mean number of excitons at time zero, and $\gamma$ is the radiative decay rate of each individual exciton. If we further assume that the emission intensity from the corresponding spectral line is proportional to the number of ways the $n$-exciton state can decay, we obtain

$$I_n(t) = I_0 n P_n(t) \,, \tag{9.4}$$

where $I_n(t)$ is the intensity of the emission line corresponding to the $n$-exciton state and $I_0$ is a constant proportional to the efficiency with which the emitted photons are collected and detected.

Fitted curves from this model are shown as solid lines in Fig. 9.7. The same $\mu$ and $I_0$ are used for all emission lines at given excitation power. $I_0$ varies between streak camera images, due to sample drift and streak-camera gain drift. The fitting parameters $\mu$ and $I_0$ were determined by minimizing the total sum-of-squares error between the model and both lines for a given streak camera image. The model is seen to provide an excellent fit to the data, supporting the assignment of single-exciton, biexciton peaks and tri-exciton (not shown here).

## 9.3   Generation of Single Photons

Since peak 1 corresponds to emission from a quantum dot containing a single and last exciton, there should be only one photon emitted at this energy for each excitation laser pulse, regardless of the number of electron-hole pairs originally created inside the dot [35]. Spectral filtering was used to select this last photon, resulting in only one photon per pulse. As shown in Fig. 9.4, a monochromator-type configuration was used to define a 2 nm-wide measurement bandwidth, with the center wavelength determined by the detector position. Additional rejection of unwanted light (scattered pump light and stray room light) was achieved with a 10 nm bandpass filter attached to each detector.

The Hanbury Brown and Twiss-type configuration was used to measure the second-order correlation function. A beamsplitter sends photons towards one of two single-photon detectors. The detectors are EG&G "SPCM" avalanche photodiodes, which have efficiencies of 40% at 877 nm and 0.2 mm-wide active areas. The electronic pulses from the photon counters were used as start $(t_1)$ and stop $(t_2)$ signals for a time-interval counter, which recorded a histogram of delays $\tau = t_2 - t_1$. Normalized histograms are shown in Fig. 9.8. In the limit of low collection and detection efficiency, these histograms approximate the second-order correlation function [36]. The $\tau = 0$ peak shows a large reduction in area, indicating strong anti-bunching. The numbers printed above the peaks indicate the peak areas, properly normalized by dividing the histogram areas by the photon count rate at each detector, the laser repetition period, and the measurement time. For the numbers shown, the only background counts subtracted were those due to the known dark count rates of the photon counters (130 s$^{-1}$   and 180 s$^{-1}$), which are small compared to the total count rates (19800 s$^{-1}$   and 14000 s$^{-1}$) for the two counters at 0.88 mW pump power. When only counts within 2.8 ns of $\tau = 0$ were included, a normalized $g^{(2)}(\tau = 0)$ peak area of 0.03 was obtained at 0.88 mW. Subtracting the constant background floor seen in the data gave an even lower value of 0.01.

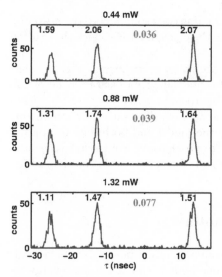

**Fig. 9.8.** Photon-photon correlation functions for emission from a single InAs quantum dot under pulsed, resonant excitation. The numbers above the peaks represent the normalized peak area $g_\tau^2$.

An additional cause of antibunching in the experiment is suppression of the probability for the dot to absorb a second photon after a first photon has been absorbed. If one collects emission from both the single-exciton and the multi-exciton lines, the $g^{(2)}(\tau = 0)$ peak area is still reduced to about 0.32. We believe that, once a single exciton has been created in the dot, the resonant energy for absorption is shifted, due to the electrostatic potential of the trapped carriers. This moves the absorption peak out of resonance with the exciting laser, reducing the probability of absorbing a second photon and creating a second exciton.

The residual non-zero probability of having more than one photon per pulse is believed to be primarily due to imperfect spectral filtering. As well, there is a broadband emission background that contributes some spurious photons. By reducing the filter bandwidth, we believe another fivefold reduction in the $g^{(2)}(0)$ peak should be possible.

## 9.4 Coupling Single Quantum Dots to Micropost Microcavities

A single quantum dot has been shown to be a good source of single photons. However, the usefulness of the source for quantum communication or other quantum information applications is limited by its efficiency. Only one out of approximately every 3000 photons emitted from the dot was ultimately detected by the single-photon detectors. The largest cause of this inefficiency is the fact that the dot emits primarily into the semiconductor substrate, and only 0.6% of the emission is collected by the aspheric lens in front of the cryostat window.

Lenses, mirrors, and other linear optical elements cannot change the brightness of a source, and are thus limited in their ability to increase the collection efficiency. However, by placing the quantum dot inside a microscopic optical cavity, its spontaneous emission characteristics can be changed, and it can be made to radiate primarily into the cavity modes. This is due to the fact that spontaneous emission is not an inherent property of the emitter, but is the result of interaction between the dipole and the surrounding electromagnetic vacuum.

The radiative transition rate of an emitter from an excited, initial state $|i\rangle$ to a lower energy, final state $|f\rangle$ depends on the available photon field density of states $\rho(\lambda)$ at the transition wavelength. In the weak-coupling regime of cavity quantum electrodynamics (QED) [37], where the atomic excitation is irreversibly lost to the field, this rate is expressed by Fermi's golden rule as $(2\pi/\hbar)\rho(\lambda)|\langle f|H|i\rangle|^2$, where $H$ is the atom-field interaction Hamiltonian, $|f\rangle$ and $|i\rangle$ are the ground and excited states of an atom. Thus, by altering $\rho(\lambda)$ using an optical cavity, the spontaneous emission can be enhanced or suppressed. Looked at another way, the spontaneous emission is analagous to stimulated emission, where the "stimulating" field is not real photons, but vacuum fluctuations. By localizing the vacuum field in a resonant cavity mode, the spontaneous emission rate can be enhanced. For a localized atom with a negligible linewidth that is on resonance at the antinode of the standing wave, the enhancement factor (known as the Purcell factor) is $3Q\lambda^3/4\pi^2 V$, where $Q$ is the cavity quality factor, $\lambda$ is the transition wavelength, and $V$ is the cavity mode volume. In order to couple the majority of spontaneous emission into the cavity mode, we require a relatively high-finesse cavity with a small mode volume.

Microscopic planar cavities can be grown by the same MBE process used to create the quantum dots. The microcavity is formed by two distributed-Bragg reflectors (DBR's) separated by a spacer layer whose thickness is equal to $\lambda$, the wavelength of the light in the material. The DBR's are dielectric mirrors consisting of alternating quarter-wavelength-thick layers of AlAs and GaAs. The Fresnel reflections from the AlAs / GaAs interfaces add up in phase, resulting in high overall reflectivity within an angular and spectral stopband. There are 29.5 mirror pairs below the spacer layer and 15 mirror pairs above. The electric field intensity has a maximum in the center of the spacer layer, where the quantum dot layer is grown.

Interaction between the ensemble of quantum dots and the planar microcavity allows for a limited enhancement of the spontaneous emission rate. This is similar to effects that have been seen for quantum well excitons in planar DBR microcavities [38]. Fig. 9.9 shows the wavelength dependence of spontaneous emission lifetime for quantum dots in a planar microcavity

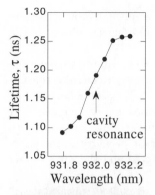

Fig. 9.9. Photoluminescence lifetime vs. wavelength for quantum dots in a planar microcavity.

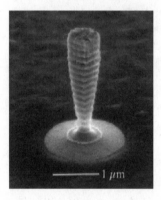

Fig. 9.10. Scanning-electron-microscope image of a micropost microcavity.

with a resonance wavelength of 932 nm and a quality factor $Q$ of 2300. (See below for an explanation of the measurement technique.) Here, there is a continuous distribution of cavity modes from the resonant cut-off wavelength $\lambda_c$ to a wavelength corresponding to the stopband edge. Since the number of modes increases with decreasing wavelength from the cutoff wavelength, the spontaneous emission lifetime is correspondingly decreased from 1.3 ns above the cutoff wavelength to 1.1 ns at shorter wavelengths. In order to achieve a larger enhancement of emission rates, both the field and the exciton must be confined in all three dimensions.

Three-dimensional confinement of the field is realized at the same time as isolation of single quantum dots by etching microposts into the MBE-grown sample. Following a similar process to that described above for etching mesas, posts are etched through the top DBR mirror, through the spacer layer containing the quantum dots, and through a small portion of the lower DBR. Designed post diameters vary from 6 $\mu$m down to 0.5 $\mu$m. A scanning-electron microscope image of an example post with nominal diameter of 0.5 $\mu$m is shown in Fig. 9.10. A taper in the post diameter arises as a result of the etch, and helps reduce the cross-sectional area at the location of the quantum dots. This results in a mode volume close to $(\lambda/n)^3$.

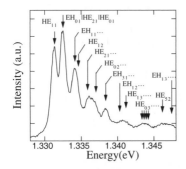

**Fig. 9.11.** Photoluminescence from a micropost microcavity with diameter of 6 $\mu$m which contains a large number of InAs/GaAs quantum dots. Peaks correspond to different resonances of the microcavity, labeled according to the corresponding transverse waveguide mode. The arrows show calculated mode energies.

Due to the large index contrast between the semiconductor material and the surrounding air or vacuum, light is trapped inside the post by total internal reflection. That is, the post acts as a small waveguide for light in the transverse directions. Together with the mirrors in the longitudinal direction, this leads to three-dimensional confinement of light in the micropost microcavity. The structure exhibits discrete resonances, to which the quantum-dot emission can be efficiently coupled. This effect can be seen in the PL spectrum from dots in a 6 $\mu$m-diameter microcavity post, as shown in Fig. 9.11.

The inhomogenously-broadened quantum dot emission is filtered by the cavity modes into a series of discrete peaks. Note that we do not consider polarization in our treatment; the lowest-energy fundamental mode, for example, is actually composed of a pair of degenerate modes of opposite polarizations.

The cavity resonances can be modeled approximately following Panzarini and Andreani [39]. We assume that the electromagnetic field can be factored into a part that depend only on the transverse coordinates and a part that depends only on the longitudinal coordinate. The longitudinal field dependence is calculated using a transfer-matrix method, using layer thicknesses from the crystal growth, and known GaAs and AlAs refractive indices[40]. The calculated longitudinal profile gives an effective penetration depth of the cavity mode into the DBR's. In addition, an average refractive index $n_{\mathrm{eff}}$ is obtained by averaging over the longitudinal direction, using the field intensity as a weighting factor. The effective index is then used in a standard waveguide model to determine the transverse field dependence. Using cylindrical dielectric boundary conditions, a characteristic equation for the modal wavenumber can be solved to provide the blue shift of the cavity resonance [41]. The waveguide equations also give a transverse field dependence. By integrating the field in three dimensions, an effective mode volume is obtained.

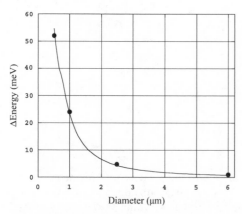

**Fig. 9.12.** Blueshift of the lowest-energy mode of micropost microcavities, as compared to the cutoff wavelength of the corresponding planar microcavity. The *points* are experimental data, while the solid line is the calculated result.

The arrows in Fig. 9.11 are the resonance energies calculated by this method for a post with 6 $\mu$m diameter; they are labeled according to the corresponding transverse waveguide modes. Also, Fig. 9.12 shows agreement between measured blueshifts for the fundamental ($HE_{11}$) mode and blueshifts calculated by this method. Note that this agreement is obtained without any fitting parameters.

Figure 9.13 shows the filtered PL spectrum for a post with a top diameter of 1 $\mu$m. If this is compared to Fig. 9.11, we can see a blueshift and an increased mode spacing, due to the increased transverse confinement. However, the modal linewidth is also increased. Figure 9.14 shows the measured values of quality factor for different post diameters, determined from the measured PL linewidths. The degradation of $Q$ with decreasing post diameter is believed to be primarily due to diffraction in the unetched portion of the lower DBR. The field spreads as it penetrates into the lower DBR and on its trip back to the post, and the amount of light that is recaptured by the mode is determined by the overlap of the diffracted field with the transverse mode

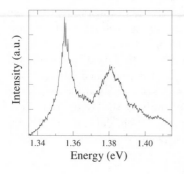

**Fig. 9.13.** Photoluminescence from a micropost microcavity with diameter of 1 $\mu$m which contains several InAs/GaAs quantum dots.

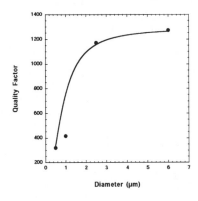

**Fig. 9.14.** Quality factor of the fundamental mode of micropost microcavities as a function of post diameter. The *points* are experimental data, while the *solid line* is the predicted result.

profile of the post. A simple model of diffraction in the lower DBR is used to calculate quality factors. As can be seen in Fig. 9.14, this model can account well for the observed loss of $Q$. This indicates that other loss mechanisms, such as scattering loss due to roughness of the post sidewalls, play a secondary role.

Time- and frequency-dependent emission from the quantum dots was measured using a streak camera with a spectrometer attachment, as described above. The decay at longer times is determined by a single exponential time constant, the spontaneous emission time for excitons in the quantum dot. This part of the curve was thus fitted in order to determine spontaneous decay times. The validity of this approach was confirmed using a detailed rate-equation model. By repeating this process at different wavelengths, the frequency-dependent emission rate across the fundamental cavity mode can be determined.

Figure 9.15 shows the results for a post with 2 $\mu$m top diameter. The lifetime is a minimum on resonance with the cavity mode, a clear signature of the discrete, three-dimensional cavity resonances. As well, the lifetimes off-resonant with the mode are nearly the same as in the absence of the cavity. This indicates that the lifetime modification is due to interaction with the cavity mode, and not due to some other effect, such as non-radiative recombination.

The expected spontaneous emission rate can be calculated according to the following equation:

$$\frac{\gamma}{\gamma_0} = \frac{Q\lambda_c^3}{2\pi^2 n_{\text{eff}}^3 V_0} \frac{\Delta\lambda_c^2}{\Delta\lambda_c^2 + 4(\lambda - \lambda_c)^2} + f \,, \tag{9.5}$$

where $\gamma_0$ is the (experimentally-determined) spontaneous emission rate of a quantum dot without a cavity, $\lambda_c$ is the cavity resonant wavelength, $\Delta\lambda_c$ is the cavity linewidth, and $(\lambda - \lambda_c)$ is the detuning of the dot emission wavelength from the cavity resonance. The cavity quality factor and mode volume are calculated as described above. $f\gamma_0$ is the spontaneous decay rate into leaky modes (i.e., emission that is incident on the post edges at an angle

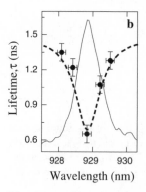

**Fig. 9.15.** Photoluminescence lifetime *vs.* wavelength for quantum dots in a micropost microcavity with a top diameter of 2 μm. The points are measured values, while the *dashed line* is the predicted result. The thin, *solid line* represents the photoluminescence intensity at the same wavelengths.

larger than the critical angle for total internal reflection). For an ensemble of dots in the post, the calculated spontaneous emission rate must be averaged radially across the post to give an expected decay rate. This rate is also plotted in Fig. 9.15, and shows good agreement with experiment. Note again that no fitting parameters are used in the theory.

Figure 9.16 shows similar results for a cavity with top diameter of 0.5 μm. Similar characteristics can be seen. However, there is now a single quantum dot on resonance with the cavity mode. A significant reduction in the spontaneous emission time, to 0.28 ns, is seen. The dashed line is the theoretical result. This is in good agreement with the measurement, considering

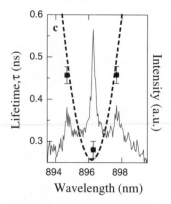

**Fig. 9.16.** Photoluminescence lifetime *vs.* wavelength for isolated quantum dots in a micropost microcavity with a top diameter of 0.5 μm. The points are measured values, while the dashed line is the predicted result. The thin, solid line represents the photoluminescence intensity at the same wavelengths.

only diffractive losses through the bottom DBR mirror have been included in determining $Q$. As well, the factorization of fields into longitudinal and transverse components is inexact.

The spontaneous emission lifetime can be converted into a spontaneous emission coupling coefficient $\beta$, the fraction of light that is captured in the fundamental cavity mode:

$$\beta = \frac{\gamma - \gamma_0 - \gamma_c}{\gamma} , \qquad (9.6)$$

where $\gamma$ is the enhanced spontaneous emission decay rate into the fundamental mode of the cavity, $\gamma_0$ is the spontaneous emission decay rate in the absence of a cavity, and $\gamma_c$ is the fractional spontaneous emission decay rate into the solid angle of the cavity mode in the limit that the mirror reflectivity, $R \to 0$. Since this solid angle is at most a few degrees in our case, $\gamma_c \ll \gamma_0$. Using this formula, we determine that 78% of the light from a single quantum dot is collected by a single mode.

Similar lifetime modifications have been reported for an ensemble of quantum dots. However, systems employing an ensemble of inhomogenously-broadened dots coupled to discrete optical modes may have large $\beta$ and short $\tau$ values at the on-resonant condition, but off-resonant $\beta$ and $\tau$ values are small and long, respectively. Averaging over the ensemble of dots coupled to cavity mode, the overall $\beta$ and $\tau$ values are poor. More recently, modified spontaneous emission has been seen for a single quantum dot coupled to a whispering-gallery mode of a microdisk cavity [42]. However, it is a challenge to efficiently couple light out of such a mode. Micropost microcavities, on the other hand, can be designed so that virtually all light escapes from the mode in a Gaussian-like wave propagating normal to the sample surface. This allows for efficient collection and coupling to downstream components.

The coupling coefficient of 78% is limited by the relatively poor cavity quality factor for small posts. Since this is mostly due to diffraction loss in the lower DBR, the most apparent way to increase the quality factor is to etch completely through the lower mirror stack. This involves reducing the taper in the cross-section, while maintaining smooth sidewalls in order to avoid scattering loss. This is done for a micropost structure shown in Fig. 9.17, for which $Q$ and $\beta$ are measured to be $\sim 800$ and $\sim 0.9$, respectively.

## 9.5   Theoretical Analysis of a Micropost DBR Cavity

The only way to accurately model optical modes in such a complex structure as described above is to numerically solve Maxwell's equations. One of the most popular ways of doing so is the finite-difference time-domain (FDTD) method, proposed by Yee in 1966 [43]. This method is based on a discretization of the differential form of Maxwell's equations. More explicitly,

**Fig. 9.17.** A new micropost DBR microcavity structure.

the following two Maxwell's equations are solved numerically:

$$\mu_0 \frac{\partial \mathbf{H}}{\partial t} = -\nabla \times \mathbf{E} \tag{9.7}$$

$$\epsilon \frac{\partial \mathbf{E}}{\partial t} = \nabla \times \mathbf{H},$$

where $\mathbf{E}$ and $\mathbf{H}$ are the electric and magnetic fields, respectively, and $\epsilon$ is the dielectric constant of the medium. A Cartesian spatial grid is defined with increments $\Delta x$, $\Delta y$, and $\Delta z$, and a time increment $\Delta t$ is also defined. Any field $F(x, y, z, t)$ can then be written in discretized form as $F^n(i, j, k) = F(i\Delta x, j\Delta y, k\Delta z, n\Delta t)$. The electric and magnetic fields are defined on two grids that are offset by half a step in time and space. The fields can then be alternately advanced in time, using a leapfrog method. Six coupled finite-difference equations are solved, one for each of the components of the electromagnetic fields. For example, the $z$-component of the electric field is given by

$$E_z^{n+1}\left(i, j, k + \frac{1}{2}\right) = E_z^n\left(i, j, k + \frac{1}{2}\right) + \frac{\Delta t}{\epsilon(i, j, k + \frac{1}{2})} \tag{9.8}$$

$$\times \left[ \frac{H_y^{n+\frac{1}{2}}(i + \frac{1}{2}, j, k + \frac{1}{2}) - H_y^{n+\frac{1}{2}}(i - \frac{1}{2}, j, k + \frac{1}{2})}{\Delta x} \right.$$

$$\left. - \frac{H_x^{n+\frac{1}{2}}(i + \frac{1}{2}, j, k + \frac{1}{2}) - H_x^{n+\frac{1}{2}}(i, j - \frac{1}{2}, k + \frac{1}{2})}{\Delta y} \right].$$

The computational mesh is truncated by placing a nonreflecting absorber at all boundaries [44].

We use this method to determine the modes of a micropost microcavity. An initial field distribution is applied to the analyzed structure, and the fields are subsequently evolved in time. We record the time evolution of the field at a point of low symmetry, and take the fast Fourier transform of the resulting time series to get the cavity mode spectrum. We then identify a mode of interest with frequency $\omega_0$. The mode is isolated by convolving the field in time (at each point of the computational domain) with an oscillating function of frequency $\omega_0$. In the frequency domain, this convolution corresponds to the application of a band-pass filter with central frequency $\omega_0$ and with a linewidth determined by the boundaries of the convolution integral. We ignore mode polarization in this analysis, so that any pair of polarization-degenerate modes will be treated as one mode.

We can take advantage of the rotational symmetry of the cavities to make the calculations more efficient, reducing the order of the computer memory requirements from $N^3$ to $N^2$, where $N$ represents a linear dimension of the computational domain [45]. In this case, each simulation performed by the cylindrical FDTD algorithm is for a particular value of the azimuthal mode number $m$.

The flexibility of the FDTD method allows simulation of both ideal structures with straight walls and realistic structures with posts that are undercut as a result of etching. We take the refractive indices of GaAs and AlAs to be 3.57 and 2.94, respectively, and the thicknesses of GaAs and AlAs mirror layers to be 70 nm and 85 nm, respectively. The central GaAs spacer layer is 280 nm thick and is sandwiched between 15 mirror pairs on top and 30 mirror pairs on bottom. The entire structure rests on a GaAs substrate. Absorption losses are neglected. The spatial discretization is performed with a 5 nm mesh. For ideal structures, we assume straight walls and etching through the entire bottom mirror. For tapered structures, the etch extends through only the top two pairs of the bottom mirror. As well, in these realistic posts the cavity diameter is constant for only 550 nm on top, after which it changes linearly with a slope of approximately 4° with respect to the micropost axis.

Figure 9.18 shows the calculated electric field intensity of the lowest-order mode in structures with a cavity diameter of 0.5 $\mu$m at the top of the post. Part (a) shows an ideal post, while part (c) shows a realistic post. A significant difference in the field distribution can be seen between the two posts.

The quality factor of a mode is determined by two independent methods: (1) the decay time constant for energy stored in the mode; and (2) the ratio of the energy stored in the cavity, multiplied by the mode frequency, to the power lost by radiation outside the cavity. Table 9.1 lists the values of $Q$ calculated for the fundamental modes of ideal posts with three different diameters. It can be seen that the quality factor decreases as the posts get smaller. This is due to the blue shift in the fundamental mode, also indicated in Table 9.1. As the post diameter decreases, the mode becomes more tightly confined, and its wavelength decreases. The resonance wavelength is thus no longer the wavelength for which the DBR's were designed. At the same time,

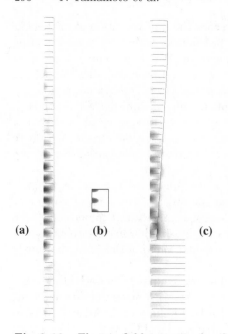

**Fig. 9.18.** Electric field intensity for the fundamental mode of a micropost microcavity with a top diameter of 0.5 $\mu$m. The intensity is represented by a grey scale. Half a longitudinal cross-section of the post is shown. The three-dimensional intensity distribution can be obtained by rotating around an axis running down the center of the post (left-hand side of the figures). Lines represent interfaces between different materials. The field in (**a**) is calculated by the finite-difference time-domain method for an ideal post, where the sidewalls are straight and the etch extends completely through the lower mirror. (**b**) shows the corresponding field near the center of the post calculated by the approximate method. (**c**) shows the field calculated by the finite-difference time-domain method for realistic posts, where the sidewalls are tapered and the etch only extends partially through the lower mirror.

the effective incident angle of light on the DBR's increases. Both these effects cause the effective mirror reflectivity to decrease slightly.

A much more dramatic decrease in quality factor is seen when we consider posts of realistic shape, as indicated in Table 9.1 and in Fig. 9.19. As can also be seen, the blue shift is similar to that for ideal posts; the wavelength shift thus cannot account for the degradation of $Q$. The additional loss in these posts is due to diffraction in the lower DBR. Light making a longitudinal round-trip in the cavity penetrates a certain distance into the unetched portion of the lower DBR before being reflected back. As it does so, it diffracts outwards, so that only a certain fraction of the light is recaptured in the post. This is also manifest in the significant field intensity outside of the lower portion of the realistic post in Fig. 9.18c.

**Table 9.1.** Quality factor $Q$ and resonance wavelength $\lambda$ for the fundamental modes of ideal and realistic micropost microcavities, as calculated by the finite-difference time-domain method.

| Diameter ($\mu$m) | Ideal $Q$ | Ideal $\lambda$ (nm) | Realistic $Q$ | Realistic $\lambda$ (nm) |
|---|---|---|---|---|
| 0.5 | 5000 | 920 | 166 | 915 |
| 1.0 | 7000 | 976 | 440 | 962 |
| 2.0 | 11 500 | 993 | 2500 | 992 |
| $\infty$ | 12 000 | 1000 | – | – |

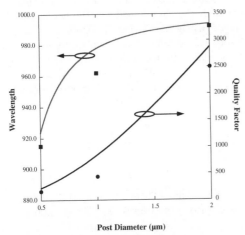

Post Diameter (μm)

**Fig. 9.19.** Quality factor and resonance wavelength for the fundamental modes of realistic micropost microcavities. Points indicate values calculated by the finite-difference time-domain method, and solid lines indicate values calculated by the approximate method.

The analysis of Purcell factors is based on a method presented in [46]. The total radiated power is calculated for a classical dipole in the cavity and for a dipole in bulk material. The ratio of these powers gives the Purcell factor. Figure 9.20a shows the Purcell factors calculated for single quantum dots on resonance with the fundamental mode of cavities with realistic shapes. The quantum dots are assumed to be located near the center of the posts. The degradation in $Q$ is seen to overtake the reduction in mode volume, reducing the Purcell factor as the post diameter decreases. Nonetheless, significant enhancement of spontaneous emission rate is seen for posts of all sizes. Figure 9.18b shows the ensemble averaged Purcell factor over all different locations of a quantum dot inside a post, which is in a fairy good agreement with the measured results shown by the dots.

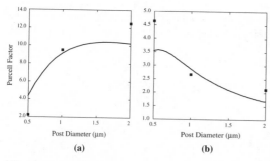

**Fig. 9.20.** Purcell factors for quantum dots resonant with the fundamental modes of realistic micropost microcavities. The Purcell factor is the ratio of the spontaneous emission rate for a dot resonant with the cavity mode to the spontaneous emission rate in the absence of the cavity. (**a**) shows calculated results for a single dot at the center of the post. The points are values calculated by the finite-difference time-domain method, and the line is calculated by the approximate method. (**b**) compares experimentally-measured values (points) with the average values calculated by the approximate method for a large number of dots distributed radially across the post (line).

## 9.6    Entangled Photon-Pairs from a Single Quantum Dot

A fundamental nonlinear effect in a QD is the saturation of a single energy level by two electrons and holes of opposite spins due to the Pauli exclusion principle. We propose a device that produces regulated pairs of entangled photons using this effect [19]. We assumed a typical InAs/GaAs dot diameter of 20 nm and height af 4 nm. The QD is embedded in a GaAs p-i-n junction, where the tunnel barriers are 190 Å (112 Å) thick in the $n$ and $p$ side. For a qualitative discussion of the device operation, the Coulomb blockade energy can be estimated in a single particle picture [47,48], for simplicity, with strain and piezoelectric effects [49] neglected. We assumed that the one and two electron ground state energy levels are 210 meV and 190 meV below the conduction band edge of GaAs, respectively, and that the one and two hole ground state energy levels are 100 meV and 80 meV above the valence band edge of GaAs, respectively. The first excited electron ($p$-like) state is about 70 meV above the ground state [50,51]. These values are consistent with experimental observations [47,50] and calculations [52,53]. If the junction voltage $V_j$ is well below the built-in potential, the carrier transport takes place by resonant tunneling of electrons and holes.

Figure 9.21 shows the calculated resonant tunneling rates for electron and hole *versus* the applied bias voltage. The calculation uses the WKB approximation with an effective electron and hole mass of 0.067 $m_0$ and 0.082 $m_0$, respectively, and a temperature of 4 K. The different lines correspond to the following (from left to right): Electron tunneling into the dot containing zero

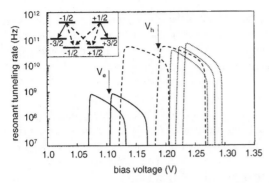

**Fig. 9.21.** Calculated resonant tunneling rates at 4 K into the QD ground state for electrons (*solid lines*) and holes (*dashed lines*) versus the applied bias voltage. Tunneling into the first excited electron state is indicated by *dotted lines*. In turnstile operation, the bias voltage is modulated between $V_e$ and $V_h$. The *inset* illustrates optical transitions in a cubic lattice. The numbers indicate the projection of the total angular momentum $J_z$ for the electrons and holes.

or one electron (solid lines) and hole tunneling into the dot containing two electrons and zero or one hole (dashed lines). Tunneling into the first excited electron state is indicated by dotted lines, where the three lines correspond to two electrons and two, one, or zero holes. The difference in the widths of electron and hole tunneling resonances is due to the asymmetric tunnel barriers and different doping levels, $n = 10^{19} \text{cm}^{-3}$ and $p = 10^{18} \text{cm}^{-3}$. We chose the position of the QD within the GaAs layer in order to have the first hole resonant tunneling condition fulfilled at a junction voltage above the second electron tunneling resonance. In this situation, we can switch on and off hole and electron tunneling by switching between different bias voltages.

Two-photon turnstile operation is achieved as follows: At a low bias voltage $V_e$ (indicated in Fig. 9.21), two electrons can tunnel into the initially empty QD. Further electron tunneling is now completely suppressed due to the Pauli exclusion principle, since the ground state is filled and the next available electron state, the first excited state, is far off of resonance. Then, we switch up to a higher voltage $V_h$ (indicated in Fig. 9.21), where two holes tunnel. Again, further hole tunneling is suppressed due to the Pauli exclusion principle since the hole ground state is filled and the first excited hole state (not shown) is off resonance. The first excited electron state shifts by typically 7 mV [50] to lower voltages when a hole tunnels. This is indicated by the three dotted lines in Fig. 9.21. However, even after two holes have tunneled into the QD, electron tunneling is inhibited. Once the holes have tunneled, radiative recombination annihilates two electron-hole pairs and produces exactly two photons. Thus, modulating the bias voltage between $V_e$ and $V_h$ produces a regulated stream of photons, where two photons are emitted per modulation cycle.

The two photons arise from the decay of the biexcitonic ground state of the QD, where the correlated electrons and holes have opposite spins. For quantum wells in direct-gap materials with a cubic lattice, any photons emitted are circularly polarized, because the $J_z = \pm 1/2$ electron recombines with the $J_z = \pm 3/2$ heavy hole [54]. This is illustrated in the inset in Fig. 9.21, where solid arrows indicate the $\sigma^+$ and $\sigma^-$ ground state transitions. In the case of a QD, the strong confinement introduces level mixing and the hole ground state may have contributions from the $J_z = \pm 1/2$ hole states. Possible transitions to the $J_z = \pm 1/2$ states are indicated by dashed arrows in Fig. 9.21. Accordingly, when a $J_z = +1/2$ electron radiatively recombines with a hole in a QD, the emitted light is predominately $\sigma^+$ polarized, but may also have a $\sigma^-$ component. Thus, the two photons that arise from the decay of the biexcitonic ground state are not necessarily perfectly anticorrelated with respect to $\sigma^+$ and $\sigma^-$ polarization. An asymmetric dot shape, strain, and piezoelectric effects [55] further reduce the anticorrelation. However, there is experimental evidence from polarized photoluminescence [56] and two photon absorption measurements [57] that the anticorrelation in $\sigma^+$ and $\sigma^-$ polarization is preserved in QD's. An exact calculation of the energy levels and oscillator strength including spin for the system discussed here would be desirable (so far optical and electronic properties of self-assembled InAs QD's have been calculated neglecting spin [55]).

We point out that a previous single photon turnstile device relies on the relatively small Coulomb splitting [10]. This limits the operation of this device to very low temperatures (40 mK) in order to guarantee that thermal energy fluctuations are negligible. In the proposed device, the turnstile operation is maintained up to much higher temperatures due to the very large splitting between the electron and hole ground and excited states. Electron and hole tunneling could be controlled merely by the Pauli exclusion principle, even if the Coulomb blockade effect were absent. For the parameters we assumed here, an operation at above 20 K should be possible. At higher temperatures, the electron and hole tunneling curves are broadened, mainly due to the thermal energy distribution of the electrons and holes in the $n-$ and $p-$doped layers. The broadening leads to a significant hole (electron) tunneling rate at lower (higher) bias voltage $V_e(V_h)$, and photon emission can no longer be controlled. With a smaller QD and a larger splitting between ground and excited states, a larger broadening could be tolerated and thus a higher temperature operation is possible. We calculated that, up to a temperature of 50 K, thermionic emission can be neglected in the proposed structure.

We now focus on the production of pairs of entangled photons at well-defined time intervals. Starting from the biexcitonic ground state of the QD, a first electron can recombine with a hole and emit a $\sigma^+$ or a $\sigma^-$ photon. Then, the second electron of opposite spin recombines with a hole, and a photon of opposite polarization is emitted. This situation is very similar to a two-photon cascade decay in an atom [58]. The two-photon state has the same form in any basis and is a maximally entangled (Bell) state: $|\psi\rangle = \frac{1}{\sqrt{2}}(|\sigma^+\rangle_1|\sigma^-\rangle_2 + |\sigma^-\rangle_1|\sigma^+\rangle_2)$. Because of additional binding energy, the biexcitonic ground state has a smaller energy than twice the excitonic ground state. Therefore, the first emitted photon 1 and the second emitted photon 2 have different energies as shown in Fig. 9.5

The advantage of the proposed structure compared to other sources of entangled photons, such as two-photon cascade decay in atoms or parametric down-conversion in nonlinear crystals, is that entangled photon pairs are provided one by one with a tunable repetition rate of up to 1 GHz by a compact semiconductor device. The source is electrically pumped and the photons are emitted in resonant modes of an optical resonator, which greatly improves, e.g., the efficiency of subsequent fiber coupling.

The inset in Fig. 9.22 sketches the setup of a possible experiment, where the nonlocal quantum correlation between photons 1 and 2 leads to a violation of Bell's inequality. The two photons are separated with the help of a dichroic mirror (DM) and analyzed by a combination of quarter-wave plates (Q1, Q2), polarizing beam splitters (P1, P2), and detectors (D1, D2). Bell's inequality in the version of Ref. [59] is

$$S = |E(\alpha,\beta) - E(\alpha^{'},\beta)| + |E(\alpha^{'},\beta) + E(\alpha^{'},\beta^{'})| \leq 2 \quad , \tag{9.9}$$

where

$$E(\alpha,\beta) = C_{++}(\alpha,\beta) + C_{--}(\alpha,\beta) - C_{+-}(\alpha,\beta) - C_{-+}(\alpha,\beta) \quad . \tag{9.10}$$

Each photon is subject to a measurement of linear polarization along an arbitrary angle $\alpha$ or $\beta$ with two-channel polarizers whose outputs are $+$ and $-$. Then, e.g., $C_{++}(\alpha,\beta)$ is the number of coincidences between the $+$ output

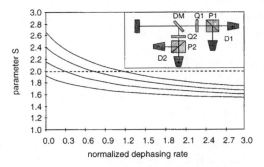

**Fig. 9.22.** The *inset* shows the setup for a proposed experiment. Photons are separated with a dichroic mirror (DM) and analyzed with the combination of quarter-wave plates (Q), polarizing beam splitters (P), and detectors (D). The figure shows the calculated left side of (9.9) (parameter $S$) versus the dephasing rate $R_d$. $R_d$ is normalized to the radiative recombination rate $R_p$ and $R_h = 10R_p$. Values above 2 (*dashed line*) are a violation of Bell's inequality. From *top* to *bottom* $\Delta_{\mathrm{corr}} = 1$, 0.9, 0.8 and 0.7.

of the polarization measurement of photon 1 along $\alpha$ and + output of the polarization measurement of photon 2 along $\beta$. Maximal violation of Bell's inequality is observed for a particular set of angles of the two polarizers: $\alpha = 0$, $\alpha' = -\pi/4$, $\beta = 3\pi4$, $\beta' = \pi/8$. For this set, quantum mechanics predicts $S = 2\sqrt{2}$, although the local hidden-variables theory is constrained by 2.

In the proposed device, several processes may degrade the entanglement and cause an evolution of the pure state into a statistical mixture of anticorrelated photons. For example, the QD initially contains two electrons, and then the bias voltage is changed to allow hole tunneling. It is possible that a first photon is emitted right after the first hole has tunneled, before the biexcitonic ground state has formed. A second photon can be emitted after the second hole has tunneled, but the final state is then a statistical mixture. Alternatively, even if the QD is in the biexcitonic ground state, spin dephasing may occur between the photon emission events. If the dephasing rate $R_d$ is much larger than the radiative recombination rate $R_p$, then the final photon state is again a statistical mixture.

In order to demonstrate that it is possible to measure a violation of Bell's inequality we performed a numerical calculation. A rate-equation model of the tunneling and radiative recombination processes in the QD, similar to that presented in Ref. [60], is used. In order to account for the above mentioned problem of imperfect correlation, we define the degree of anticorrelation $\Delta_{\text{corr}}$ in the following way:

$$\Delta_{\text{corr}} = \frac{R_p^+}{R_p^+ + R_p^-} \ . \tag{9.11}$$

In this equation $R_p^+ (R_p^-)$ denotes the radiative recombination rate of a $J_z = +1/2$ electron with a hole in the biexcitonic ground state of the QD through to the emission of $\sigma^+ (\sigma^-)$ photons. In this notation, $\Delta_{\text{corr}} = 0.5$ corresponds to no anticorrelation and $\Delta_{\text{corr}} = 1$ to perfect anticorrelation.

Figure 9.22 shows the left side of (9.9) v.s. the dephasing rate $R_d$ for a hole tunneling rate of 10 times the radiative recombination rate, in agreement with the calculated hole tunneling rate of 10 GHz and radiative recombination rate greater than 1 GHz in the proposed device. Values above 2 (dashed line) are a violation of Bell's inequality. The different curves correspond to different values of the degree of anticorrelation $\Delta_{\text{corr}}$; from top to bottom, $\Delta_{\text{corr}} = 1, 0.9, 0.8$ and $0.7$. Clearly, a violation of Bell's inequality can be measured even with imperfect anticorrelation if the dephasing rate is small enough. Recent experiments in QD's indicate that the spin dephasing rate of conduction band electrons is much lower than 0.3 GHz [61], and thus much lower than the radiative recombination rate and tunneling rates.

## 9.7   Conclusions

Quantum dots can be used as sources of non-classical light. We have grown InAs/GaAs quantum dots by a self-assembly process in molecular-beam epitaxy, and have isolated single dots in microposts by using electron-beam lithography and plasma etching. Spectroscopy on single dots shows an energy shift between emission from single excitons and biexcitons in the quantum dot. This means that pulsed excitation and spectral filtering can be used to isolate a single emitted photon per pump pulse. Using this method, we demonstrated a hundred-fold reduction in the multi-photon probability as compared to attenuated coherent light. Further reduction should be possible by improved spectral filtering.

As well, we have incorporated single quantum dots into microscopic optical cavities. By modifying the spontaneous emission rate, this technique results in 90% of the light emitted from the dot being collected in a single cavity mode. The efficiency of coupling can be further improved by improving the fabrication process.

By combining these two elements, it should be possible to create an efficient source of triggered single photons. As well, an extension of the technique should make it possible to create photon pairs that are at least partially entangled in polarization. These two sources could then be incorporated into quantum key distribution systems in order to improve the transmission rate of secure information, as well as linear optical quantum computation.

## References

1. C.H. Bennett and G. Brassard, in Proceedings of IEEE International Conference on Computers, Systems, and Signal Processing, Bangalore, India (IEEE, New York, 1984), p. 175.
2. C.H. Bennett, F. Bessette, G. Brassard, L. Salvail, and J. Smolin, J. Crypto. **5**, 3 (1992).
3. N. Lütkenhaus, Phys. Rev. A **61**, 2304 (2000).
4. E. Waks, A. Zeevi, and Y. Yamamoto, e-print quant-ph/0012078, Dec. 16, 2000.
5. H. J. Kimble, M. Dagenais, and L. Mandel, Phys. Rev. Lett. **39**, 691 (1977); F. Diedrich and H. Walther, Phys. Rev. Lett. **58**, 203 (1987).
6. T. Basché, W. E. Moerner, M. Orrit, and H. Talon, Phys. Rev. Lett. **69**, 1516 (1992); L. Fleury, J.-M. Segura, G. Zumofen, B, Hecht, and U. P. Wild, Phys. Rev. Lett. **84**, 1148 (2000).
7. C. Kurtsiefer, S. Mayer, P. Zarda, and H. Weinfurter, Phys. Rev. Lett. **85**, 290 (2000); R. Brouri, A. Beveratos, J.-P. Poizat, and P. Grangier, Opt. Lett. **25**, 1294 (2000).
8. F. De Martini, G. Di Giuseppe, and M. Marrocco, Phys. Rev. Lett. **76**, 900 (1996); B. Lounis and W. E. Moerner, Nature **407**, 491 (2000).
9. C. Brunel, B. Lounis, P. Tamarat, and M. Orrit, Phys. Rev. Lett. **83**, 2722 (1999).

10. J. Kim, O. Benson, H. Kan, and Y. Yamamoto, Nature **397**, 500 (1999); A. Imamoglu and Y. Yamamoto, Phys. Rev. Lett. **72**, 210 (1994).

11. M. Heinrich, T. Legero, A. Kuhn, and G. Rempe, Phys. Rev. Lett. **85**, 4872 (2000).

12. P. Michler et al, Nature **406**, 968 (2000); B. Lounis, H. A. Bechtel, D. Gerion, P. Alivisatos, and W. E. Moerner, Chem. Phys. Lett. **329**, 399 (2000).

13. C. Santori, M. Pelton, G .S. Solomon, Y. Dale, and Y. Yamamoto, Phys. Rev. Lett. **86**, 1502 (2001).

14. P. Michler et al, Science **290**, 2282 (2000); V. Zwiller et al, Appl. Phys. Lett. **78**, 2476 (2001).

15. E. M. Purcell, Phys. Rev. **69,** 681 (1946); K. H. Drexhage, in Progress in Optics Vol. XII, edited by E. Wolfe (North-Holland, Amsterdam, 1974), p. 165.

16. P. Goy, J.-M. Raimond, M. Gross, and S. Haroche, Phys. Rev. Lett. **50,** 1903 (1983); G. Gabrielse and H. Dehmelt, Phys. Rev. Lett. **55,** 67 (1985); R. G. Hulet, E. S. Hilfer, and D. Kleppner, Phys. Rev. Lett. **55,** 2137 (1985); D. H. Heinzen, J. J. Childs, J. E. Thomas, and M. S. Feld, Phys. Rev. Lett. **58**, 1320 (1987).

17. B. Ohnesorge et al., Phys. Rev. B **56,** R4367 (1997); J.-M. Gérard et al., Phys. Rev. Lett. **81**, 1110 (1998); L. A. Graham, D. L. Huffaker, and D. G. Deppe, Appl. Phys. Lett. **74,** 2408 (1999).

18. G. S. Solomon, M. Pelton, and Y. Yamamoto,  Phys. Rev. Lett. **86**, 3903 (2001); Y. Yamamoto, S. Machida and G. Björk, Phys. Rev. A **44**, 657 (1991).

19. O. Benson, C. Santori, M. Pelton, and Y. Yamamoto, Phys. Rev. Lett. **84**, 2513 (2000).

20. A. K. Ekert, Phys. Rev. Lett, **67**, 661 (1991).

21. C. H. Bennett, G. Brassard and N. D. Mermin, Phys. Rev. Lett. **68**, 557 (1992).

22. Y. Yamamoto, M. Kitagawa and K. Igeta, Proc. 3rd Asia Pacific Phys. Conf. (World Scientific, Singapore, 1988) p. 777.

23. G. J. Milburn, Phys. Rev. Lett. **62**, 2124 (1989).

24. I. Chuang and Y. Yamamoto, Phys. Rev. A **52**, 3489 (1995).

25. E. Knill, R. Laflamme and G. Milburn, Natuer **409**, 46 (2001).

26. M. Tabuchi, S. Noda, and A. Sasaki, in  Science and Technology of Mesoscopic Structures (Springer-Verlag, Tokyo, 1992), p. 375; D. Leonard et al, Appl. Phys. Lett. **63**, 3203 (1993); J.-Y. Marzin et al, Phys. Rev. Lett. **73**, 716 (1994); M. Grundmann et al., Phys. Stat. Sol. (b), **188**, 249 (1995).

27. Zh. I. Alferov et al., Semiconductors **30**, 194 (1996); D. Bimberg et al., Quantum Dot Heterostructures (John Wiley & Sons, Chichester, 1999).

28. M. A. Hreman and H. Sitter, Molecular Beam Epitaxy: Fundamentals and Current Status, 2nd Ed. (Springer, Berlin, 1996).

29. G. S. Solomon, J. A. Trezza, and J. S. Harris, Jr. Appl. Phys. Lett. **66**, 991 (1995); G. S. Solomon, J. A. Trezza, and J. S. Harris, Jr. Appl. Phys. Lett. **66**, 3161 (1995).

30. S .A. Campbell, The Science and Engineering of Microelectronic Fabrication (Oxford University Press, Oxford, 1996).

31. B. Ohnesorge, M. Albrecht, J. Oshinowo, A. Forchel, and Y. Arakawa, Phys. Rev. B **54**, 11532 (1996).

32. L. Landin, M. S. Miller, M.-E. Pistol, C. E. Pryor, and L. Samuelson, Science **280**, 262 (1998); L. Landin et al, Phys. Rev. B **60**, 16640 (1999); M. Bayer, O. Stern, P. Hawrylak, S. Fafard, and A. Forchel, Nature **405**, 923 (2000).

33. J. J. Finley et al, Phys. Rev. B **63**, 073307 (2001).
34. C. Santori, M. Pelton, G. S. Solomon, and Y. Yamamoto (to be published).
35. J.-M. Gérard and B. Gayral, J. Lightwave Technol. **17**, 2089 (1999).
36. S. Reynaud, Annales de Physique **8**, 315 (1983).
37. P. R. Berman ad., Cavity Quantum Electrodynamics (Academic Press, Boston, 1994).
38. Y. Yamamoto, S. Machida, K. Igeta, and Y. Horikoshi, Coherence and Quantum Optics, Vol. VI, edited by E. H. Eberly, et. al. (Plenum Press, New York, 1989), p. 1249; H. Yokoyama et al., Appl. Phys. Lett. **57**, 2814 (1990).
39. G. Panzarini and L. C. Andreani, Phys. Rev. B **60**, 16799 (1999).
40. G. Björk, H. Heitmann, and Y. Yamamoto, Phys. Rev. A **47**, 4451 (1993).
41. A. W. Snyder and J. D. Love, Optical Waveguide Theory (Chapman & Hall, New York, 1983).
42. A. Kiraz et al, Appl. Phys. Lett. **78**, 3932 (2001).
43. K. S. Yee, IEEE T. Antenn. and Propag. **AP-14**, 302 (1966).
44. A. Taflove, Computational Electrodynamics: (Artech House, Norwood, Massachusetts, 1995).
45. Y. Chen, R. Mittra, and P. Harms, IEEE T. Microw. Theory **44**, 832 (1996).
46. Y. Xu, J. Vučković, R. Lee, O. Painter, A. Scherer, and A. Yariv, IEEE T. Microw. Theory **16**, 465 (1999).
47. H. Drexler, D. Leonard, W. Hansen, J. P. Korthaus, and P. M. Petroff, Phys. Rev. Lett. **73** 2252 (1994).
48. E. Dekel, D. Gershoni, E. Ehrenfreund, D. Spektor, J. M. Garcia, and P. M. Petroff, Phys. Rev. Lett. **80** 4991 (1998).
49. M. Grundmann, O. Stier, and D. Bimberg, Phys. Rev. B **52**, 11969 (1995).
50. M. Fricke, A. Lorke, J. P. Kotthaus, G. Medeiros-Ribeiro, and P. M. Petroff, Europhys. Lett. **36**, 197 (1996).
51. B. T. Miller, W. Hansen, S. Manus, R. J. Luyken, A. Lorke, J. P. Kottehaus, S. Huant, G. Medeiros-Ribeiro, and P. M. Petroff, Phys. Rev. B **56**, 6764 (1997).
52. A. Wojs and P. Hawrylak, Phys. Rev. B **53**, 10841 (1996).
53. F. M. Peeters and V. A. Schweigert, Phys. Rev. B **53**, 1468 (1996).
54. C. Weisbuch and B. Winter, Quantum Semiconductor Structures (Academic Press, San Diego, 1991).
55. O. Stier, M. Grundmann, and D. Bimberg, Phys. Rev. B. **59**, 5688 (1999).
56. Y. Toda, S. Shinomori, K. Suzuki, and Y. Arakawa, Phys. Rev. B **58**, R10147 (1998).
57. K. Brunner, G. Abstreiter, G. Böhm, G. Tränkle, and G. Weimann, Phys. Rev. Lett. **73**, 1138 (1994).
58. A. Aspect, J. Dalibard, and G. Roger, Phys. Rev. Lett. **49**, 1804 (1982).
59. J. F. Clauser, M. A. Horne, A. Shimony, and R. A. Holt, Phys. Rev. Lett. **23**, 880 (1969).
60. O. Benson and Y. Yamamoto, Phys. Rev. A **59**, 4756 (1999).
61. J, A. Gupta, D. D. Awschalom, X. Peng, and A. P. Alivisatos, Phys. Rev. B. **59**, R10421 (1991).

# Index

Printing: Mercedes-Druck, Berlin
Binding: Stein+Lehmann, Berlin

MICHIGAN MOLECULAR INSTITUTE
1910 WEST ST. ANDREWS ROAD
MIDLAND, MICHIGAN 48640